Palgrave Studies in World Environmental History

The widespread perception of a global environmental crisis has stimulated the burgeoning interest in environmental studies and has encouraged a range of scholars, including historians, to place the environment at the heart of their analytical and conceptual explorations. An understanding of the history of human interactions with all parts of the cultivated and non-cultivated surface of the earth and with living organisms and other physical phenomena is increasingly seen as an essential aspect both of historical scholarship and in adjacent fields, such as the history of science, anthropology, geography and sociology. Environmental history can be of considerable assistance in efforts to comprehend the traumatic environmental difficulties facing us today, while making us reconsider the bounds of possibility open to humans over time and space in their interaction with different environments. This series explores these interactions in studies that together touch on all parts of the globe and all manner of environments including the built environment. Books in the series come from a wide range of fields of scholarship, from the sciences, social sciences and humanities. The series particularly encourages interdisciplinary projects that emphasize historical engagement with science and other fields of study.

More information about this series at
http://www.palgrave.com/gp/series/14570

Carolyn M. King

Invasive Predators in New Zealand

Disaster on Four Small Paws

palgrave
macmillan

Carolyn M. King
Environmental Research Institute
University of Waikato
Hamilton, New Zealand

Palgrave Studies in World Environmental History
ISBN 978-3-030-32137-6 ISBN 978-3-030-32138-3 (eBook)
https://doi.org/10.1007/978-3-030-32138-3

Cover illustration: Nga Manu Images, courtesy of David Mudge

This Palgrave Macmillan imprint is published by the registered company Springer Nature Switzerland AG.
The registered company address is: Gewerbestrasse 11, 6330 Cham, Switzerland

In memoriam: For Kenneth Robert Ayers, with love and thanks.

Foreword

Human introductions of organisms into new environments began when people walked to a new gathering place with weed seeds on their clothes, tossing their fruit pits beside the trail. As farmers, they continued, more purposefully. After 1500, introductions became an important part of the world's biotic history as Europeans took their crops and stock to lands they conquered and moved other species to new fields around the globe. In the nineteenth century, steamships and the acclimatisation movement—an enthusiasm for new species that swept Europe and the European-settled lands—produced some of the most rapid, dramatic changes ever seen. The disasters that followed produced some caution but the flow continued, particularly of accidental introductions. Jet cargo planes now make it ridiculously easy to bring an alien like the brown tree snake to even the remotest oceanic island. Meanwhile, the consequences of earlier good ideas, mistakes, misfortunes and follies continue to unfold.

Early control measures relied on simple ideas about reducing populations and tended to simple methods—poison, guns, and fences. Ecology's insights into the interactions of species on the land allowed more sophisticated controls, but the science could not, unfortunately, demand them. In the half century since Charles Elton wrote the first scientific survey of introductions, *The Ecology of Invasions by Animals and Plants*, we have moved towards better methods and found more tools for both study and control, as witness the stories in this volume.

The reasons for control have changed as well. Early efforts focused on protecting crops and grazing land, but while money remains a key factor, nationalism has become important too. Programmes now seek to save native species and pre-European landscapes as part of the national heritage. In the fields of both introductions and controls, Australia and New Zealand hold pride of place, for it was there that new species most visibly changed the land, and there that attitudes towards native species changed most quickly.

North Americans and Europeans had as much enthusiasm, energy and ignorance as their cousins to the south, as well as more money, but most new species found it harder to get a foothold among continental biotas (preventing many but, alas, not every disaster). Americans made the native landscape a source of national pride in the early nineteenth century (not that that saved the bison) and Europeans already regarded themselves as rooted in their lands. But even into the twentieth century, the antipodeans saw their native species as somewhat embarrassing primitives who could not compete with superior, European species.

For drama, nothing matched the story of the rabbit in Australia—the 'grey blanket', hecatombs of dead rabbits, fences hundreds of miles long and the apocalypse of myxomatosis—but drama is not a particularly useful measure, and New Zealand, which this book treats, has its own interest. Humans came late to the islands and found no resident mammals beyond bats and seals. Birds occupied all sorts of ecological niches. The Māori reshaped the country and its fauna, but it was the Europeans, arriving in force in the early nineteenth century, who made the modern landscape with their crops and stock, the species they brought for sport, beauty and nostalgia, and others introduced to check some of the mistakes and accidents. Volumes like this one are vital in showing the history of some of those species. Introductions continue to adapt and change, even as they shape the land, and we need this kind of thorough analysis, done here in a detail unimaginable a generation ago, to understand where we are and to suggest what we must or can do.

While the data here are fascinating to any student of natural history, the book also suggests how much remains unknown, and it offers some useful cautions. Most obviously, humility is—still—in order. Earlier generations told, often with great certainty, stories we now know were incomplete or wrong (at times, disastrously so), and in addition to documenting their errors, this book also shows how much we still do not understand. Difficult as some will find it to believe, we too may be wrong. On the other hand, the accounts of unexpected success in clearing some of New Zealand's offshore islands give cause for hope. Programmes seen as difficult or impossible have turned out to be possible, given public commitment for the long term, dedicated people on the ground and some bold imagination. We will need those qualities, as well as hope and humility, for we have set in motion processes that will find new forms in the Anthropocene but will not vanish.

Texas A&M University
College Station, TX, USA

Thomas R. Dunlap

Preface

There is a story about an elderly English academic who was stopped in an Oxford street by a tourist and asked for directions to Beckley.

'Ah', he said, fishing out his pipe and slowly filling it (he was a great believer in the old saying, attributed to an apocryphal rural Irishman, that lighting a pipe gives a wise man time to think, and a fool something to stick in his mouth). 'Well', he said at last, 'well, if I wanted to go there, I wouldn't start from here'. In other words, to know how to get to where you want to be, it helps to know where you have come from.

The whole complicated relationship between pest animals and the humans whose interests they exploit is rather like that. Few scientists of the time understood that interaction better than the pipe-smoking academic, my own doctoral supervisor, Dr. H. N. Southern; I heard this story directly from him in about 1970. He had spent the years 1939–45 labouring to find ways to control rodent damage in Britain's essential wartime food stores. For centuries before that, ship owners, settlers and farmers of all cultures around the world have battled against the rats and mice that foul stores of food and damage almost any kind of chewable materials.

In New Zealand, we not only have that ancient problem in spades, but we also urgently need to get rid of the mustelids (stoats, weasels and ferrets) that help rats to kill and maim our native birds and lizards, and the Australian brushtail possums that damage our forests and birdlife and pass bovine tuberculosis to our cattle. We struggle to understand why the organised introductions of thousands of mustelids (stoats, weasels and ferrets, all specialised predators of small animals and birds) was accepted as an official policy that seems to us completely irrational, but was, at the time, seen to be the only possible way to save the trade in wool, New Zealand's principal export product, from ruination by rabbits. We kill all these pests in thousands every year and win many local and temporary battles, yet we never seem to win the war.

History Is Important

All of the species described here are mammals that have been introduced by human agency, deliberately or accidentally (i.e. not including those counted as native, because they got here by flying or swimming under their own steam), but that does not automatically make them invasive. Not all mammals transported to New Zealand established self-sustaining populations, and not all of those that did have become invasive pests. As a broad rule of thumb, invasive species can be recognised as the ones that behave quite differently from their ancestors. Their wider distributions and higher numbers in their new countries can have unexpected and frequently damaging consequences. New Zealand is a textbook example of this selective process, so to protect our native biota, we have become a world leader in invasive species management.

History is important in helping us to think about the past, not because the past has any magical quality, but because we cannot study the future, and yet, still need something to set against the present. Historical information reminds us that the basic assumptions of any given period can be quite different from those of any other. Ideas that were obvious to the Victorians often seem irrational to us, and statements that seem certainly true to us now may not seem so to our descendants. Likewise, those who have travelled much or lived in different places do not assume that everywhere is the same, so are unlikely to be surprised by local variations in custom. They are, to some extent, immune to the great cataract of nonsense that pours from local vested interests, sensational journalists and nimbyists.

History is even more important in helping us to think about the future, because a detailed knowledge of how proposals of the past worked out, or didn't, is needed to inform our much-needed critical judgement on which current proposals might eventually work, or not. History reminds us forcibly that nature is dynamic and that plans assuming that the future will be some minor variation of the present are at risk of encountering some unwelcome surprises.

The ongoing declines of New Zealand's endemic birds, bats, lizards and frogs can be measured in distressing detail by every repeat survey and in every habitat. The groundswell of community concern for what has been lost—and worse, what is still at immediate risk of future loss—has now moved far beyond the decades-long protests of conservationists. Of course, there are multiple other contributing causes of environmental change, and we are increasingly recognising the importance of these, but much of the strongest conservationist focus has been on the alien introduced predators—especially rodents, cats and mustelids—which have had a devastating and pervasive effect on the numbers and distributions of the unique native fauna of New Zealand.

What This Book Is, and Is Not

This book is a critical review of the historical background required to understand one of New Zealand's longest-continuing and very expensive national dilemmas. In order to make the point that this part of the world has seen many previous and even more drastic environmental changes and species extinctions, it begins by painting a brief picture of the New Zealand of the remote, pre-human, land mammal-free past. Since then, there have been immense shifts in climate, land area and native fauna, from before the demise of the dinosaurs, through the ice ages, to the richly forested, uninhabited, bird-thronged archipelago of islands of less than a thousand years ago (Chap. 1). Then, it describes the arrival of the first Polynesian colonists and the unintended but drastic consequences of their settlement, accompanied by two small predators of Pacific origin, neither now common (Chap. 2).

Without minimising that critically important event, or the huge significance of Māori culture for conservation history, the rest of the book concentrates on summarising recent research documenting the additional effects of the arrival of hundreds of thousands of European colonists and their diverse retinue of mammalian companions. Together, these alien animals are the main focus of anti-predator anguish in the present conservation debate. All of the introduced species described here are well equipped with teeth and claws, and all have caused untold damage to New Zealand's native fauna over a turbulent few centuries. While there are many good books and papers that touch on the different species and circumstances contributing to this sad story, I have aimed to bring all the threads of the saga together in one connected narrative.

Of course, we have developed a substantial arsenal of long-tried and tested traps and poisons that can knock down the numbers of pests in some times and places, but on the inhabited main islands, none of them ever gets the last one. A carefully planned toxic operation can reduce the number of rats in a forest from a wall-to-wall rampage down to almost zero in a week, but in six months, you would never know they had been away. Diseases such as myxomatosis and cat flu can be fatal to more than 95% of the individuals infected, but the last few to escape and develop immunity can re-establish the population in a few generations. House mice, ship rats and rabbits are paradigm pests: whatever killing devices we throw at them, of however many sorts and however often we repeat the known successful lethal techniques or try new ones, the pests always come back. Even their natural enemies, imported into New Zealand from their home country, where they seldom cause as much trouble as they do in New Zealand, could not effectively control them, for reasons connected with the deep network of animal community interactions that did not become clear until much later (Chap. 12).

The nineteenth century was a pivotal period of rapid environmental and social changes in New Zealand, so the story described here concentrates on that arbitrarily defined hundred years. After five centuries of changes precipitated by the arrival of Polynesian settlers and their Pacific rats and dogs (Chap. 2), European activities facilitating the invasions of more small mammalian predators from the northern hemisphere set off further and faster significant environmental changes. Not all of them were confined to that period, because house mice and rats started arriving in the mid-to-late 1700s; domestic cats and Australian brushtail possums were already present in the mid-1850s (although full documentation of their potential for causing damage to native wildlife appeared only much later). But it makes a neat story to focus our attention on that period.

This book is also limited in another sense: it concentrates only on describing a particular group of invading alien small mammal predators and their prey, which interact with each other and directly affect each other's populations in a particular environment unlike any other in the world. All are small and furry, foraging silently on their four small paws, famously unfussy eaters and widely disliked by people, so seldom seen by day or far from escape cover. Other extremely serious pest invasions into New Zealand are also important, but are not part of this narrative because they belong to different animal communities (deer and goats in forests, chamois and tahr on the alpine tops, trout in the rivers and lakes). Their stories are for a different book.

Further again outside my brief are other tragedies enacted by invasive species on the global stage, skilfully summarised in previous classic books by Charles Elton, Alfred Crosby and Thomas Dunlap. But nowhere else has that particular combination of historical, geographical, biological and human circumstances that make the New Zealand story unique. Cats, mice and Norway rats are common problems on islands around the world, but no other country has such large populations of black rats, feral ferrets and wild (non-commensal) house mice as New Zealand has, or such huge numbers of brushtail possums (some 20 times more abundant here than in Australia, with very different relationships to forest trees and their animal communities). So, our particular combination of players places some limits on the idea that we could transfer much specific advice to elsewhere in the world about how native species might interact with their invaders. In the very unusual and artificial communities assembled over the history of human interference in New Zealand, many native populations have not only been invaded, they have been simply swept off the map.

Therefore, predator-prey interactions cannot be interpreted in isolation, as in many older models; they are always part of a community, a web of interactions specific to a time, place and particular combination of species. Even though the key players in such communities in New Zealand are not always the same as those of anywhere else, the basic principles still apply here. So, the most useful branches of the massive theoretical base of mammalian predation biology for understanding events in New Zealand are those taking account of location-specific community-level interactions, habitat requirements and climate.

We have discovered that the penalties of failing to do that are high. For example, the old idea of 'the balance of nature' has long been discredited, but is still widely believed. It led directly to the tragic outcomes of Benjamin Bayly's confidence in the ability of imported 'natural enemies' to control rabbits (Chap. 7). The theories of Paul Errington (that predators take only the 'doomed surplus') inhibited senior managers from taking prompt action to save the birds of Big South Cape Island from rats (Chap. 4).

On the other hand, our findings certainly do have some wider resonance, and others can learn much from the widely recognised expertise in pest control and habitat restoration that we have developed. New Zealand cases are frequently cited as paradigm examples prominent in the global conservation literature and are well represented in big international conferences such as the latest one organised by IUCN, *Island Invasives: scaling up to meet the challenge*, in Dundee, 2017.

The painful lessons of our past offer the world one overwhelmingly important cautionary tale, concerning the unpredictable dangers of implementing any self-perpetuating weapon. The possible gene drives of the future are generating both excitement and concern among pest managers, just as stoats and ferrets did in their time. But once released, neither can ever be recalled. No one can be fully confident of the ultimate outcome of such a policy, however carefully planned; and fast-moving current events would soon overtake any predictions from current knowledge. Still, the most obvious take home messages visible at present are clearly articulated at the end of Chap. 12.

Predator-Free New Zealand 2050

In New Zealand, as everywhere else, the conservation of biodiversity is a matter of great and enduring public interest with complicated historical roots, and the global losses of irreplaceable species are accelerating. Deforestation and commercial farming have always played a hugely significant role in this crisis, by contributing directly or indirectly to all major drivers of biodiversity loss, in complex and interrelated ways. These difficult matters are the topic of extensive and often polarising public debate, all too often conducted without reference to their important historical and ecological backgrounds.

For the 250 years since the first Europeans landed in New Zealand, there has been no large-scale solution in sight, and not nearly enough money or political will to find one. Now, the agencies responsible for conservation policy have been stimulated to take real, expensive, government-sponsored action. New Zealand is facing the challenges posed by small predatory pests more seriously than ever before. In 2016, the national government committed 28 million dollars to support an aspirational programme, ambitiously titled Predator-Free New Zealand 2050 (PFNZ2050), aiming to banish rats, stoats and possums from the whole country by 2050. In the light of this proposal, now is a good moment to ask how and why so many alien rodents and carnivores have colonised New Zealand, exactly what their arrival has meant for the native fauna, and what can be done about them now. Above all, surely it would be useful to know something about how to avoid repeating the mistakes we have made during past attempts to deal with them (Chap. 12).

We know from hard experience that continuing with the pest control business as usual, by searching for more and better ways to kill individual pests in the hope of eradicating whole populations, can work spectacularly well on smaller offshore islands, but not wherever reinvasion is likely. So, the first lesson of intelligent pest control is, learn from history. We have to do that because it is not reasonable to keep on doing the same thing and expect a different outcome. Especially, we cannot continue to forget that introduced predators are not solely to blame for keeping native species out of radically modified landscapes where few natives can *live*.

History can explain why PFNZ2050 cannot get to where it wants to be from where it is now, but it also offers reasons to hope that new developments in understanding biodiversity loss, aided by as yet unknown advances in technology and radical increases in public engagement, could make that journey possible in the future. To do that, we need to know what did—and did not—work in the past. That is what this book is about.

Hamilton, New Zealand

Carolyn M. King

Acknowledgements

It is a truism, but still true after many repetitions, that authors of academic works never work alone. We all build on what others have done before us; in the words of Isaac Newton, we can only stand on the shoulders of giants. Therefore, my first acknowledgements go with heartfelt gratitude to the many people and institutions with whom I have worked since first starting as a newbie graduate student in Charles Elton's Bureau of Animal Population in Oxford, a fortnight before he retired in October 1967. All the work described in these pages is based on studies begun then—my own and others'—and continued ever since I moved to New Zealand four years later.

This particular book owes its origin to the initiative and support of Dr. James Beattie and the Board of Palgrave Macmillan and to their editors Molly Beck and Oliver Dyer in London and to Megan Laddusaw and Christine Pardue in New York. No question was too inconsequential to require their full and helpful attention. It has been a pleasure to work with them. I am also indebted to the anonymous reviewers who examined my proposal and to Austin Gee and Charles Daugherty, who patiently reviewed late drafts of the whole text.

A long list of people supplied reviews of one or more chapters and/or provided information or illustrations or many other forms of encouragement: Peter Banks, Fred Brook, Ellen Clarkson, Mick Clout, Miriama Cribb, John Dyer, Neil Fitzgerald, John Flux, Patrick Garvey, Keith Giles, Dave Hansford, Ray Harlow, Alastair Henshaw, Peter Holland, John Innes, Richard Jennings, Wayne Johnson, Daphne and Bill Lee, David Miller, Maxine Moerbe, Pauline Morse, Pat Morris, Grant Morriss, David

Mudge, Dorothy Nash, Grant Norbury, Max Oulton, Bruce Patty, Ray Richards, James Russell, Janice Salmon, Linda Smith, Andrew Thompson, Andrew Veale, Dave Ward, Cheryl Ward, Janet Wilmshurst, Kerry-Jane Wilson and Trevor Worthy. I am especially grateful to Thomas Dunlap for adding a thoughtful Foreword. All applied their individual expertise with great kindness and made many perceptive suggestions for improvement. Several potentially embarrassing errors were thereby caught in time, and of course, those that remain are entirely my responsibility.

Sources of data and ideas in the published literature, and of key quotations, are given in the endnotes to each chapter and in the captions to figures and tables. My own papers cited all list in the original versions my thanks to the people and institutions who assisted those particular projects. All chapters comprise current reviews of each topic at the time of writing, formulated for readers without experience of biological research. The result is far from a comprehensive survey of this huge subject, but contains enough clues to lead to sources of further information.

Chapters 3 and 5 are condensed from two papers published in 2016 by the journal *Biological Invasions*, with permission from Springer. Chapters 7, 8, 9 and 10 are abbreviated and modified versions of three papers published in 2017 by the *International Review of Environmental History* and the *New Zealand Journal of Ecology*, with permission from The Australian National University and the New Zealand Ecological Society, respectively. The last part of Chap. 8 is based on a note published in 2017 in *Notornis*, with permission from the Ornithological Society of New Zealand. For more lengthy and detailed citations to primary historical sources underlying these chapters, please refer to the original papers, all available online and most open-access. Acknowledgements due to other institutions are listed in captions and endnotes.

By far the most reliable and accessible sources of primary historical information are the official New Zealand Government publications—the reports of Parliamentary debates in Hansard, and the *Appendices to the Journals of the House of Representatives*, both searchable online. Equally reliable deposits of official documents are held in Archives NZ, most freely accessible on site. Non-governmental documents such as letters, farm diaries, and unpublished analyses are held in the Alexander Turnbull Library (Wellington), the Hocken Library (Dunedin), the Auckland Institute and Museum; the city libraries of Auckland and Dunedin; the Maritime Museums of Auckland, Port Chalmers and Bluff (regrettably, the Lyttleton Maritime Museum was closed in 2011 by earthquake damage). All of these I have searched in person. In addition, professional archivists searched several local museums on my behalf.

Historical news items—preferably those written by eye-witnesses, especially if citing reputable authorities or government officials—often recorded and discussed invaluable contemporary observations of the natural world. Modern newspapers are notoriously unreliable as sources of factual data, but historical news reports, if handled and cross-checked carefully, can supply the only accounts now available of how particular people thought and behaved in the past. We are fortunate that by the early-to-mid-nineteenth century, there were dozens of provincial newspapers in New Zealand, and reading them over several years gave me the strong impression that the standard of journalism then was higher than it is now. Each of the old newspapers was a rich source of information about local events and people, widely syndicated between regions, and they often reprinted significant items from correspondents and news publications overseas. Much of this massive resource is now digitised and easily searchable via Papers Past (NZ) and the British Newspaper Archive. See their online newspaper databases in PDF format at https://paperspast.natlib.govt.nz/ and http://www.britishnewspaperarchive.co.uk/ for easily accessible and valuable accounts of shipping movements, and reports of meetings, letters and editorials showing how people reacted to the first arrivals of alien animals.

Dates of named ship departures from England and arrivals in New Zealand are taken from the Alexander Turnbull Library's copy of Comber's exhaustive documentation of shipping arrivals and departures, which ends in December 1889. From 1890 onwards, confirmation of shipping data is more difficult, and passenger lists become swamped with thousands of names.

The names of nineteenth-century New Zealand provinces varied: Southland seceded from Otago in 1861, but rejoined in 1870. The provinces were abolished as separate regional governments in 1876, but the names remain as geographic descriptions. The name given by the Māori to their own country, at least since the early nineteenth century, was and is Aotearoa and is used here in that context. The name New Zealand has been used overseas since the mid-seventeenth century and is applied to it here after it became a colony of the British Empire in 1840.

Financial records are cited in the currency of New Zealand used in the nineteenth century, the British pound sterling (£), divided into 20 shillings (s), each of 12 pence (d). The Reserve Bank of New Zealand provides an online calculator by which sterling amounts back to 1862 can be converted into contemporary NZ$ (http://www.rbnz.govt.nz/monetary-policy/inflation-calculator).

To assist overseas readers, a map of New Zealand common place names (Fig. 1.1) is offered as a general background (more detailed maps are provided in context in Chaps. 3, 5, 8, 9, 10 and 11), and the text uses the English names of animals, if available, while their scientific and Māori equivalents are listed in the Appendix.

Māori is an official language of New Zealand, so in this text, Māori words are not italicised, as would be usual for non-English words in an English text. Māori vowels are pronounced separately, so a word like "kiore", which a European might expect to sound as two syllables 'kee-aw', is pronounced with three, as 'kee-or-eh'. Māori has five vowels and ten consonants: h, k, m, n, p, t, w, r, n, g, all pronounced as in English, plus wh (pronounced f).Māori has no equivalents to the English b, d, j, s, v, x, or z. Māori words do not add a suffix in the plural, so 'moa' and 'kiwi' are spelled the same, whether they refer to one or many individuals, just as are 'deer' and 'sheep' in English. Long vowels in written Māori are distinguished by means of a macron, a bar inserted above the vowel, because the length of vowels can sometimes distinguish words different in meaning.

Contents

About the Author

Carolyn M. King FRSNZ (DPhil Oxon 1971, PhD Waikato 1999) is recognised as an international authority on the biology of small mammals, especially mustelids and rodents. Born in Tunbridge Wells, England, Carolyn (known to friends as Kim) emigrated to New Zealand in 1971, straight from Oxford, where she completed the first ever full field study of the ecology of weasels. From 1971 to 1977, she worked with the Department of Scientific and Industrial Research (DSIR) Ecology Division in Lower Hutt as a scientist specialising on introduced carnivores, especially stoats. After she left DSIR to concentrate on her family, she continued to write papers and books part-time. In 1983, she was appointed Scientific Editor to the Royal Society of New Zealand, and edited *The Journal of the Royal Society of New Zealand* until 2001 and *The New Zealand Journal of Zoology* from 1991 to 2009. In 1995, she joined the Department of Biological Sciences at Waikato University (Hamilton, New Zealand), until she retired as a teaching professor in 2018, but has stayed on part-time to continue writing. She received the annual Award from the NZ Ecological Society in 1999, the Silver Medal of the Mammal Society in 2005, the Kudos Lifetime Achievement Award in 2010, and was elected a Fellow of the Royal Society of New Zealand in 2018.

Carolyn's other books include:

Immigrant Killers: Introduced predators and the conservation of New Zealand birds (1984).

The Natural History of Weasels and Stoats (1ed 1989; 2ed (*with R.A Powell*) 2007).

The Handbook of New Zealand Mammals (Ed) (1ed 1990; 2ed 2005; 3ed (*with D. M. Forsyth*) due from CSIRO Publishing in 2020).
The European Rabbit: History and Biology of a Successful Coloniser (Ed with H.V. Thompson) (1994).
The Drama of Conservation: History of Pureora Forest Park (Springer Verlag/Department of Conservation 2015, with DJ Gaukrodger and NA Ritchie).
Habitat of Grace (Australian Theological Forum 2002).

List of Figures

List of Tables

PART I

The Years Before Cook

CHAPTER 1

Introduction (82 Million Years Ago to AD 1280)

Travelling to New Zealand is the nearest any ordinary folk can get to arriving on another planet. Jared Diamond (1990)[1]

Life on a Lonely Island

Charles Darwin, arriving on the Galapagos Islands after long expeditions to observe the continental fauna of South America, was astonished at the variety of animals and plants he found on these strange and remote specks of land so far from the nearest mainland coast. He wrote:

> by far the most remarkable feature in the natural history of this archipelago … is that the different islands to a considerable extent are inhabited by a different set of beings … I never dreamed that islands, about fifty or sixty miles apart, and most of them in sight of each other, formed of precisely the same rocks, placed under a quite similar climate, rising to a nearly equal height, would have been differently tenanted.[2]

Islands are often believed to have a near-magical quality, places to escape enemies, havens of peace and freedom from conflict, protected by open seas that invaders cannot cross. Darwin discovered that island species are just as competitive as any other, but they interact within an environment

C. M. King, *Invasive Predators in New Zealand*, Palgrave Studies in World Environmental History, https://doi.org/10.1007/978-3-030-32138-3_1

and community different from those on the mainland, and with a more limited cast of characters.[3] The more isolated they are, the less easily they can be influenced by other populations, and so, the more they differ from their relatives.

As island populations accumulate characteristics advantageous to their new environment without being able to exchange genes with their continental relatives, they gradually become distinct from each other over time. Many biologists since Darwin have used islands as ideal natural laboratories for demonstrating how the dynamic nature of life, and the consequences of natural selection in isolation, can produce a range of new species with characters all their own.

Like the Galapagos, the islands of modern New Zealand also comprise an archipelago ideally placed to encourage the evolution of new species (Fig. 1.1). They span 2813 km of the southwest Pacific Ocean—from the subtropical Kermadec group, Raoul and Macauley Islands (29° South) to Campbell Island (52° South). They include the North and South Islands (the 'mainland'), 114,740 km^2 and 151,120 km^2, respectively, sprawled across 1475 km between latitudes 34.4 and 46.6 decimal degrees South, through the Chatham Islands (800 km east of Christchurch) and Stewart Island (30 km south of the South Island) to the subantarctic groups (Snares, Auckland, Campbell, Antipodes and Bounty Islands). These are only the largest of 735 inshore and outlying islands larger than 1 hectare (ha), ranging from fragments separated from the mainland coast by narrow channels to larger chunks of land in distant, splendid isolation.

The total land area of New Zealand adds up to 270,000 km^2, about the same as the UK but less than 4% of that of Australia, and much more variable in geography and climate than either. All this wonderful diversity means that the habitats offered to animals in the New Zealand region range from temperate coastal bush to high mountains (23 of them reaching to over 3000 m) covered in permanent snow; from rich subtropical forests through near monocultures of hardy mountain beeches to (since only the last couple of million years) the alpine zone; from improved farmland to huge plantations of exotic radiata pine; from limestone caves with constant temperatures to semi-arid tussock-lands with severe winters and hot summers; from swamps to massive freshwater lakes; from small island refuges to modern cities; and in the far south, a scattering of subantarctic islands, and a large slice of Antarctica. For most of the long period from the disappearance of the dinosaurs to the cooling climate of the later

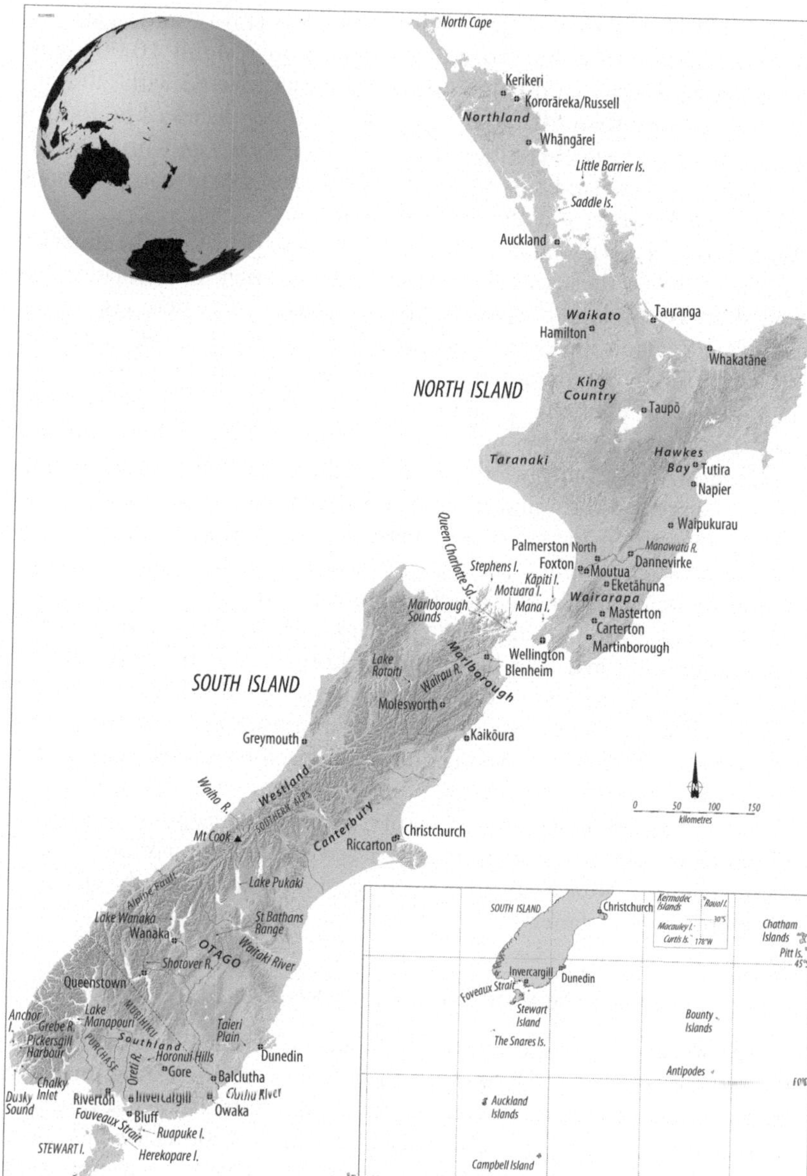

Fig. 1.1 Map of New Zealand showing places mentioned in the text. (Cartography by Max Oulton)

Tertiary, that is, from 66 to 10 million years ago (mya), and again, for most of the period since the retreat of the last ice age roughly 10,000 years ago, almost all the dry land below the treeline was covered with tall evergreen forest. More than 80% of it comprised unique species of podocarps (native conifers) or southern beeches (Nothogafaceae) and broadleaved trees, with tree-ferns and shrubs in the under-storey, and native tussock grassland in drier areas and above the treeline.

Animals of one sort or another live in all of these places, so the diversity of wildlife native to these islands is impressive, and their long isolation has compiled an assemblage of species different from everywhere else in the world. New Zealand illustrates brilliantly well how the marvels and perils of life on remote islands are the products of diversification, successive immigrations and extinctions, adaptive radiations and evolutionary interactions. But it also illustrates more clearly than almost anywhere else how the deep-down instability of the restless earth has enforced irresistible dynamic changes on the animals and plants that inhabit its surface. Written in the rocks is the evidence of an unimaginably long and unpredictable history of geological upheavals, whose consequences have sifted through an unending series of variable subsets of animals and plants. The characters in this long saga are part-resident, part-immigrant, thrown together in combinations unknown elsewhere. Those whose lineages have not adapted to each new challenge and to each other have simply disappeared.

The result is an assemblage of biotic communities that has evolved on the New Zealand islands, surviving and changing through tectonic rifting, mountain building, varying sea levels, storms, down-wind drifts, immigrations, extinctions and shipping accidents, surprising enough to northern-hemisphere visitors to have prompted Jared Diamond's famous remark that 'Travelling to New Zealand is the nearest any ordinary folk can get to arriving on another planet'.[4]

The Story of New Zealand Wildlife

New Zealand is a large and remote archipelago, protected since the time of the dinosaurs by up to 2000 km of turbulent ocean, deterring intrusion by all but occasional windblown or marine immigrants. In such severe solitude, priceless remnants of early experiments in evolutionary adaptation, which have been extinguished everywhere else, have been preserved alongside new lineages arising from scattered populations stranded by varying sea levels or tectonic uplifts (Fig. 1.2). Together, they formed an

Fig. 1.2 Relics of Gondwana: two living animals whose ancestry goes back to the time of the dinosaurs. (**a**) the tuatara; (**b**) the peripatus or velvet worm, *Peripatoides novaezealandiae*. (Photographs by Author (**a**) and Neil Fitzgerald (**b**))

entirely new, globally unique feather-based-cum-cold-blooded economy, in which all the ground-dwelling ecological niches filled elsewhere in the world by mammals were occupied by invertebrates, avian and non-avian reptiles and birds. The English, Māori and scientific names of those mentioned in this book are listed in the Appendix.

The absence in modern New Zealand of the terrestrial mammals that have flourished in other countries—not to mention the snakes, dinosaurs and crocodiles, and mammals resident in nearby Australia—is usually explained on the assumption that their ancestors all 'missed the boat'. But, in fact, the earliest members of all these groups (and their archaic predecessors) were already well established in Australia long before 80 mya.[5] There must be some other explanation for their absence in New Zealand. There is, but it is not simple.

J.R.R. Tolkien's masterful trilogy, *The Lord of the Rings*, is a classic story in three volumes, magically brought to life in Sir Peter Jackson's three films, all shot in New Zealand. One after another, the three films introduce us to totally unknown creatures that we have never met in the flesh (or would ever want to), but are so believable that thousands of people visit New Zealand to see the places where they were supposed to have lived. The contemporary New Zealand landscape was convincingly portrayed as the Middle Earth of the mythical past, and some of the star-struck visitors are almost surprised to find that orcs and elves no longer live there.

The history of New Zealand's native fauna is a real-life parallel to Tolkien's vision. Like Middle Earth, it has been populated by hundreds of creatures totally unfamiliar to us, which played their successive parts over the eras of geological time, the equivalent of Tolkien's three volumes, against a landscape that has changed out of their ancestors' recognition and no longer supports any of them. Most of us know nothing of this vanished cast of real creatures, so our understanding of the present tends to be distorted by our unfamiliarity with the past.

This is unfortunate, because the very long and complicated three-stage history of the New Zealand landscape[6] cannot be understood simply by looking at the species now living there. Fossils tell us of ancestral forms that have never been seen alive by any human eyes and would have been unrecognisable if they had. The latest descendant species often look quite different from the earliest members of their genetic lineage—for example, the extinct giant Haast's eagle was, at 10–15 kg, one of the largest flying birds that ever lived. Yet, this monstrous predator was descended from a small Australian eagle only a tenth of its size, which arrived in New Zealand a mere 0.7–1.8 mya.[7]

Neither are the living species a random sample of the previous residents, because the winnowing processes set off by environmental changes were strongly biased against the most different and highly specialised species that were once well adapted to long-gone conditions. We know of them only because they left messages to us in their remains, now readable after decades of detailed analyses of fossils and genetic lineages. To appreciate the potential changes we might face in the future, and to put the present conservation crisis into the context of a very, very long history of dynamic changes in native fauna, and not just in New Zealand, it is worth taking some trouble to understand the reasons for the dramatic changes of the past.

The Gondwanan Fauna (82–c.55 Million Years Ago)

The story began about 180 mya, at a time when the southern continents were united into one single land mass, the supercontinent of Gondwana.[8] Its original cast of animals and plants spread freely across a vast, continuous land surface now divided into chunks we label as Antarctica, South America, Africa, Madagascar, Australia, the Arabian Peninsula and the subcontinent of India.[9] Forests, mountains, coasts and inland waters were inhabited by an immense variety of animals, all adapted to a warm, ice-free world (Fig. 1.3), but now vanished into pre-history. There were many predatory reptiles, including dinosaurs, crocodiles, tuatara (an ancient lizard-like reptile whose ancestors once coexisted with dinosaurs

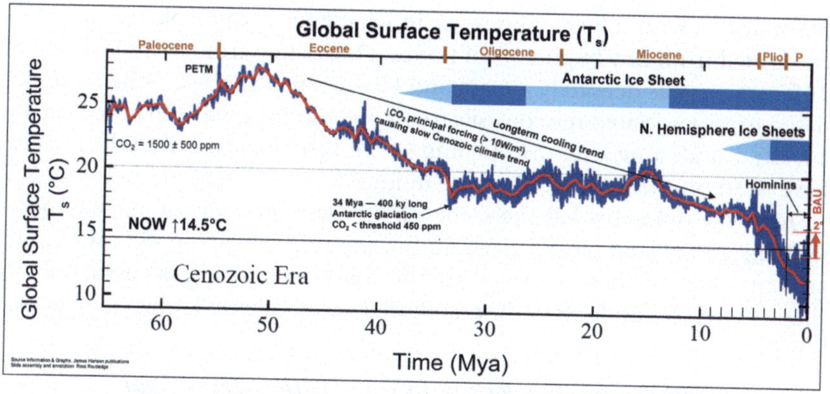

Fig. 1.3 Global climate change over the Tertiary period, since New Zealand became isolated. (Steven Earle, http://www.alpineanalytics.com)

worldwide, whose descendants now survive only in New Zealand) and the earliest marsupials, monotremes and constrictor snakes. Frogs, small lizards and the earliest mammals kept a low profile on the forest floor; birds thronged the skies, the freshwaters and the shores; in the seas, huge coiled ammonites drifted with the currents, and mosasaurs hunted the early diving birds.

The first of three distinct volumes in the story of wildlife in New Zealand, the Gondwanan phase, started about 82 mya, when deep earth movements opened a rift in the surface of the southeastern edge of Gondwana. A large fragment of continental shelf the size of India (known to science as Zealandia, but ten times the size of New Zealand today) began to split off from the rest of Gondwana. A narrow seaway opened up a gap between Australia and Antarctica (still ice-free and joined to Australia until about 40 mya) and gradually widened from the south as the continental fragment drifted away. Intermittent land connections with Australia remained in the north, direct or via chains of stepping-stone islands, for many millions of years.

All or any of the animals and plants typical of eastern Gondwana could have been carried along, and at first, they were indistinguishable from many that remained on the continental landscape that later became divided into the separate southern continents. They included at least six different types of land dinosaurs, not identifiable to known species (but they included two forms of large carnivores, three of herbivores and one pterosaur[10]), whose former presence in what later became New Zealand is confirmed from tiny fragments of bones. One comes from rocks laid down 145 mya, so was definitely a mainland Gondwanan species. Others can be dated from associated microfossils to 75 mya, so must have survived on its ancestral lands long after the opening of the Tasman Sea, and could already have been on the way to becoming unique.

After the dinosaurs left the scene, the earliest ancestors of many important lineages of land birds arrived during the Palaeocene (66–55 mya), and at least some of them survived all the following geological upheavals.[11] Plus, of course, hundreds of thousands of invertebrates unknown to science.

The Zealandian Fauna (55–25 Million Years Ago)

The second volume of the story began when the last land bridge between Zealandia and Australia was lost by about 55 mya, around the beginning of the Eocene period.[12] The gap completed the formation of the Tasman Sea, which widened the minimum water crossing to Australia/Antarctica to more than 1000 km by 45 mya, and has been stable ever since. The climate was much warmer then (Fig. 1.3), until the opening of the Drake Passage, about 30 mya, cut the last link between the southern tip of South America and the Antarctic peninsula. After that, there was no land to stand in the way of the near-continuous westerly wind and cold currents roaring clockwise around Antarctica, cutting the southern continent off from all contact with warmer waters from the north. Antarctica and its surrounding waters began to freeze, and its last surviving Gondwanan residents slowly perished.

Zealandia, the separated continental fragment whose emergent parts later became New Zealand and New Caledonia, was far enough north to escape freezing, but as its isolation lengthened, its residents lost contact with the rest of the world. It began to develop its own independent identity as a completely separate world. Earth movements continued to stretch and thin its crustal rocks, and over time, the land surface lost buoyancy and gradually sank. Space for land species diminished, but the cool shallow seas surrounding the remaining coasts provided rich feeding for a long succession of endemic seabirds and marine mammals. At least three species of huge penguins (and others known, but still undescribed), plus many other seabirds, also evolved in this area, illustrating the importance of coastal Zealandia to breeding marine species, including some that were much larger than any known today.[13]

The bits of Zealandia that remained above sea level have varied in extent over time, but they were always thickly forested. Diminishing numbers of surviving representatives of the ancient Gondwanan fauna and flora were joined by increasing proportions of immigrants. Zealandia, then as now, lies down-wind from the teeming continent of Australia/Antarctica, so most were borne across the Tasman by the prevailing westerly winds and currents—including the ancestral moa and kiwi.

Over the next 30 million years, the continuing dynamic processes of evolution in isolation welded together the ancient and immigrant elements into a Zealandian fauna distinctly different from that of Gondwana. Many of the most ancient elements are hugely important, because they have survived alone in Zealandia as living representatives of the early stages of their evolutionary lineages no longer surviving elsewhere. For example, the New Zealand wrens and frogs are the most ancient members of their groups known.[14]

But by about 25 mya, the land area available was not only totally isolated, but also getting smaller and smaller as the sea encroached onto it. If there had not been enough land still above water to support a variety of representatives of the Zealandian land fauna, we would never have known of them. They included both small vertebrates (frogs, tuatara, wrens, and short-tailed bats), a thriving community of invertebrates in leaf-litter and soil, and some unique fluorescent freshwater limpets with South American relatives.

The Late Tertiary Fauna (25 Million–11,700 Years Ago)

The third volume of the story began after the end of the Oligocene period, when renewed earth movements along the Alpine Fault (Fig. 1.1) and volcanic activity started a mountain-building phase, which reversed the drowning trend and created not only the Southern Alps, but also most of what we now recognise as the modern New Zealand archipelago. The result of those processes—still ongoing—is an unstable land surface still comprising only a fraction of the total area of Zealandia, and the only visible part, because most of it remains under water.

Over the last 20 million years, the re-emerging land has been colonised with more new immigrants, even as changing sea-levels and tectonic movements have frequently reorganised the geography of the islands. The most distinct vertebrate members of faunas of modern New Zealand are descended from either a few dogged Gondwanan survivors released from the Oligocene bottleneck, or from ancestors that have flown or swum across the Tasman since then.

For example, old assumptions that the whole known range of moa, kiwi, wrens and other strongly endemic birds must have lived aboard 'moa's ark' since it separated from Australia have been blown apart by recent DNA analyses (For more details on how this technique works, see Chap. 3. But note that molecular divergence data are often debatable and subject to modification under new techniques). These studies show that the Holocene diversity of moa dates back not more than 25 million years[15]—that is, since our islands began to emerge from their drowning episode. Over time, the dynamic processes of natural selection and of long-term species turnover have continued to sift through both the oldest survivors and the newest arrivals in ways that we are only just beginning to understand.

During the Miocene period, roughly in the middle of the third phase at some 16–19 mya, the southern South Island supported tropical forests filled with a huge variety of wildlife. New excavations by Trevor Worthy and his colleagues in an old lake bed near St Bathans, in Otago, have found fossil fragments of a rich assembly of invertebrates, fish, frogs, reptiles (including the tuatara, a tortoise and a freshwater crocodile), mammals (several representatives of the known short-tailed bats, a newly described extinct bat, plus several others still undescribed). Plant fossils are also abundant, showing that eucalypts, acacias and many other species we think of as typical of Australia were once at home here. The most significant finds were three fragments of a hitherto unknown archaic mammal (not an ordinary mouse, or a marsupial; indeed, it is apparently unrelated to any known group). Most wonderful of all, there were at least 40 species of birds, including a tiny kiwi, small enough to represent the first flying ancestors of kiwi to reach Zealandia,[16] and many waterfowl. The riotous scene on and around the ancient lake is recreated in a detailed painting reproduced in the New Zealand Geographic.[17] Abundant vegetation covered the land surrounding the lake, supporting flocks of warmth-loving parrots (one, a real giant), and fruit-eating pigeons, all now extinct.

All the ancient endemic New Zealand groups are represented at St Bathans, all endemic at very high levels and recognisable but still not fully modern species, showing that they were not recent arrivals. As Trevor Worthy points out, if all these had arrived on a naked land surface newly emerged since the end of the Oligocene drowning, the St Bathans fauna must have been assembled and adapted to local conditions within 3–4 million years, and yet, had added no more significant groups in the following 16 million years. This seems less likely than the simpler explanation, that the Oligocene drowning did not quite cover the whole of Zealandia. Nevertheless, when the land surface reached its minimal area, there was much less room for the most ancient of terrestrial survivors, especially the largest individuals. The genuine Gondwanan ghosts that remain are often identifiable only from their genetic lineages (fossils are simply too few)—but there are probably more than we know of now since most New Zealand species have never been genetically sequenced.[18]

Twenty million years later, during the Pleistocene ice ages, the rich tropical habitat represented at St Bathans had become a cold, arid grassland. With no northern refuges within reach, the tropical rainforest was swept away by freezing winds, and the hills were dissected by bone-chilling rivers fed from mountain glaciers. These conditions were fatal for warmth-loving birds and reptiles, but at the same time, favoured the appearance of hardier forms, including former forest lineages that became adapted to wide treeless grasslands and alpine tops, such as the takahē, the rock wren and the kea. Fossils and genetic analyses can now tell us much about the surprising number of warm-weather inhabitants of old Zealandia that have died out and about their tougher replacements.

The Australian mainland has not suffered such large-scale fragmentation and isolation, so the huge variety of extinct animals represented there can give us a backwards glimpse of some of the species that were probably also once part of the Gondwanan phase of New Zealand's wildlife. Like the Miocene-age community at St Bathans, a rich fossil site of about the same age at Riversleigh, a remote cattle station in Queensland, has yielded beds of fossils preserving the remains of a tropical forest fauna. Both sites supply abundant evidence of a diverse assemblage of birds dependent on a now long-vanished lush forest environment.[19]

The Holocene Fauna (11,700 Years Ago to the Present)

These successive changes of scenery through the sequence of acts of this long-running drama, and the associated entries and exits of the players, remind us of an important principle relevant to modern conservation policy. There is not, and never has been, a fixed 'native fauna of New Zealand', a past assemblage of species that could remain stable in its isolation from human activity until less than a thousand years ago. On the contrary, the wildlife of New Zealand has continuously evolved as a dynamic mixture of survivors from the Gondwanan or Zealandian years, plus immigrants arriving throughout those years and since, all adjusting to each other and to the changing geography of life on the shaky edges of the planet.

Restless changes in sea levels have caused intermittent joinings and separations of the islands of the archipelago. The number of surviving pre-Holocene species was probably cut down by periodic rises in sea level reducing the areas of their terrestrial refuges, since the smaller an island, the fewer species can survive on it. At the same time, isolated populations soon become distinct, so over the long term, these changes in sea level tended to help create new species. Rising sea levels encouraged the splitting of previously widespread populations into daughter species that could no longer interbreed if they met again. Falling sea levels often allowed dispersal across dry land bridges—that was how moa got to the North Island from their centre of origin in the South Island.[20]

Those few land animals representing ancient lineages that have maintained a foothold for generations on Zealandia's unstable land surface are mostly small (including the tiny endemic frogs, lizards and dozens of unique invertebrates) because even the smallest remnants of land were enough to let them survive the most extensive drowning episodes. Likewise, the tropical species that inhabited Otago during the Miocene were the first to be affected as the world cooled down towards the glacial periods, but even then, the hardier forest species and their resident animals could survive if they could reach the cool temperate forest that remained in the far north of the North Island, a long finger of land reaching towards the Equator.

Conversely, new immigrants have constantly appeared to fill the gaps, mostly from Australia. Those most likely to arrive were those that could swim or fly across the Tasman Sea (insects, birds, bats, seals and sea lions, whales and dolphins). Those most likely to settle were the ones that lived in mixed-gender groups, and which could compete for living space with the established native species. That was hard for Australian land birds unused to the thick native forests of New Zealand already stocked with full complements of forest-specialists. Australian birds often cross the Tasman as vagrants, but for long stretches of the Tertiary period, few of them managed to settle in New Zealand.

Then, at the onset of the Pleistocene glacial cycles around 2.5 mya, much of the forest was cleared away by massive upheavals in landscape, vegetation and climate. Suddenly, New Zealand offered any passing vagrants an attractive new home on vast swathes of open country, largely deserted by their former forest-adapted inhabitants. These newly created niches suited Australian birds well, so the rate of establishment of new bird species, arriving as already pre-adapted to treeless habitats, increased until the postglacial recovery of the forests. Ironically, history is now repeating itself. Human destruction of at least 30% of the Holocene forest that formerly covered 85–90% of New Zealand has been followed by a new influx of Australian birds of open country.[21]

Even so, none of the immigrants included representatives of the non-flying, non-swimming terrestrial mammals and reptiles that were evolving in the rest of the world. Among the most significant absentees were three families of true carnivores (the dogs, cats and weasels) and the venomous snakes. Constrictor snakes were present in Australia before Zealandia cast off, so in theory, their fossils could yet be found here. Crocodiles were in Otago during the Miocene.

The main features of the modern landscape of New Zealand are surprisingly recent, and on a geological time scale, frequently very violent. The Southern Alps, reaching to 3724 m at the present peak of Mount Cook-Aoraki, did not exist 5 mya. Uplift along the Alpine Fault (Fig. 1.1) has totalled about 20 vertical kilometres so far, and is still continuing, at about 10 mm a year: even now, the ranges are so young and fragile, their tops are being worn down by torrential rainstorms even as they are lifted up.[22] Occasional ruptures along a network of faults along the boundary between two tectonic plates, marked by the Alpine Fault and its many parallels, can trigger sometimes catastrophic earthquakes—most recently at Kaikoura in 2016.[23]

One of the most formidable volcanoes in the southern hemisphere, the Taupo caldera, produced the most violent volcanic eruption known in the world over the last 5000 years, in 232 AD. It devastated the central and eastern North Island with ignimbrite and ash fall, and remains a major threat because it is well overdue for another eruption.[24] Ruapehu, the currently most active volcano in the central North Island, began erupting less than 0.5 mya, and has produced at least minor performances some 60 times since 1945 alone.[25] Living and fossil genetic representatives of ancient groups, including many insects plus kiwi and frogs, which once lived side by side in the southern South Island, now live up to 400 km apart, separated by massive differential horizontal movements of the landscape under them.[26]

The surprising and necessary conclusion is that most of the present land forms in such geologically active areas, with all their multitude of inhabitants, are younger than the forests that clothe them. Whatever ancient, unknown, four-footed land mammals were here in geological time, none of them, of any kind (obviously, not counting bats), were still alive by the Holocene era. Instead, the emergent surface of Zealandia became, alone among all substantial land masses on earth, a kingdom dominated by birds, lizards and invertebrates—and among them, more than anywhere else, huge flightless birds and insects.

Advantages and Perils of Flightlessness

As generations of Pacific explorers have noticed, the birds native to islands free of ground predators tend to become larger than their continental relatives, and to become flightless. The explanation for this nearly universal trend is a simple matter of cost-benefit analysis as calculated, not deliberately but by the inescapable consequences of natural selection. Individuals are in constant competition with each other for reproductive success, and those breeding adults that have even a slightly greater probability of leaving surviving offspring will be the ones whose genes will be represented in the next generation.

Flight is an effective, necessary but very expensive escape mechanism for birds living in habitats patrolled by ground predators. But on an island where flight is not needed as a safety device, parent birds can save the energy required to develop and maintain large flight muscles. Those that invest their savings into, say, better equipment for attracting mates, fighting off rivals or larger and more nutritious eggs, are the ones most likely to succeed in reproductive competition. Flightless birds can also disregard the strict weight limits that govern the body size of flying birds, so are also free to become larger and heavier—an advantage for keeping warm in a cool climate, and in competition for mates. So, those genes for smaller wing size carried by flightless parents will be the ones most likely to be passed on, to construct the same characteristic in their offspring, and in theirs. It follows inevitably that the individuals we see today are the descendants of the winners, the ones that inherited the genes providing the right combination of characteristics for that environment.

The re-emerging surface of Zealandia did indeed provide exactly the right conditions offering the energy-saving advantages of flightlessness. The transition from flying ancestors took time, of course, which is why the option has in effect been limited only to those birds whose lineages have been resident for the longest. But given enough time, remarkable transformations become possible, as is demonstrated better in Zealandia's endemic fauna than anywhere else on earth. The ancestors of some, including the moa and the kiwi, arrived many millions of years ago, and now retain few clues as to what their flighted ancestors looked like. By contrast, the ancestry of the takahē is pretty well represented by both the agile eastern swamp hen of Australia and by the latest arrivals in New Zealand from the same source, the pūkeko. These slim, fully flighted trans-Tasman adventurers have arrived several times during the recent glacial periods, and took less than a million years to transition into two species of larger, fatter, flightless versions of themselves, restricted to waddling across the open grasslands.

The energy-saving strategy of flightlessness was a huge advantage to the long-resident birds of Zealandia, but it had two serious drawbacks. First, it offered no protection against the *native* predators— the harriers, falcons, eagles and owls that hunted adult birds from the air, and tuatara, geckoes and skinks that could easily clean up a nest of small eggs. So, archaic Zealandia was certainly *not* predator-free: the native birds were well used to the risk of sudden death or the loss of their young, but most of the predators they knew hunted by sight during the day. So, their best anti-predator defences were, besides their large body size, to make themselves invisible by day, or to be active only at night. The kākāpō has mossy green plumage that makes it nearly impossible to see when sitting motionless even in plain sight; the semi-flighted North Island kōkākō builds its treetop nests hidden under a thick canopy of vegetation; the kākāpō and kiwi nest on the ground and are both nocturnal and very noisy, but are far too large for any owl to tackle. All had become dangerously naïve in the face of unexpected encounters with sharp-eared, four-footed competitors and predators who followed their noses by night.

Second, flightlessness and large body size together encouraged the endemic species of Zealandia to become specialists in a slow, low-energy lifestyle that minimised both fertility and mortality rates. The energy equations controlling life on remote, secure islands that gave the long-resident native birds such valuable savings remained the best option in a live-long, breed slowly environment. The largest of the islanders simply saved the energy that, in continental species, is invested in producing large numbers of short-lived young, and used it to produce fewer, long-lived and well-educated young. They had no experience of having to defend themselves against organised, well-armed human hunters and their dogs, so remained tame and approachable, completely unaware of the potential danger.

So, isolation is the engine of variety, but it is also very risky. The longer it continues, the greater the loss in preparedness for change, and the more drastic the consequences when isolation breaks down. Island faunas are tragically vulnerable to invasion from outside. Wherever exotic animals have burst in upon previously untouched offshore islands, they have done terrible damage in the past. Untold numbers of Pacific island species have simply disappeared when their solitude was breached,[27] like cargo ships boarded by pirates.

We need to understand the histories of these invasions as a background to appreciating how alien furry predators (but not the native feathered ones) have so very recently changed the history of New Zealand's wildlife. Descriptive, factual accounts make possible comparisons with similar patterns, and speculations on the underlying causes, as this book aims to show. On the other hand, although history, like science, must be built upon facts, in much in the same way that a house is built with bricks,[28] the mere collection of facts is no closer to real history than a pile of bricks is a house. The bricks need to be built up, in the right places and in the right order, and with all the right accompanying architectural information and clearly identified assumptions. Hence, we need to reconstruct the whole story from the beginning, including all the previous acts in the long saga of New Zealand's wildlife, in order to reveal in full context the dramatic impact of the first human actors to burst upon the stage.

Summary

New Zealand is the only emergent part of Zealandia, an isolated fragment of Gondwanaland, on which a historic series of three endemic faunas can be distinguished: (1) From 82 to 55 mya, a subset of Gondwana species survived, including at least six species of dinosaurs, at least one crocodile and proto-mammals. (2) From 55 to 25 mya, a distinctly new endemic fauna was formed from new arrivals integrated with ancient survivors, as Zealandia rafted eastwards and gradually sank into an archipelago of low-lying islands. (3) From 25 mya to the end of the Pleistocene, earth movements and volcanic activity created the mountainous geography of modern New Zealand. Key fossil sites of Miocene age (16 mya) preserve samples of tropical forest and fauna now long vanished. New alpine habitats <5 million years old are populated with endemic species evolved from forest ancestors as the land uplifted beneath them. In the absence of any terrestrial mammals or snakes, flightlessness and large size evolved several times independently, associated with slow breeding and low mortality. The history of New Zealand explains why the endemic fauna were so uniquely vulnerable to the arrival of human and mammalian predators.

Notes

1. Diamond, J. M. (1990). New Zealand as an archipelago: An international perspective. In Ecological restoration in New Zealand islands (pp. 3–8). Auckland: Department of Conservation.
2. Darwin, C. (1845). Journal of Researches into the Natural History and Geology of the Countries Visited during the Voyage Round the World of HMS Beagle under the Command of Captain Fitzroy RN. London: John Murray (Reprinted by Folio Society, London, 2003). Pp 392–93.
3. King, C., Lomolino, M., Roemer, G., & Godley, B. (2013). Habitat case studies: Islands. In D. W. Macdonald & K. J. Willis (Eds.), Key Topics in Conservation Biology 2 (pp. 204–221). Oxford: John Wiley & Sons.
4. Diamond, J. M. (1990).
5. Archer, M., Hand, S. J., & Godthelp, H. (1991). Riversleigh: The story of animals in ancient rainforests of Australia. Balgowiah, NSW: Reed Books Pty.
6. Campbell, H., & Hutching, G. (2011). In search of ancient New Zealand. Auckland: Penguin and GNS Science.
7. Bunce, M., Szulkin, M., Lerner, H. R. L., Barnes, I., Shapiro, B., Cooper, A., & Holdaway, R. N. (2005). Ancient DNA provides new insights into the evolutionary history of New Zealand's extinct giant eagle. PLOS Biologia, 3, e9.
8. Campbell, H., & Hutching, G. (2011).
9. Gibbs, G. W. (2006). Ghosts of Gondwana. Nelson: Craig Potton Publishing.
10. Campbell & Hutching, 2011:124.
11. Worthy, T. H., de Pietri, V. L., & Schofield, R. P. (2017). Recent advances in avian palaeobiology in New Zealand with implications for understanding New Zealand's geological, climatic and evolutionary histories. New Zealand Journal of Zoology, 44, 177–211.
12. Worthy et al. (2017).
13. Mayr, G., Scofield, R. P., De Pietri, V. L., & Tennyson, A. J. D. (2017). A Paleocene penguin from New Zealand substantiates multiple origins of gigantism in fossil Sphenisciformes. Nature Communications, 8(1), 1927.
14. Worthy, T. H., Hand, S. J., Nguyen, J. M. T., Tennyson, A. J. D., et al. (2010). Biogeographical and phylogenetic implications of an early Miocene wren (Aves: Passeriformes: Acanthisittidae) from New Zealand. Journal of Vertebrate Paleontology, 30(2), 479–498.
15. Bunce, M., Worthy, T. H., Phillips, M. J., Holdaway, R. N., et al. (2009). The evolutionary history of the extinct ratite moa and New Zealand Neogene paleogeography. Proceedings of the National Academy of Sciences, 106, 20646–20651.
16. Worthy, T. H., de Pietri, V. L., & Schofield, R. P. (2017).

17. Worthy, T. H., & Jones, J. (2011). Deep Secrets: Discovering New Zealand's tropical past. New Zealand Geographic, 107, 44–55.
18. Gibbs (2006) p.199.
19. Archer et al. (1991).
20. Bunce et al. (2009).
21. Rawlence, N. J., Scofield, R. P., McGlone, M. S., & Knapp M. (2019). History Repeats: Large Scale Synchronous Biological Turnover in Avifauna From the Plio-Pleistocene and Late Holocene of New Zealand. Frontiers in Ecology and Evolution, 7, 158. doi: https://doi.org/10.3389/fevo.2019.00158
22. Coates, G., & Cox, G. J. (2002). The rise and fall of the Southern Alps. Christchurch: Canterbury University Press.
23. Duputel, Z., & Rivera, L. (2017). Long-period analysis of the 2016 Kaikoura earthquake. Physics of the Earth and Planetary Interiors, 265, 62–66.
24. Wilson, C. J. N., & Walker, G. P. L. (1985). The Taupo eruption, New Zealand. I. General aspects. Philosophical Transactions of the Royal Society, London A, 314, 199–228.
25. Donoghue, S., Neall, V., Palmer, A., & Stewart, R. (1997). The volcanic history of Ruapehu during the past 2 millennia based on the record of Tufa Trig tephras. Bulletin of Volcanology, 59, 136–146.
26. Gibbs (2006).
27. King, W. B. (1973). Conservation status of birds of central Pacific islands. Wilson Bulletin, 85, 89–103.
28. Poincaré, H. (1952). Science and Hypothesis. New York: Dover Publications.

CHAPTER 2

The Māori Era (1280–1769)

The Māori, with great skill, made river, sea, and bush minister to his necessities. He lived well where the average European would have starved. T. W. Downes (1928)[1]

Aotearoa Discovered by People

At the beginning of the thirteenth century, a mere 800 years ago, there were no people in this part of the Southwest Pacific. The New Zealand archipelago was the last substantial habitable land mass in the world unknown to any human explorers, because its isolation from inhabited lands (1635 km east of the closest point on mainland Australia, 1485 km from the nearest coast of Tasmania, 1500 km south of New Caledonia, and 9300 km west of South America) had shielded it from human discovery for so long. It also included the largest by far of the Pacific islands, and the coldest, but their sandy shores were untouched by any human footprint.

When eventually the first people arrived from eastern Polynesia[2] in about AD 1280,[3] birds and lizards were still the dominant form of life in or around New Zealand. The only land mammals to be seen were the thousands of bats pouring out of caves and tree roosts, while seasonal

C. M. King, *Invasive Predators in New Zealand*,
Palgrave Studies in World Environmental History,
https://doi.org/10.1007/978-3-030-32138-3_2

assemblages of breeding fur seals and sea lions gathered in noisy crowds on rocky shores and beaches, even venturing far inland, and dolphins and whales romped offshore in their thousands.

The Polynesian voyagers came in huge double-hulled canoes, carrying all the equipment needed to colonise a new land, including plants and seeds, tools and fishing gear, their women, children and wider families, and two of their familiar animals—rats and dogs. In Rarotonga, the largest island of the southern Cooks group, there is a memorial to the pioneering group of canoes, which tradition says left from there. They named their new home Aotearoa.

The Pacific rats and dogs they brought were the very first four-footed animals to set paw on New Zealand shores since long before the ice ages, and they were the only two resident land-breeding mammal species for the next 500 years other than bats and seals. Such a paucity of furred inhabitants astonished the early European observers, used as they were to the variety of animals native to their home countries. As Joseph Banks commented in his journal, written in early 1770 during James Cook's first visit to New Zealand,

> In all the times that we have landed in this Countrey, we have seen, I had almost said, no Quadrupeds really original natives of it: Dogs & Rats indeed there are; the former, as in other Countries, companions of the Men, & the latter, probably brought hither by the Men, especially as they are so scarce, that I myself have not had an opportunity of seeing even one … It appears not improbable that there really are no other species of Quadrupeds in the Countrey, for the Natives, whose cheif luxury of dress consists in the skins & hair of Dogs, & the Skins of divers Birds … would probably have made use of some part of any other animal they were acquainted with; a circumstance which tho' we carefully sought after, we never saw the least signs of.[4]

Inevitably, the arrival of human hunters and their predatory companions onto this previously untouched land disrupted the settled communities of resident fauna. The detailed consequences went undocumented at the time, but in recent years, have been deduced from multiple clues provided by radiocarbon dating, genetics, archaeological digs and biogeographic comparisons.

Pacific Rat, Kiore

The Pacific rat (kiore in Māori: see p. xxii) originally came from Southeast Asia, but has been nearly ubiquitous around Polynesian settlements for many centuries. Polynesian explorers preparing for a long journey brought their large double-hulled voyaging canoes close inshore while stocking them with provisions. Opinions are still divided as to whether the rats were deliberately loaded onto the canoes or simply boarded them as stowaways. These little rats are not great swimmers, but it seems unlikely that canoes being loaded with stacks of edible supplies while anchored close to a rat-infested island could always have been kept clear of uninvited boarders. Either way, rats have made the journey across vast stretches of ocean often enough, and in enough numbers, to establish new populations on many islands, including the main and more than 50 offshore islands of New Zealand.[5]

The Māori thought of themselves as part of nature, siblings with the native species as the offspring of Tane, the spirit of the forest. One beautiful illustration of this attitude can be found as a carving at the entrance of the Whanganui National Park (Fig. 2.1). The carver, Maramarua Cribb, has shown Tane cuddling the kiore affectionately under his arm, with the kiore's front paw resting on his finger.

Fig. 2.1 Carving at the entrance to Whanganui National Park, showing Tane, the god of the forest, with one of his children, a kiore. Carved in the early 1980s by Maramarua Cribb. (Photograph by Author, reproduced by permission of Maramarua's granddaughter Miriama)

In addition, to some Māori tribes (though not all), the kiore is or was valued as a taonga (tribal treasure) because it was one of very few sources of fresh red meat available in precolonial times. Whenever the beech forest trees clothing vast areas of mountain country staged a mass heavy seed fall, kiore populations irrupted, just as wild house mice now do the same (Chap. 5). Māori hunters were quick to harvest a bonanza. One such event in 1884–85 was described in detail:

> The magnitude of the plague is the subject of ordinary conversation … Living rats are sneaking in every corner, scuttling across every path; their dead bodies … strew the roads, field and gardens, pollute the wells and streams in all directions.[6]

Nowadays, the kiore offers a different sort of treasure; it has become a mine of information for deducing the history of human colonisation of the Pacific islands. Kiore have proved an invaluable tool for the analysis of the many different and informative signs of their presence. With their help, archaeologists and geneticists armed with the latest analytical methods have finally settled long-standing arguments about the prehistory of the Pacific.

Kiore are especially useful for this purpose, because they certainly could not have travelled across the Pacific independently. More importantly, they were not only the earliest by far of the three species of rats to reach New Zealand with their human companions, but also, they do not interbreed with other rats, so (unlike the dogs, chickens and pigs also transported by people around the Pacific) their genetic footprints are not confused with those of any later-imported European species. DNA analyses (Chap. 3) show that they came from somewhere in eastern Polynesia, and the kiore living today are the direct descendants of those ancestral colonists.[7]

The technique of radiocarbon dating applied to organic materials excavated from well-dated sites shows without any doubt that there were no rats available to leave bones, chewed seed cases or broken snail shells anywhere around New Zealand before the late thirteenth century. Rat bones are sometimes preserved in cave deposits and middens, but the samples are rather small and not always reliable. By contrast, plants with dateable woody seed cases large enough to preserve rat tooth marks are widespread around the Pacific, and have become an especially valuable and sensitive method of detecting the earliest arrival of rats on a previously rat-free

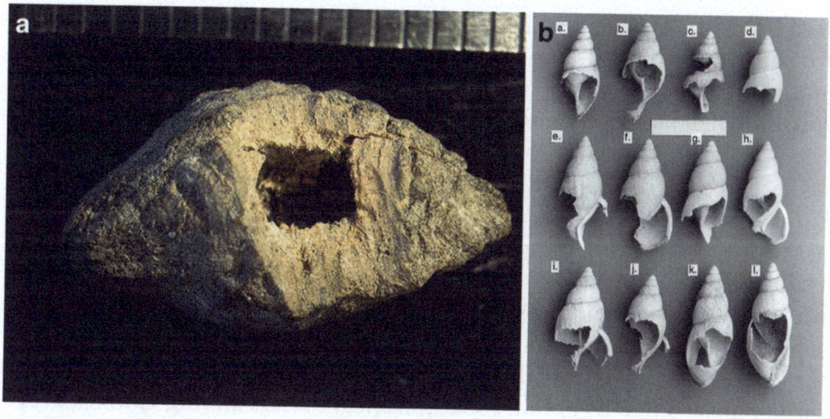

Fig. 2.2 Rat-feeding signs, broken moa egg and snail shells, and pollen deposits all agree on dating the first arrival of human colonists to 1280–1282 AD. Shown here are (**a**), the distinctive kiore incisor bite marks on seeds of hinau (length 16 mm) and (**b**), shells of flax snails, (d) and (h) damaged by kiore, others by predatory birds. (Photographs by Janet Wilmshurst (**a**) and Fred Brook (**b**))

island.[8] In New Zealand, these tooth marks can be found only in deposits laid down later than about 700 years ago. Snail shells with characteristic rat-gnawing marks date from about the same time (Fig. 2.2).

The Polynesians and their companion animals certainly arrived in New Zealand together at about the same time. The dates derived from archaic Polynesian graves, and occupation levels at the oldest archaeological sites, provide firm evidence of the first human settlements, all in close agreement with the dates from natural sites signalling the first arrival of rats.

Like other rats, kiore have had a massive effect on the original fauna and flora of all the islands they visited, although that was not obvious for a long while.[9] They were once assumed to be harmless vegetarians, but there is no doubt now that their arrival was disastrous for dozens of small native species that once lived in safety on the ground. The effect is best seen now by comparing the wildlife of offshore islands with or without kiore, and the enormous benefits of removing them, as illustrated in startling detail by recent eradication experiments.

For example, on islands that are or have been occupied by kiore, the destruction of fruits and seedlings of forest trees removes vital foods of birds, and changes the composition of the canopy in ways that disadvantage birds. David Towns lists at least 15 species of invertebrates, two species of frogs, juvenile tuatara, and 11 species of lizards (geckoes and skinks) as among the species most badly affected by kiore predation, and which recover when kiore are removed. At least ten species of small seabirds have become locally or totally extinct on islands invaded by kiore, and at least 13 species of terrestrial birds disappeared from the mainland soon after the kiore arrived.[10]

An even more significant story emerged on Little Barrier Island-Te Hauturu o Toi, a relatively undisturbed island of 2817 ha, and one of the very first national wildlife sanctuaries. Kiore had been on the island for centuries, and feral cats since about 1867 (Chap. 6). Both contributed to the local extinction of North Island saddlebacks and the total extinction of Little Barrier snipe by about 1882, plus heavy losses among Cook's, grey-faced and black petrels, as well as bush birds, tuatara and lizards. Removing the cats in 1980 showed only, to everyone's surprise, that the cats had been actually protecting the birds' nests from kiore. With cats gone, seabird fledging success dropped dramatically, until the kiore were eradicated too—then, and only then, did fledging success increase.[11] (The eradication programme was complicated by cultural differences in opinion over whether the kiore should be eradicated, like any other rat, from this and other offshore islands that are needed to be clear of all introduced predators in order to create sanctuaries for threatened native birds. With goodwill and sensitive negotiation, a solution was found.) The disappearance of the kiore also benefitted the local giant wētā, whose population doubled over the next five years. And since then, an hitherto invisible population of the New Zealand storm petrel, first and last seen by Captain Cook's crew in 1769, has suddenly bred up enough to become visible and is now present in the hundreds of individuals.

Polynesian Dog, Kuri

Of the three domestic animals found on almost all the Polynesian islands (pigs, chickens and dogs), only dogs were successfully brought to Aotearoa/New Zealand. The Polynesian dog, the kuri, was described as a small, short-legged breed somewhat resembling the 'fox in shape, but variable in the colour',[12] with a broad head, pricked ears and a bushy tail. In its time, the kuri was much valued by Māori; they were kept as pets and fed on human food, including a lot of fish. Kuri were much larger than rats, so were a welcome source of (non-human) meat, but it was reserved for very high-status diners on special occasions.

Kuri were useful watchdogs; they could not bark, but had no difficulty in announcing the approach of strangers. Ferdinand von Hochstetter, an Austrian geologist who travelled widely in the North Island in 1859, described the reaction as he entered a remote Māori village late one night: 'The dogs hailed our arrival with a perfect jackal's howl'.[13] The skins and fur of kuri were used to make cloaks and to decorate weapons,[14] and their bones and teeth were used for fish hooks and necklaces.[15]

Kuri were also trained as hunting dogs. Richard Cruise, a military officer aboard *HMS Dromedary* sent to the Bay of Islands in 1822 to collect kauri spars, met a Māori chief wearing an 'emu-feathered mat'. Understandably, Cruise did not recognise the kiwi, which is much smaller than an emu but has the same loose, shaggy plumage. Unlike the emu, the kiwi is nocturnal, which accounts for Cruise's comment in a footnote that 'the natives go out after dusk, with lights, which attract their attention, and they kill them with dogs'.[16] The German naturalist Ernst Dieffenbach thought the kuri relatively harmless, not a sufficiently powerful animal to do harm to domestic sheep, so 'In want of better sport they hunt young birds, and to this cause the scarcity of many indigenous birds must be ascribed'.[17]

By contrast, Dieffenbach added, the dogs needed to help in hunting for larger game were 'introduced and mongrel dogs, mostly bull-terriers or bloodhounds, which are savage pig-dogs'. The artist Georg Angas observed in 1844 that 'Every pah abounds with dogs, which are used principally to hunt the wild pigs that run loose in the woods'.[18]

The kuri was fully inter-fertile with all of the many breeds of dogs that accompanied the European settlers, so the swamping of purebred kuri genes has been both rapid and irresistible. Further, the original genetic diversity of kuri was very low, showing that only a few dogs had been brought from the Pacific islands, or the source population also was limited in diversity, or those that came were closely related. By the 1840s, even so acute an observer as Dieffenbach commented sadly that 'It [the kuri] is now rarely met with, as almost the whole race of the island has become a mongrel breed'.[19]

The Fourteenth-Century Transformation

Polynesian settlement set off a disturbance unlike any other, an extraordinary, immediate and rapid transformation of the late pre-human environment of Aotearoa, whose effects were felt—first and worst—by the irreplaceable endemic species. Of course, non-endemic native species are valuable too, because they belong to New Zealand, by ancestry or independent colonisation, and damage to them is certainly to be prevented wherever possible, but they also live in other countries.

Extinction in such circumstances is not random: it acts rather like a catastrophic epidemic of plague, which first removes the oldest and most vulnerable. The only small consolation is that many of the uninhabited, remoter islands of the archipelago and their fragile cargoes completely escaped the most serious consequences of the Polynesian transformation.

Endemic species are found nowhere else in the world, so are ranked at the highest step in conservation value. A species can be endemic to various extents. If its members are only slightly different from their nearest relatives elsewhere, they may be endemic only at the subspecies or species level. The more different they are (which usually means, the longer they have lived isolated from their relatives), the higher up the taxonomic scale their species is placed, through genera to family level or above.

The highest level of distinction possible is that of an endemic Order, and New Zealand has three of these, represented by the five species of kiwi, the nine extinct moa and the two living tuatara. The wrens (rock wren and rifleman) have recently moved up from Family to Subordinal level. Below them are nine groups of vertebrates and five of invertebrates classed as endemic at Family level, including the wattlebirds (kōkākō and saddlebacks), the mohua, the four living native frogs plus three others now extinct, and the two short-tailed bats. The tūī and Hooker's sea lion are endemic at Genus level. The black stilt, the white heron, the NZ long-tailed bat, the NZ fur seal, and the Hector's and Maui's dolphins are among more than 200 vertebrates and 20,000+ invertebrates endemic at Species level.[20] More recent arrivals, such as the kingfisher and the fantail, have few differences from their (mostly Australian) relatives, so qualify as endemic only at Subspecies level. The most recent, such as the welcome swallow and the spur-winged plover—established and spread within the last 50 years—are identical with their relatives overseas. Some of the endemic species listed here are shown in a reconstruction of the pre-human forest in Fig. 2.3. (For species lists, see Appendix).

Differences in degree of endemic status are important, because those species that are endemic at a higher level are nearly always the most vulnerable to a new disturbance, compared with species barely distinguishable from their relatives overseas. The longest residents had the advantage over their competitors, but only for as long as the islands remained remote and inviolate. Such isolation could not last, and eventually, the outside world broke in.

Fig. 2.3 An imaginary scene in a North Island forest a few centuries before humans arrived. Left to right, above: red-crowned parakeet, morepork, NZ pigeon (kererū), **piopio (native thrush), *North Island saddleback, *North Island kōkākō, grey warbler, short-tailed bat, harrier hawk, kākā. Left to right, below: brown kiwi, **bush wren, *Hochstetter's frog, *kākāpō, *snipe, *tuatara, *blue duck, **giant moa, weka, forest gecko, **laughing owl, **small moa, **stout-legged wren, *little spotted kiwi, **owlet-nightjar. Species marked with one asterisk are now rare or threatened, and survive only under Department of Conservation protection or on offshore islands; those with two asterisks are totally extinct. (Modified by Ellen Clarkson from a painting by Pauline Morse for Wilson, K.-J. (2004). Flight of the Huia. Christchurch: Canterbury University Press. For scientific names, see Appendix)

By the mid-1700s, the Zealandian fauna had been through 500 years of massive changes (Chap. 1). The total land area of native forest had been reduced from roughly 21 million hectares to about 14 million—mainly the drier, eastern regions—by huge, uncontrollable fires set off by human hunters.[21] (The actual areas, like the numbers of native species, depend on definitions, which vary between sources.) Scorched stumps and layers of charcoal under the open grasslands of Canterbury and Otago still speak of widespread burning of the formerly huge forests that could not later regenerate. Within a few centuries, the charred remains had been converted into vast, natural-looking open tussock grasslands, the last large unclaimed pastoral resource in the world (Chap. 7).

Hunted by Māori, then by whalers and sealers, many coastal breeding rookeries of fur seals, sea lions and penguins,[22] and formerly teeming colonies of seabirds breeding on inland mountains, became deserted. Tuatara and lizards, which once abounded under every mainland rock or bush, could survive in their former numbers only on inaccessible offshore islands. And, most extraordinary of all, there were the sudden mass extinctions of all the largest and meatiest flightless birds.

Losses of Endemic Birds

How many birds were present on the New Zealand islands at the climax of their evolution in isolation, just before the first human contact, and how many of them were unique to those islands? Certainly, the scene would have been very different from that which can be reconstructed from Miocene-era fossils. But from later subfossils and genetic analyses, some idea of what a forest community might have looked like by about 20,000 years ago is illustrated in Fig. 2.3.

On the other hand, asking for the actual numbers of extinct versus living birds is a simple question to which there is no simple answer, because the numbers depend on which islands are included, the dates covered by the count and, most contentious of all, the allocation of taxonomic names and endemic status to the whole range of recognisable birds differing to debatable degrees from their relatives. Furthermore, authors differ as to whether to include the (at least) 16 species of European-era unaided colonists, while the known extinctions over the millions of years since the Miocene era (Fig. 1.3) would have been caused by many different factors, especially changes in climate and habitat. Compare the following:

> In 2001, Richard Holdaway and colleagues came up with a total of 245 species in 110 genera, representing 46 families spread across the entire archipelago from the subtropical Kermadec group to the subantarctic islands, including Norfolk I. as part of the NZ geographic region, counting only those species present during the Holocene (the present geological epoch, starting after the end of the most recent glaciation roughly 10,000 years

ago). Of the 245 species, only 169 survive (69% extinct); 171 of the total were ranked as breeding residents endemic at least at subspecies level.[23]

In 2010, Alan Tennyson counted an original total of at least 242 birds native to New Zealand (not including Norfolk I.) during the Holocene era – not all alive at the same time – of which at least 59 (24%) are now extinct.[24]

In 2017, Trevor Worthy and colleagues, allowing for various taxonomic updates, counted 217 native breeding species (145, or 67%, of them endemic) to New Zealand excluding Norfolk I., of which 58 (25%) are extinct.[25]

The differences between these estimates do not disguise the non-random pattern of the extinctions, which becomes starkly obvious when we break them down by degree of endemism. More than three-quarters of species endemic at the ordinal level have gone, and more than half of those endemic at family or genus level; the losses decline down to a third of endemic species and a fifth of endemic subspecies. The dominance of endemic units among the extinctions reflects both the great age and size of the surviving fragments of the Gondwana continental landmass, and the distance they are from outside additions.

On the inhabited islands, most of the large endemic bird species, known (from their fossils) to have been present at first human contact, have disappeared. They included all nine species of moa, a swan, two geese, at least three ducks, a merganser, a coot, a harrier hawk, a giant eagle, a giant rail or adzebill, four or more smaller rails, five wrens, an owlet-nightjar and a raven. Others such as the takahē and some kiwi and frogs survived, but only just—they are now confined to drastically reduced distributions on the main islands, or like the tuatara, snipe and saddlebacks, on uninhabited predator-free islands only. Fossil bones preserved in caves and empty snail shells show where ancient residents were once common. Fortunately for us, their remains have stayed untouched for thousands of years, thanks to the pre-human absence of small mammalian scavengers.

All the most strongly endemic species were representatives of either the last survivors of Gondwana, or the earliest immigrants of the Zealandian era, so all were unique and, by definition, irreplaceable. As is usual for species adapted to undisturbed islands, the adults were long-lived, naïve and unwary, and bred slowly. The females usually had to survive several years before breeding for the first time, and then, they laid small clutches spread over many seasons of a long lifetime (Chap. 1).

It could be argued that many of these birds would eventually have been lost even if they were never hunted, along with huge areas of their forest habitat, because those forests were the only places that offered the living conditions to which they were so long adapted. That is not necessarily true for slow transitions like the glacial-interglacial cycles, but it is certainly true for sudden, catastrophic events like those of the fourteenth century. Even then, it is not easy to disentangle separate but interrelated triggers of a series of historical events. But a rough rule of thumb can be derived from a well-established correlation between the number of species and the area of the island or patch of habitat they live on.[26]

Plotting species number against declining area predicts that about 10% of resident forest birds will be lost for every 50% cut in forest habitat. One unpublished analysis by Phil Millener of the non-marine birds of the North Island alone reckoned that at the time the Polynesians arrived, they found about 73 species and 10.9 million hectares of forest. By the time the Europeans followed them centuries later, the birds were down by 29% to 52 species, and fire had reduced the forest by 23% to 8.4 million hectares. But according the species/area theory, a loss of 23% of forest alone should have reduced the number of birds by around 5%.[27] As calculations by rule of thumb go, this one is old and pretty rough and cannot explain exceptions like Stewart Island. (Stewart Island is still almost entirely forested, but has lost many vulnerable species. Perhaps the presence of all three invasive rats, a deadly combination of predatory talent unmatched elsewhere in New Zealand, has something to do with it.) The underlying theory has many flaws and assumptions, but no recent updates of the numbers could be large enough to challenge the general conclusion that many more unique birds were lost during the Polynesian period than can be accounted for by deforestation alone. The difference is a stark illustration of the vulnerability of the slow-breeding endemic birds to suddenly accelerated mortality.

Moa Hunting

The ancestors of the Māori came from Pacific islands where chickens were common domestic animals. 'Moa' is the usual Polynesian name for a chicken. For some reason, the colonists did not bring live chickens with them to Aotearoa, or they brought some that did not survive. The Māori had no chickens until European stocks were brought from Port Jackson by

Samuel Marsden, and even then, they were valued mostly for their feathers.[28] So, it is entirely understandable that when Polynesian hunters first encountered the huge moa of the South Island, they thought of them as super-normal chickens, and applied the same name to them. The same logic is not so obvious to a contemporary biologist: hence, I was startled to spot a roadside notice some years ago in Samoa advertising 'Moa, 1.65 per kilo'.

Joseph Banks and the many other naturalists who described in careful detail the wildlife they saw in Aotearoa from the late 1700s to the 1850s knew nothing of the moa. But in the Bay of Islands in the 1830s, there were people who preserved ancient traditions that their ancestors hunted the moa 'until it became extinct'—including even some details that suggest visual memories such that 'it had a plume of feathers on its head'.[29] The survival of this memory confirms the long-held opinion that cultural traditions can be preserved over very long periods—in this case, 300–400 years. An even more specific report, never followed up, was given by the long-resident Bay of Islands trader Joel Polack:

> That a species of the emu, or a bird of the genus Struthio, formerly existed on the [northern] island, I feel well assured, as several large fossil ossifications were shown to me when I was residing in the vicinity of the East Cape ... The natives added that, in times long past, they received the tradition, that very large birds had existed, but the scarcity of animal food, as well as the easy method of entrapping them, had caused their extermination.[30]

James Cook had formed a bad impression of that area, naming it Poverty Bay, which might explain why he did not stay long and never talked to anyone who knew of the moa.

Other Europeans to visit that area had better luck. John Harris, a young man sent by a Sydney trader in the mid-1830s to establish a supply of New Zealand flax for export, arrived at a village near what is now Gisborne (not far from the East Cape). Harris collected from local Māori a series of interesting artefacts, including a strange bone, which they told him had come from an eagle. He took it to his uncle John Rule in Sydney in 1837, and two years later, Rule took it to England to be identified by Richard Owen, the leading anatomist at the British Museum of Natural History. Owen, sceptical at first, saw that it was part of the thigh bone of a large animal. He compared it with the bones of cattle, and then of the ostrich, and finally announced in 1839 the startling news that the bone had belonged to a hitherto unknown, extremely large flightless bird.[31]

Excited palaeontologists dug up many more moa bones from both main islands over the next decades, and named up to 64 new species. Cooler and more comprehensive analyses by Trevor Worthy and others, using the sharper tools of DNA analyses, have now reduced these to nine species, belonging to six genera in three families.[32] Tens of thousands of these huge birds once lived all over the main islands, ranging from the smaller turkey-sized species to the larger versions with massive strength in their legs and long necks tipped with a sharp beak. They were formidable targets even for hunters armed with effective spears and axes. But they offered an abundant and apparently inexhaustible supply of fresh meat, easily harvested by well-organised hunting parties and cooked in underground ovens (Fig. 2.4).

Statistical models simulating the decline in number of adult female moa depended, as do all models, on some primary assumptions. These were: the initial populations of moa taken as about 160,000; clutch size 1–2, age at first breeding five years. Then, some variable assumptions were added to test what would happen if the model took the number of human colonists as 100 or 200; the rate of human population increase as 1%/yr or 2%/yr; and the number of moa killed per week as 1 kill per 10 or 20 people. The results predicted that the maximum period to extinction was no more than 160 years; the realistic period, including the independent effects on moa of forest fires and habitat loss, was less than 100 years. It is safe to assume that

Fig. 2.4 Leg bone of a moa cooked underground for an ancient feast. (From Buist, A. G., & Yaldwyn, J. C. (1960). An 'articulated' moa leg from an oven excavated at Waingongoro, south Taranaki. The Journal of the Polynesian Society, 69, 76–88)

the moa were all gone by the mid-fifteenth century.[33] In the oldest archaeological sites, layers of moa remains were gradually replaced with those of shellfish, which then became overgrown with trees dated to the 1600s.

Some authors confidently reckon that the Polynesian population could not have numbered more than a couple of thousand during the time it took to exterminate the moa.[34] How could such a comprehensive slaughter over such a large land area and in such a short time been possible for small groups of hunters with hand-held weapons? Perhaps the moa were already declining for other reasons? Maybe they were becoming inbred, or had accumulated too many deleterious mutations? But the gene pools of 281 moa belonging to the four best preserved species were large and stable right up to their demise, with no loss of genetic health over the previous 2000 years.[35] Neither had they been much affected by the comings and goings of the glaciers, except that climate changes assisted their dispersal between islands temporarily joined by periodic low sea levels. An alternative explanation is that the number of people arriving has been grossly underestimated. Rather than a few colonists who took time to build up their numbers from small beginnings, Walter et al.[36] reckon that there was a deliberate, well-planned mass migration.

It does seems certain that any widespread dispersal of large numbers of hungry hunters would have had an immediate and irreversible effect on the moa populations—not because Polynesians were any more ruthless than any other hunters, but because the birds were unable to sustain even moderate harvesting. The vast forests of the hunters' new homeland were stocked with supplies of game that must have seemed inexhaustible by comparison with the limited resources of much smaller islands. But any disturbance that increases the number of birds dying faster than they can be replaced will quickly knock down the population, especially those acutely vulnerable birds that normally live long and breed slowly. The arrival of the first human hunters had exactly that effect, not only because they killed and butchered the adult birds and chicks, but also because they took the eggs to clean out and use as water bottles—a valuable possession in a culture with no pottery. Persistent over-harvesting is also, of course, the primary aim of modern pest control (Chap. 12), but unfortunately, does not have the same effect on short-lived, fast-breeding rats as it had on the moa.

By 1450,[37] the moa were extinct in their previous strongholds in the South Island, so the increasingly hungry southern hunters turned to other game. Coastal colonies of the extinct endemic waitaha penguin, that once nested from Stewart Island all the way up to the southern end of the North Island, had all been exterminated by the late 1480s. As Māori shifted northwards, and their economy made a forced transition from hunting to kumara-growing and fishing, the related yellow-eyed penguin, previously confined to subantarctic islands, moved in to occupy the vacated southern penguin roosts from 1515 onwards.[38]

Māori hunters also made the traditional rookeries of fur seals and sea lions around the mainland coasts much less safe for breeding females and their pups than the inaccessible offshore islands. As the mainland rookeries were deserted by the surviving animals, Māori moved northwards to warmer climates more likely to support horticulture, and the human hunting pressures on the abandoned rookeries eased. In time, the original mainland sea lion lineage was replaced by a related but genetically distinct subantarctic lineage.[39] Much later in 1804–1830, these, in turn, were nearly exterminated by European sealers, until the sealing trade collapsed (Chap. 3). The surviving sea lions are protected but still vulnerable; fur seals have recovered much more rapidly since 1840 than have sea lions, and their rookeries are crowded again.

One conspicuous absentee from the roll call of Māori food supplies is the domestic pig. Pigs had provided a nearly universal staple of the diet of the Polynesian islanders since they arrived in the Pacific with the Lapita people, but pigs apparently did not reach New Zealand with the original Polynesian settlers. What a difference they would have made to Māori history! Especially as pigs were known to the ancestors of the Māori, as Cook's companion Tupaia discovered when talking to local people in December 1769. He asked them:

> if they knew of any Countreys besides this or ever went to any, they said no, but that their ancestors had told them to the NWbN or NNW was a large Countrey, to which some People had sailed in a very large Canoe which passage took them up a Month from this Expedition a part only returned who told their countreymen that they had seen a Countrey where People Eat Hogs, for which animal they used the same name (Booah) as is used in the Islands, & have you no Hogs among you said Tupia? no; & did your ancestors bring none back with them? no; you must be a parcel of Liars then said he, & your Story a great Lye, for your ancestors would have never been such fools as to come back without them.[40]

Cook, however, decided to make good this loss. By September 1769, some months into the voyage, Banks reports that Cook still had on board

'17 Sheep, 4 or 5 fowls, the same number of South Sea Hogs, 4 or 5 Ducks, and an English Boar & Sow with a litter of Pigs'. The pigs were important reserve supplies of fresh meat for the ship's company, but Banks remarks that 'in the use of these we are rather sparing, as the time of our getting a supply is rather precarious'. So, it says a lot for Cook's generosity if he was prepared to part with any of them to establish a reliable and more abundant source of meat for the Māori than the only animals they knew of, which were rats, dogs or—and then, only after a victorious battle—human enemies. All the same, Cook's pigs were less likely to be the ancestors of any later populations than were the 56 pigs sent in three shipments to the northern chief Te Pahi from Norfolk Island in 1804 and 1805. By 1847, Angas could comment that pigs brought by European settlers were being bred in great numbers throughout the country.[41]

Aotearoa at the Height of the Classic Māori Era

Over the centuries during which Māori had New Zealand to themselves, their culture evolved from its Polynesian roots into a vibrant community of fiercely independent tribes, closely adapted to the resources of their vast new land. Every corner of it—orders of magnitude more extensive and varied than any of the Pacific islands—was explored and claimed as communal property by a local tribe. They knew of every accessible source of all useful mineral deposits, such as greenstone for carving and obsidian for tools (but not gold or coal, which meant nothing to them). Locations of important fruiting trees and the seasons that these provided food for forest creatures, and the best places to gather forest fruits and to hunt birds and rats with beautifully made snares and traps, were essential parts of local knowledge passed down from one generation to the next.[42] Ropes and fishing nets were woven from natural materials gathered in the forest, fish hooks from bones, and personal ornaments from the feathers of birds and the teeth of marine mammals. In short, 'the Māori, with great skill, made river, sea, and bush minister to his necessities. He lived well where the average European would have starved'.[43]

Over the 500 years of mutual adaptation between the Māori and the native fauna, the long processes of species turnover produced a very different natural environment for the Māori from the one first encountered by their Polynesian ancestors. Endemic species that could not survive the accelerated mortality imposed by human and rat predators were gone. Those that could survive were supplemented by new arrivals, who became in turn subject to the same sifting effects of selective extinction and immigration. This new, post-human fauna was full of variety and abundance. That is very clear from the observations of the first European naturalists to see it in its fully formed state, and to record many detailed descriptions published in words that bring that lost world vividly to life.

Among the best of those records, and the most significant for the story outlined in this book, were written in March 1773 by members of James Cook's crew when they arrived, cold and exhausted after weeks in the freezing Southern Ocean, at the entrance of Dusky Sound.[44] The Fiordland coast was and is wild and remote, clothed in thick forest from the timberline to the water's edge, and indented with multiple inlets offering calm refuges from the heaving waves of the Tasman Sea. As the *Resolution* glided in to anchor, the ship's naturalist Georg Forster observed flocks of water birds on every shore, fur seals basking on the rocks, and 'the whole country resounded with the wild notes of the feathered tribe'. Lieutenant Richard Pickersgill described their position as in 'one of the most enchanting little Harbours I ever saw'… surrounded by mountains clothed with tall trees where the 'sweet swelling Notes of a number of Birds made the finest Harmony'.[45]

Everyone on board was keen to explore the fragrant forest and its abounding wildlife. The deep water of the glacier-carved Dusky Sound allowed *Resolution* to be tied up so close to the shore that its yard-arms tangled in the trees (Fig. 4.2). Cook and his surveyors spent weeks mapping the sound and its islands, creating a reliable guide to the most important and best-known anchorage in this stormy corner of the Pacific. The existence of Cook's maps and detailed descriptions guaranteed a long succession of visits by other ships.[46] The cove where they anchored was later named Pickersgill Harbour.

The first ecstatic written descriptions of the beautiful and abundant native bird life of the mid-eighteenth century, so different to what any of us can observe in the same areas today, help to emphasise an important point.

The first European explorers believed they were observing New Zealand's wildlife in its pristine state, and they assumed it had always been as they saw it then, but they were quite wrong. There had already been millions of years of intermittent transformations, driven by the restless scrutiny of natural selection interacting with a geologically young and dynamic environment. The losses of all nine species of moa, and at least 19 other large flightless birds that followed them into oblivion during the

Holocene, represented only the latest round of extinctions (admittedly, also by far the fastest). The departures of the moa and their companions did not destroy the native fauna, but they certainly precipitated further massive changes in it—on top of millions of years of previous changes. It helps to understand the context and the dynamic background of the world that was encountered by the first Europeans in the late eighteenth century, the better to appreciate the events of the nineteenth century, the focal period of our story.

The nineteenth century saw even greater changes in human culture. In 1800, the entire land surface of the main islands and many of its inshore attendants were claimed by one or other Māori territorial tribe. They had a well-established and successful material culture based on lithic tools, fire, human muscle power, gardening, wood carving, canoeing and hunting of native birds, all held under common ownership and all fiercely defended by skilled warriors led by powerful chiefs. The human population of the islands comprised some 100,000 full-blooded Māori[47] plus a tiny handful of European traders, and two-thirds of the original forest still stood.

By 1901, much of the country had been taken over by European settlers and divided into parcels of private property. They had steel tools, steam engines, horse and bullock power, heavy horticultural, agricultural and logging equipment, improved crop species, libraries, railways, cities and many introduced companion animals and livestock. Only 45,549 Māori were left, out of a total New Zealand population of 815,862, and another third of the forest was gone, with consequences that will be explored in the following chapters.

Summary

New Zealand was the last substantial land mass discovered by humans. Polynesian voyagers arrived in about AD 1280, and evolved into the Māori—a unique indigenous culture. They were accompanied by the Polynesian rat (*Rattus exulans*, Māori name kiore) and domestic dog (in Māori, kuri)—the first four-footed mammals to encounter the rich endemic fauna of pre-human times. Genetic analyses of the kiore have been instrumental in determining the history of Polynesian settlement throughout the Pacific. Human hunters assisted by fire and dogs targeted the large, meaty native birds and their eggs and chicks; kiore destroyed the fruits and seedlings of forest trees, and the smaller birds, lizards and invertebrates. The Māori settled on both the main and many offshore islands, unintentionally setting off a massive transformation of the New Zealand environment. Of 217 known species of native breeding birds (not all living at the same time), 145 (67%) are listed as endemic, and 57 (25%) as extinct. Offshore islands provided refuges for many other forest birds, penguins and pinnipeds. The most immediate, inevitable losses rapidly removed those most vulnerable to over-harvesting, including all nine species of moa in <160 years. Extinctions during Māori times must be understood against this biological background.

Notes

1. Downes, T. W. (1928). Bird-snaring, etc., in the Whanganui River District. Journal of the Polynesian Society, 37, 1–29.
2. Wilmshurst, J. M., Hunt, T. L., Lipo, C. P., & Anderson, A. J. (2011). High-precision radiocarbon dating shows recent and rapid initial human colonization of East Polynesia. Proceedings of the National Academy of Sciences, 108(5), 1815–1820.
3. Wilmshurst, J., Anderson, A. J., Higham, T. F. G., & Worthy, T. H. (2008). Dating the late prehistoric dispersal of Polynesians to New Zealand using the commensal Pacific rat. Proceedings of the National Academy of Sciences, 105, 7676–7680.
4. Morrell, W. P. (Ed.) (1958). Sir Joseph Banks in New Zealand, from his journal. Wellington: A.H and A. W. Reed, and www.enzb.auckland.ac.nz. P. 123.
5. Atkinson, I. A. E., & Towns, D. R. (2005). Kiore. In C. M. King (Ed.), The Handbook of New Zealand Mammals, 2nd edition (pp. 159–174). Melbourne: Oxford University Press. Third edition in press, Eds C. M. King and D. M. Forsyth, CSIRO Publishing 2020.
6. Meeson, J. (1885). The plague of rats in Nelson and Marlborough. Transactions & Proceedings of the New Zealand Institute, 17, 199–207.
7. West, K.., Collins, C., Kardilsky, O. et al. (2017). The Pacific Rat Race to Easter Island: Tracking the Prehistoric Dispersal of *Rattus exulans* Using Ancient Mitochondrial Genomes. Frontiers in Ecology and Evolution. https://doi.org/10.3389/fevo.2017.00052
8. Wilmshurst, J., Anderson, A. J., Higham, T. F. G., & Worthy, T. H. (2008). Also Wilmshurst, J. M., & Higham, T. F. (2004). Using rat-gnawed seeds to independently date the arrival of Pacific rats and humans in New Zealand. The Holocene, 14(6), 801–806.
9. Matisoo-Smith, E. (2017). The Human Landscape: Population Origins, Settlement and Impact of Human Arrival in Aotearoa/New Zealand. In J. Shulmeister (Ed.), Landscape and Quaternary Environmental Change in

New Zealand (pp. https://doi.org/10.2991/2978-2994-6239-2237-2993_2998). Paris: Atlantis Press, Atlantis Advances in Quaternary Science, vol 3.
10. Towns, D. R. (2009). Eradications as reverse invasions: lessons from Pacific rat (*Rattus exulans*) removals on New Zealand islands. Biological Invasions, 11, 1719–1733.
11. Rayner, M. J., Hauber, M. E., Imber, M. J., Stamp, R. K., & Clout, M. N. (2007). Spatial heterogeneity of mesopredator release within an oceanic island system. Proceedings of the National Academy of Sciences, 104(52), 20862–20865.
12. Cruise, R. A. (1823). Journal of a Ten Months' Residence in New Zealand [1820]. Christchurch: Republished 1957 by Bagnall, A.G. (Ed.), Pegasus Press. P. 270
13. Hochstetter, F. von (1867). New Zealand, its physical geography, geology and natural history. Stuttgart: J. G. Cotta. P. 351.
14. Colenso, W. (1887). Notes, chiefly historical, on the ancient Dog of the New Zealanders. Christchurch: Kiwi Publishers. Wellington: New Zealand Electronic Text Collection, Victoria University of Wellington Library.
15. Greig, K., Boocock, J., Prost, S., Horsburgh, K. A., Jacomb, C., Walter, R., & Matisoo-Smith, E. (2015). Complete Mitochondrial Genomes of New Zealand's First Dogs. PLOS One, 10, e0138536.
16. Cruise (1823). Pp 201, 316.
17. Dieffenbach, E. (1843). Travels in New Zealand, Volume 2. Christchurch: Capper Reprints. P. 15.
18. Angas, G. F. (1847). Savage Life and Scenes in Australia and New Zealand (2 vols). London: Smith Elder. P. 337.
19. Dieffenbach, E. (1843). Travels in New Zealand, Volume 2. P. 184.
20. Gibbs, G. W. (2006). Ghosts of Gondwana. Nelson: Craig Potton Publishing.
21. McWethy, D. B., Wilmshurst, J. M., Whitlock, C., Wood, J. R., & McGlone, M. S. (2014). A High-Resolution Chronology of Rapid Forest Transitions following Polynesian Arrival in New Zealand. PLOS One, 9, e111328. Perry, G. L., Wilmshurst, J. M., & McGlone, M. S. (2014). Ecology and long-term history of fire in New Zealand. New Zealand Journal of Ecology, 157–176.
22. Collins, C. J., Rawlence, N. J., Prost, S., et al. (2014). Extinction and recolonization of coastal megafauna following human arrival in New Zealand. Proceedings of the Royal Society B, 281, 20140097.
23. Holdaway, R. N., Worthy, T. H., & Tennyson, A. J. D. (2001). A working list of breeding bird species of the New Zealand region at first human contact. New Zealand Journal of Zoology, 28, 119–187.
24. Tennyson, A. J. D. (2010). The origin and history of New Zealand's terrestrial vertebrates. New Zealand Journal of Ecology, 34, 6–27.

25. Worthy, T. H., de Pietri, V. L., & Schofield, R. P. (2017). Recent advances in avian palaeobiology in New Zealand with implications for understanding New Zealand's geological, climatic and evolutionary histories. New Zealand Journal of Zoology, 44, 177–211.
26. Lomolino, M. V., Riddle, B. R., Whittaker, R. J., & Brown, J. H. (2010). Biogeography, 4th Edition. Sunderland, MA: Sinauer Associates.
27. King, C. (1984). Immigrant Killers. Introduced Predators and the Conservation of Birds in New Zealand. Auckland: Oxford University Press. Pp 50, 213.
28. Cruise (1823, p. 32); Wood, J. R., Herrera, M. J., Scofield, R. P., & Wilmshurst, J. M. (2016). Origin and timing of New Zealand's earliest domestic chickens: Polynesian commensals or European introductions? Royal Society Open Science, 3(8).
29. Wilson, J. A. (1907). The Story of Te Waharoa: Sketches of ancient Maori life and history. Christchurch: Whitcombe and Tombs. P. 126.
30. Polack, S. J. (1838). New Zealand: Being a Narrative of Travels and Adventures during a Residence in that Country between the Years 1831 and 1837, Vol. 1. London: Richard Bentley. P. 303.
31. Berentson, Q. (2012). Moa: the life and death of New Zealand's legendary bird. Nelson: Craig Potton Publishing. P. 246.
32. Worthy, T. H., & Scofield, R. P. (2012). Twenty-first century advances in knowledge of the biology of moa (Aves: Dinornithiformes): a new morphological analysis and moa diagnoses revised. New Zealand Journal of Zoology, 39, 87–153.
33. Perry, G. L., Wheeler, A. B., Wood, J. R., & Wilmshurst, J. M. (2014). A high-precision chronology for the rapid extinction of New Zealand moa (Aves, Dinornithiformes). Quaternary Science Reviews, 105, 126–135.
34. Holdaway, R. N., Allentoft, M. E., Jacomb, C., Oksam, C. L., Beaven, N. R., & Bunce, M. (2014). An extremely low-density human population exterminated New Zealand moa. Nature Communications, 5, 5436.
35. Allentoft, M. E., Heller, R., & Oskam, C. L. (2014). Extinct New Zealand megafauna were not in decline before human colonisation. Proceedings of the National Academy of Sciences, 111, 4922–4927.
36. Walter, R., Buckley, H. L., Jacomb, C., & Matisso-Smith, E. (2017). Mass Migration and the Polynesian Settlement of New Zealand. Journal of World Prehistory, 30, 351–376.
37. Holdaway, Allentoft et al. (2014).
38. Rawlence, N. J., Perry, G. L. W., & Smith, I. W. G. (2015). Radiocarbon dating and ancient DNA reveal rapid replacement of extinct prehistoric penguins. Quaternary Science Reviews, 112, 59–65.
39. Collins, et al. (2014).

40. Morrell, W. P. (Ed.) (1958). Sir Joseph Banks in New Zealand, from his journal. Wellington: A.H and A. W. Reed, and www.enzb.auckland.ac.nz. Pp. 90–91.
41. Angas, G. F. (1847). Savage Life and Scenes in Australia and New Zealand (2 vols). London: Smith Elder.
42. Best, E. (1942). Forest Lore of the Maori. Wellington: Dominion Museum.
43. Downes, T. W. (1928).
44. Salmond, A. (1991). Two Worlds. First Meetings between Maori and Europeans 1642–1772. Auckland: Viking.
45. Original reports quoted by Salmond (1991:48) within the context of a full description of the Resolution's visit and the European's reactions.
46. Begg, A. C., & Begg, N. C. (1966). Dusky Bay. In the Steps of Captain Cook. Christchurch: Whitcombe & Tombs.
47. Pool, D. I. (1991). Te iwi Maori: a New Zealand Population, Past, Present & Projected. Auckland: Auckland University Press.

PART II

The Years of Accidental Invasions

CHAPTER 3

Reconstructing Invasion Histories from European Shipping Records (1769–1900) and Genetics (Since 2000)

> Commercial activity in New Zealand's frontier era between 1792 and 1840 [depended on] a number of realities that had relevance for the country's immediate and long-term future. One was that ... Māori turned out to be capable and competitive entrepreneurs who could grow, produce and harvest commodities such as flax and timber on a large scale; ... Second, there would always be some commercial and industrial activities – which in the early days included sealing and whaling – that [could prosper] only with investment from overseas. Third, in an unregulated and untaxed era, some entrepreneurs stood to make enormous profits.
> Michael King (2007)[1]

From as soon as the early Polynesian voyagers learned to build and steer ocean-going canoes capable of reaching distant islands, and sixteenth-century AD European maritime engineers learned to build square-rigged ships that could sail close to the wind for months out of sight of land, the expansion of human interest in the Pacific became inevitable. And, since all vessels venturing on long voyages had to carry supplies of food for the people and domestic animals they carried, it was equally inevitable that nearly all such voyages offered a passport to the world for commensal rats and mice. Had these familiar rodents not accompanied human travellers

C. M. King, *Invasive Predators in New Zealand*,
Palgrave Studies in World Environmental History,
https://doi.org/10.1007/978-3-030-32138-3_3

everywhere they went, and sent their descendants to scramble ashore uninvited and in droves at every opportunity, New Zealand might still be the only large landmass in the world with no small rodents.

To reconstruct a plausible invasion history of New Zealand, we need to put together two independent sources of information, about the people and about the intrepid little wanderers that accompanied them. For the Polynesian era, these sources have been discussed in Chap. 2. For European times, the task of assembling what we know from both sources, and taking account of the political decisions and market data that influenced them, is complicated and time-consuming, but at least it is made much easier by the recent timing of these events, well into the era when official records were carefully archived, and which have now been made easily accessible.

Shipping Records

European shipping records kept by colonial authorities, and stored in ships' logs, port documents and newspaper databases, can not only provide significant information about potential sources of rodent colonists, but also define some dates, routes and means of transport as possible, and rule out others as improbable.[2] Conversely, genetic data from stowaway rodents could potentially provide independent evidence of unrecorded shipping movements in New Zealand waters (Fig. 3.1).

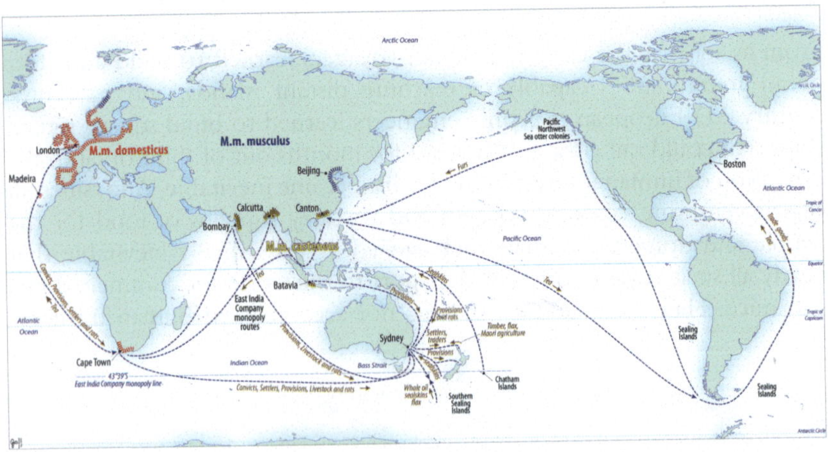

Fig. 3.1 Principal trade routes to New Zealand, 1788–1830s. Mice of three distinct subspecies originally came from three different areas; rats of two distinct species were carried on the same ships. (Cartography by Max Oulton from Author's data)

Establishment of the Port of Sydney, 1788

Historic ports on the route between Britain, India and Southeast Asia (Bombay, Calcutta and Batavia) were part of an extensive late-eighteenth-century Far East trading network. The English East India Company (EIC) had been established to pursue trade in the East Indies, under a charter granted in 1600, but over time, it concentrated on India and China. EIC ships plied standard routes from London via the Cape of Good Hope (a British possession since 1814), through the Indian Ocean and the Strait of Malacca to the South China Sea and the northern Pacific Ocean,[3] to the exclusion of all other British and colonial shipping working in the SW Pacific.[4]

The EIC had long possessed a fiercely guarded monopoly on its lucrative trade in Chinese tea, but at the time it was founded, there were no British-controlled ports east of Calcutta where its ships could call for supplies or repairs. Cook's narrow escape from the Great Barrier Reef in June 1770 emphasised the need for the development of local facilities to support the expected future increases in shipping further south—and not only because of the planned use of Botany Bay as a convict settlement. An additional part of the justification for the establishment of Port Jackson (Sydney) proposed to the British Government in 1783 was to 'enter into a commerce that would render our trade to China ... very favourable' in competition with American ships working much further from their home ports.[5]

The First Fleet ships sent to establish the penal colony arrived in Port Jackson in 1788 (six convict transports, three store ships, two naval escorts). They carried 645 people and only enough food to provide for the settlers until they could establish their own agriculture to support themselves. Such a concentration of resources and people must surely have been impossible to keep clean of rodent stowaways. Contemporary accounts of these and many other voyages of the time comment on how shipboard rats could torment the passengers and crew (Chap. 4).

A more significant problem was that the London bureaucrats who planned and supplied the First Fleet were far too optimistic in their estimate of how long it would take the new colony to become self-sufficient in food. Worse, the convict workforce, once landed, comprised too many people skilled only in the criminal arts useful in cities,[6] and too few from rural environments with any experience of farming and handling domesticated livestock. Worse again, the completely unfamiliar soils and climate of New South Wales demanded radical adaptations in agricultural tools and attitudes that took a generation to develop. Until then, the infant settlement endured a long period of semi-starvation.

Throughout the hard years of early colonial times, the EIC continued to protect its profits, especially in tea from Canton, by vigilantly enforcing its monopoly. Most British and colonial ships (other than its own and naval vessels) were prevented from trading with India or Canton direct. In addition, to insure against escape of transported convicts, shipbuilding in Sydney was at first totally prohibited. Eventually, the EIC monopoly on southwest Pacific shipping was reduced, first to the international tea trade in 1813, then to the commerce of Sydney merchants in 1820, and finally abolished in 1833. Only then did the new opportunities thereby permitted allow a huge development of trade and associated opportunities for the movement of trading goods to Sydney from other ports, including Calcutta, Bombay and Batavia. Some proportion of rodents picked up from these non-European sources were likely to represent quite different Asian genetic lineages, as did indeed prove to be the case.

For those responsible for managing the stores landed at Port Jackson, the prevention of theft and vermin damage was a constant nightmare from its earliest days. Governor Arthur Phillip reported in February 1790 that:

> The stores have been lately overrun with rats, and they are equally numerous in the gardens, where they do considerable damage; and as the loss in the stores could only be known by removing all the provisions, that was done, and many casks of flour and rice were found to be damaged or totally destroyed. The loss in those two articles by the rats since landing has been more than twelve thousand [pounds] weight.[7]

Philip does not mention mice as a separate problem, but it is a fair bet that they were there too. Not everyone can tell the difference.

Every transport ship arriving over the next decade brought more hungry convicts and proportionately fewer provisions. The new settlement went through

> a period of squalor and misery, with the threat of famine never far away and short and dismal commons the normal fare … few ships called … The chief visitors were transports bringing hungry prisoners and very inadequate provisions, [plus] East Indiamen bringing a cargo [of convicts] on contract for the Transport Board … [and] odd sealing and whaling ships.[8]

The Transport Board paid private ship owners a small commission to carry convicts to Port Jackson, but at first, there was no back cargo for the ships on their return voyage.[9] So, some British convict ships negotiated permission to drop their 'live lumber' and return home via Canton with a cargo of tea. Three of the transports of the First Fleet were under charter to take cargoes of tea from China to London after landing stores and convicts at Port Jackson.[10] Many others followed.

Imported Food Supplies and Rodent Invasion Opportunities, 1788–1820

The more convicts arrived, the more the colony's sources became insufficient, so provisions had to be bought in from anywhere they could be had. Getting them to Sydney safely was an even bigger problem. Governor Phillip had only two naval vessels at his command, of which the flagship *Sirius* was wrecked off Norfolk Island in 1789. On 17 April 1790, the *Supply* was sent to Batavia for emergency provisions, returning on 19 September, having chartered a Dutch vessel, the *Waaksamheid*, to follow with more stores. The British government sent supplies from London, but they were never enough.

For example, Henry Dundas [Secretary of State for the Home Department, London] wrote to Governor Phillip on 5 July 1791:

> You will receive by ship *Pitt* ... a proportion of salted beef and pork for 400 convicts for 12 months. The supply was confined to these articles on the idea that, with the grain produced in the settlements, the flour already sent from home, the quantity purchased from Batavia, and the supply intended to be forwarded to you from Calcutta, you would not, at least for the present, be in want of flour or rice.[11]

Dundas was way out of touch with the realities faced by the infant colony. Wheat farming in New South Wales was still barely getting established in 1800, and it was precarious for a long time.

Governor Phillip was supposed to enforce the EIC monopoly, but things got so bad by October 1792 that Major Grose, Commander of the NSW Corps guarding the convicts, wrote to Phillip proposing to commission *Britannia* to fetch supplies from the Cape of Good Hope. Grose pointed out that:

> The situation of the soldiers under my command, who at this time have scarcely shoes to their feet, and who have no other comforts than the reduced and unwholesome rations served out from the stores, has induced me to ... raise a sufficient sum to take up the *Britannia* [to the Cape].[12]

He added, concerning the need for a regular supply of provisions, and the opinion of the troops on those supplies that had been received from India,

> I request ... you will assist us to escape the miseries of that precarious existence we have hitherto constantly been exposed to.

Phillip, in defiance of his orders, did not stop the voyage, since the colony was in danger of starvation. He only forwarded Grose's letter to Dundas, and appealed for retrospective permission to overlook the proposed illegal violation of the EIC monopoly.[13]

Dundas replied, suggesting that further supplies could be got legally from Bengal—an idea first suggested by Calcutta merchants in 1791 on the grounds that the route from there to Port Jackson was shorter than from London, prices cheaper and freight charges lower. The EIC was not concerned, since it was interested only to protect its profits from tea, so it allowed trade in provisions to 'country ships' licensed individually by the EIC.[14] Thus permitted, British merchants in India were very willing to supply the Sydney authorities on credit to government accounts, which was the easiest transaction for a colony drastically short of currency. All these cargoes from India could potentially have carried new lineages of Asian stowaway rats and mice.

But supplies from India were not reliable either. In September 1793, Grose complained that:

> not a single article was received by the *Atlantic* [from Bengal, in August 1791] which in any other part of the world would have been served as a ration.[15]

The problem was not only the lack of any effective means of food preservation, but the inescapable, infuriating, massive scale of the damage inflicted by the rats and mice that inevitably accompanied cargoes of imported supplies. Food shortages had predictable consequences for the health of the colonists, so they all eagerly expected relief to arrive with the Second Fleet. When it arrived in June 1790, it provided little in the way of help, only still more convicts.

By 1803, the Port Jackson settlement was 'now 15 years old and about as non-productive as if in its second or third year'.[16] Sir Joseph Banks—as always, deeply interested in colonial affairs, and President of the Royal

Society 1778–1820—expressed the opinion in 1806 that although the colonist was still supported by 'supplies of provisions, clothes etc, sent to him from Home at no small expence. He will, however, soon be in a situation to provide for himself'.[17] But Banks was too optimistic: local supplies remained vulnerable to multiple problems of supply, bad harvests, bad weather and rodent damage in storage for years to come.

The famine of 1806 was described in a letter of October 1807 from Surgeon Lutterell to Under-Secretary Sullivan:

> last year ... we were then involved in all the miseries of famine ... the consequences of the flood [one of several suffered in the Hawkesbury Valley, an important wheat growing area] is still felt ... Pig stealing in this colony ought to be a capital offence, as it is at present the only animal food, except poultry, of the great body of settlers, as sheep are not yet numerous ... and the horn cattle is still fewer.[18]

Crop failures caused by further flooding in Hawkesbury in 1809 made things so much worse that Macquarie was on the point of commissioning another relief shipment from Bengal when the convict ship *Experiment* arrived in June 1809 with a cargo of grain.[19]

Rodent infestations were a common problem for ships' officers during the eighteenth and early nineteenth centuries. Packaging damaged by rats would also be accessible to mice, although the records of the time did not always distinguish them. Ships could be fumigated only if the cargo could be protected. On 31 May 1833, Captain McNeill of the *Lama* recorded that he parked his cargo temporarily on the *Dryad*, a sister ship, while he: 'Closed the Hatches for Smokeing having a great number of Mice on board destroying property of every kind'.[20]

In Canton, the compradors (Chinese agents), who were assigned to every visiting trading ship, provided a shack on an island outside the city where foreigners could store supplies and repair rigging, keep livestock and care for sick sailors while their ship was fumigated for rats and mice or laid on its side for careening.[21] No doubt the ships were soon re-infested along with the reloaded equipment and local stores.

Ships carrying provisions, farming equipment and livestock reaching Sydney from Britain, the Cape of Good Hope and India must also have carried forage for the stock. Robert Campbell (1769–1846) was a Sydney trader who was closely linked with India (he was a junior partner in Campbell & Co of Calcutta, founded by his uncle John Campbell). Under

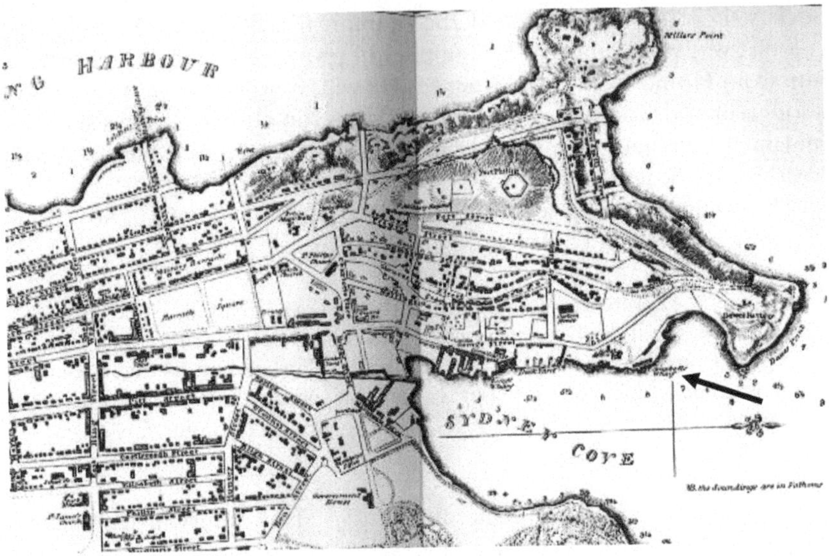

Fig. 3.2 Sydney Cove in 1831, with Campbell's Wharf (arrowed) on its western shore. (From NSW General Post Office Directory, reproduced by courtesy of Wayne Johnson)

Governor Phillip King (1800–08), Campbell had free rein to import Indian cargoes in his firm's country trade ships. Campbell's wharf, waterfront residence and warehouses were built on land on the shores of Sydney Cove, which he acquired in 1799.

Between 1800 and 1808, Campbell was responsible for importing to his own wharf (Fig. 3.2) 'upwards of two thousand Cows; the origin of those numerous herds from which our Colonists and troops and Convicts are … supplied with excellent meat'.[22] Campbell was paid more than £30,000 in Treasury bills for bringing in cattle and foodstuffs from India on contract.[23] His first warehouses date from 1800; their replacements, built in 1844–52 with the famous row of gable ends, still stand. His nephew, Robert Campbell Junior (1789–1851), joined him in Sydney in 1806. Robert junior's grandson, Hon. Robert Campbell of Otekaike (1843–1889), appears in our story as a prominent sheep-runholder in New Zealand (Chap. 7).

Cargoes of grain and cattle forage such as those imported by Robert Campbell offered by far the best opportunities for stowaway rodents. No accurate surveys of his ships were likely in his time, but some clues can be suggested from one that was done a century later. J. B. Cleland reported in

Table 3.1 Commensal rodents reported after 325 voyages by 189 ships fumigated in Sydney harbour during the years 1913–17

	Rodents found after 325 fumigation operations:		
	Norway rats	*Ship rats*	*Mice*
One species alone	3	293	53
With Norway rats	–	1	1
With ship rats	1	–	22
With mice	1	22	–
Absent	322	32	272
Total individuals	7	2968	487

Cleland, J. B. (1918). Presidential Address. Journal and Proceedings of the Royal Society of New South Wales, 52, 1–166

1918 on the numbers of rodents submitted to the Department of Public Health, collected from ships fumigated after berthing in Sydney docks (Table 3.1). The survey confirmed that stowaway rats were alarmingly common: at the end of their voyages, 90% of ships entering Sydney docks were carrying ship rats, and 16% were carrying mice. The main difference between Campbell's time and Cleland's is that in the early 1800s, the common rat on board ships was the Norway rat. By the time of Cleland's report, ship rats had replaced them (Chap. 4), and Norway rats were scarcer. The species difference does not affect the conclusion that most ships were infested with rats, and many also carried mice.

The largest numbers of mice collected by Cleland's inspectors were from vessels that often carried animal forage from the north coast of New South Wales. Baker[24] reckoned that 50% of grain/forage cargoes might carry mice. Surely Campbell's ships of a century earlier could not have been any cleaner.

Accidental introductions of rodents in landed stores is common even in the present day. Therefore, the New Zealand Department of Conservation requires containers of goods carried to islands to be inspected inside a sealed room when first landed. One report[25] lists a number of heart-stopping 'near misses' for example, a pregnant female mouse found among stores being unloaded on mouse-free Raoul Island in 1972. Another report summarising known island incursions between 1907 and 2006[26] lists 18 by mice, of which 6 were specifically recorded as mice brought ashore in stores, and 23 by rats, of which at least 6 arrived by swimming. One Norway rat reached rat-free Mana Island in 1974/75, hidden in bales of hay that landed when Mana was being used as a quarantine station for imported livestock. Such accidents must have happened at least as often in the past.

The China Fur Market, 1790–1810

The convict settlement at Port Jackson was not the only attraction for entrepreneurs keen to profit from new opportunities for trade in the Pacific. China was not only the primary source of tea, but also a rich market for other trade goods. As the New Zealand historian Robert McNab[27] commented, 'References to the China market crop up everywhere in the old records'. After all, the relative proximity of the east coast of Australia to the China market was one of the main reasons why the British Government established Port Jackson in the first place. So, one of the most obvious possible means by which the colony might earn funds to make itself financially independent of London was to take advantage of its geographical position relative to the China trade. That, in turn, could open many more opportunities for Asian rodents to reach Sydney, and thence to New Zealand.

Publication in 1784 of Cook's journals describing his third voyage[28] revealed that the Chinese were willing to pay very high prices for sea otter fur. American ship owners established 'The Golden Round', the route taken on triangular voyages, Boston-Cape Horn-NW Pacific coast-Canton-Cape Horn-Boston. From 1788 to 1825, American ships made at least 127 such voyages,[29] occasionally diverting to call at Port Jackson or New Zealand's Bay of Islands. The Boston ship owners bought up American goods, took them to the NW Pacific coast to trade with First Nation Americans for sea otter skins, and took these to Canton and traded them for tea, which was returned to Boston. Over 30 years, the North West Company sent huge but diminishing numbers of furs to China, averaging 36,822 sea otter skins/year in 1792–96, 20–25,000/year in 1804–08, and 15,000/year in 1813–23.[30]

As the numbers of sea otters crashed, supplier demands stiffened and the profits in Canton ebbed away (down from a US$1 million a year to nothing by 1834), American fur traders combed the Pacific for alternative trade goods, especially fur seal skins (Fig. 3.3). Boston traders collected seal skins from SW Atlantic and SE Pacific rookeries as they passed on their way to the NW coast. The Chinese would usually buy fur seal skins, although rarely at a good price.

Sydney-based merchants could not trade directly with India or Canton themselves, but American and Continental European ship owners were not affected by the EIC monopoly, and later, many private traders managed to evade it, illegally or legally. For example, the *Nile* (322 tons) arrived in Sydney from London on 14 December 1801 with convicts, and

Fig. 3.3 (**a**) Fur seals were once hugely abundant all around the coasts of southern New Zealand and its subantarctic islands. (Photograph by Author). (**b**) Map of seal colonies on South and Stewart Islands. (Drawn by Llewelyn Richards, from Richards, R. (2010). Sealing in the Southern Oceans, 1788–1833. Wellington: Paremata Press. P. 186)

a licence to return home via Canton. It carried a cargo of 3000 fur seal skins from Bass Strait (north of Tasmania) sent by 'Robert Campbell who arranged its shipment to China in a vessel named the *Nile*'.[31] But these voyages did not offer many opportunities for rodents to travel to Sydney because after selling seal skins in Canton, convict transport ships loaded with a cargo of tea returned to Britain.

However, EIC-licensed country ships could take commissions to load back cargoes of China goods and provisions[32] to return to Sydney merchants. After receiving the seal skins, the Chinese compradors obtained supplies from local markets, including livestock and forage for them, and supplied rattan or bamboo matting to put between layers of cargo, all offering plenty of concealment for rodents. American ships could do the same on their own account, although they usually avoided Port Jackson if they could.

Governor King strongly disliked foreigners and their collusion with local merchants to circumvent the EIC's monopoly, and in 1804, he issued a General Order forbidding clearance of foreign vessels from Port Jackson for sealing 'within the limits of this territory', which included southern New Zealand and the subantarctic islands. That did not stop the more enterprising of the Sydney merchants and sealing masters from quietly colluding with American ship owners, so there were probably many trips to and from Canton that were never recorded.

Sealing and Whaling in Southern New Zealand, 1792–1830s

Sealing was a competitive and secretive business, and many sealers were illiterate.[33] Reports of expeditions were discouraged, so documentary evidence of sealers' movements is scarce. The usual strategy was for sealing captains to land a gang in a likely area with piles of provisions, preferably near known fur seal colonies, and return later to pick up the men and the skins they had collected.[34] Fiordland was a very attractive hunting ground because the existence of seal rookeries there had been known since Cook's time, and Dusky Sound had been a well-known safe harbour since Cook charted it while resting his crew in 1773. Shipping in that area needed shelter from frequent storms, so there were many visits to Dusky Sound over the following years. Down the coast, Preservation Inlet supported a range of short-lived sealing, whaling and mining camps for more than 100 years,[35] but the rugged, roadless hinterland prevented any permanent settlements along the whole of Fiordland's coast.

The first attempt to harvest seals on the New Zealand coasts and islands started in Dusky Sound in 1792, where a gang was landed by Captain Raven of the *Britannia* (en route to the Cape of Good Hope to get supplies for Port Jackson, see p. 51) at Luncheon Cove on Anchor Island,[36] with supplies for 12 months.[37] But the number of sealskins returned to Sydney by Captain Raven (4500) was disappointing compared with the harvests then available in Bass Strait, so there was no serious seal harvesting in New Zealand while Australian rookeries were closer and more profitable. Then, in 1803, the Bass Strait sealing grounds were discovered and invaded by American sealers; the harvesting was totally indiscriminate, and rapid extermination was correctly predicted. Sealers turned their attention to New Zealand.

After Foveaux Strait was discovered 1804, all the southwestern coasts of Otago were explored for seal colonies (Fig. 3.3). Many good new sealing sites were found in Fiordland and on Stewart Island, and their locations were kept secret for intensive private harvesting by Sydney merchants for as long as possible. But by 1809, the southern New Zealand seal rookeries had been destroyed as thoroughly as the Bass Strait ones had been, and most of the harvests were sent to market via Sydney. Visits to the mainland and Stewart Island coasts were few after 1810, when sealing was transferred to the newly discovered untouched colonies on the subantarctic islands.

Unfortunately, for the sealers working the New Zealand coast, even before they started there, the prices paid at the Canton market for fur seal skins were well past the peak of the 1790s. By 1801, an influx of 200,000 fur seal skins (not all from Australasia) had reduced the Canton price to US$ 75 cents per skin. The annual demand for fur seal skins at Canton was only about 70–100,000/year, but at least 353,000 were brought in during 1800–01, 569,000 in 1801–02 and 280,000 in 1802–03.[38] Chinese import duties and commissions payable to middlemen caused the profit on skins to drop to less than the cost of transport. Chinese prices for furs at the Canton market were still worthwhile for small shipments until around 1810; back cargo included Chinese provisions, plus their potential stowaways. Larger operators needed a more reliable market.

> Happily for the Sydney sealers their industry began to burgeon at a time when sealskins had recently become a valued commodity in England. During the 1790s a Thomas Chapman had invented a technique for separating the fur of the fur seal from the coarse hairs, salt and oil ... prices rose from 13-14 shillings for medium to 30 shillings for large skins. A hat worth a couple of shillings could be sold for more than a pound if covered with seal fur.[39]

The London market for seal skins was further away, but preferable when Canton prices made illegal voyages there too unprofitable to be worth the risk. Thereafter, the best strategy was to send cargoes of skins to London, and the potential route for new cohorts of Asian rodents to reach New Zealand with sealers was closed. Sealing continued into the late 1820s,[40] but it could not be a long-term staple export industry to support the financial independence of Sydney.

Inevitably, as the China fur market fell, and as extra duties were imposed in London on colonial skins and oil, sealers turned to other trades. The emphasis of trading to and from New Zealand gradually shifted to flax, spars, timber, pork, and potatoes for the Sydney market rather than skins for export to London, and within a year, to shore whaling.

Whaling stations were established all along the south coast of the South Island between 1829 and 1840.[41] During the 1830s, whaling was the most important commercial activity in New Zealand, but the shore stations were established, manned and supplied from Sydney, and returned their products back there for forwarding to London. Unlike sealer's camps, whaling stations developed into villages, some of which became permanent settlements. In the off-season, and as a part-time occupation as the whales disappeared, the whalers turned to farming, so they needed to import livestock from Sydney. In due course, they extended their activities inland, as always accompanied by rats and mice.

In 1853, the colonial government purchased 27,923 km^2 of Māori land in the southern South Island and opened it for pastoral settlement, labelled as the Murihiku Purchase (Fig. 1.1). Immediately, whalers from Howell's cattle station, one of five in the Riverton district, started applying for leasehold pastoral runs on the newly available empty tussock grasslands. They explored inland, searching for new country and importing their stock through Bluff and Invercargill. As their prosperity increased, they built country houses and woolsheds throughout the 1850s. By the end of 1859, all accessible country for 200 km north to Lake Wakatipu had been taken up.[42]

After the first major gold strike in Otago in 1861, miners arrived from all directions.[43] Most of inland Otago and Southland had been accessed by thousands of hopeful prospectors by 1864. The already-established pastoral station owners and farmers were a significant element in that story, because they were able to supply the miners with meat, bread and beer. Their help greatly assisted the spread of mining into the remote back country, which explains how such great numbers of miners could get to work so quickly—along with their animal companions. Miners and settlers often complained about swarms of rats and mice invading their huts (Chaps. 4 and 5), disturbing their sleep and destroying their provisions.[44]

The Development of Trans-Tasman Trade, 1800–1840

The EIC monopoly and the ban on colonial shipbuilding were two important reasons for the slow development of Trans-Tasman trade. However, from at least 1800, the Sydney yards had been gradually developing the capacity to build ships of up to 200 tons burthen. At first, this extended cargo capacity had to be done illegally, but was tolerated by governors needing the services of at least small coastal vessels for local communications.

Records of European trading activity in the Bay of Islands had started, very slowly, in 1803. The number of traders (excluding whalers) entering the Bay rose to 10–12 a year in the decade to 1810. This modest but steady growth ceased abruptly after the *Boyd* massacre of 1809 and the following reprisals, and ships stayed away for several years. Not until after 1815, when the missionary leader Samuel Marsden brokered a series of reconciliations,[45] did the annual number of visiting traders get to double figures again.

By the early 1820s, Sydney had been converted from dismal prison farm to a 'bustling colony' that astonished Commissioner John Bigge during his official enquiry. Home-grown wheat production in the colony was sufficient to require the building of new flour mills in Sydney, and shipbuilding of any tonnage was now encouraged. Tariff barriers, so long a heavy disincentive to Sydney's independence, were lifted in response to Bigge's report in 1821.

Changes in regulations controlling colonial shipping encouraged a substantial increase in trading ship visits to the Bay of Islands over the next two decades. Trans-Tasman trade took off: in the 1820s, trading ship visits more than trebled to an average of 24/year.[46] Growth in the Māori capacity to supply potatoes and pork made the Bay of Islands an established provisioning depot by 1830. In 1836 alone, 151 whaling and trading vessels visited the Bay of Islands, said Joel Polack, a leading local trader who would have known better than anyone.[47] The whole Kawakawa estuary become lined with stores, marine suppliers (carpenters, blacksmiths and chandlers) and ships at anchor. The opportunities for rodents to get ashore increased in proportion to the number of trading ships arriving, adding new cohorts of eager furry colonists to those already established.

In the last decade before the Treaty of Waitangi removed control of New Zealand affairs from the governors of New South Wales, the Trans-Tasman trade to New Zealand from Port Jackson grew substantially, focussed on two centres independently: the far north of the North Island (Bay of Islands) and the far south of the South Island (Fiordland and Stewart Island). Comparison of Otago and Bay of Islands shipping records show how separate these two trading areas were.[48] These records go far to explain the curious geographical differences in the ancestry of Norway rats and house mice in New Zealand (Chaps. 4 and 5).

Outlying Islands

During the height of the sealing trade, the subantarctic Antipodes Island (discovered 1800 but not worked until 1804), Auckland Island (discovered 1806) and Macquarie Island (discovered 1810) were by far the most profitable sealing grounds, like all remote islands that had long been well out of reach of Māori hunters (Chap. 2). In all, about one million fur seal skins were taken from New Zealand and the southern islands.[49] Most came from the Antipodes Islands (37%), Macquarie Island (19%), Bounty Island (5%), Auckland Island (3%) and Campbell Island (2%). At least while the opportunity lasted, provisions had still to be sent by sealing masters to their gangs, along with any rodents they carried. Sydney was the main base and supply depot for sealing masters organising the gangs working on the subantarctic islands.

Commensal rodents can maintain independent populations on board ships in much the same way as other rodents do on land, except for one major difference: the lack of exit routes available for the dispersal of surplus numbers. Wherever such opportunities arose, on mainland coasts or offshore islands, rodents gleefully disembarked, along mooring lines or hidden among packaging, not knowing or caring where in the world they reach land. This is how New Zealand became one of the more than 80% of the world's oceanic island groups that has been invaded by one or other of the three commonest commensal European rat species.[50]

So, the previous secure isolation of even the most remote subantarctic islands was at risk of invasion by these destructive colonists during the nineteenth century. Of all the subantarctic islands visited by sealers, only the Bounty group (22 granite outcrops without landing points) presented a sufficiently hostile defence against close approach by infested vessels to evade invasion. Now, with the rapid advance of twenty-first-century eradication techniques, salvation for four of them has been achieved, and for the other three (by far, the largest) is in sight (Table 3.2).

On the other hand, historical opportunity is only one of many different factors determining the colonisation of remote islands by rodents escaping from ships: for example, ship rats arrived too late in the Pacific (Chap. 4) to have any chance of reaching most of the subantarctic islands with sealers and whalers. They got to Macquarie Island only because it has long been occupied by a research base (Chap. 6).

Interspecific competition is another important one. Rats can beat up mice, or prevent them establishing, so these distributions suggest a measure of mutual avoidance. Mice get their best chance to survive long term if they happen to land either on an island where rats are absent, or else, on a very large island where there is enough room and habitat diversity for mice to avoid rats. Stewart Island is different because it is not nearly so cold as the subantarctic islands and is the only island that supports all three rat species present in New Zealand, including the warmth-loving ship rat. Mice often arrive with people but do not establish, perhaps because their place in the community is already taken by the kiore.

The hold of a ship is a restricted space with limited opportunities for emigration, in some ways similar to a small island. On that logic, comparing

Table 3.2 Distribution of rodents on the major outlying islands as at mid-2018. Islands marked '0' have never had a permanent colony of that species

Island	*Area (ha)*	*Islands supporting permanent populations of:*		
		Norway rats	*Ship rats*	*Mice*
Disappointment	286	Eradicated 2001	0	0
Enderby	699	0	0	Eradicated 1993
Antipodes	2018	0	0	Eradicated 2016
Main Campbell	10,891	Eradicated 2001	0	Disappeared when humans left the island in 1931
Main Auckland	46,119	0	0	Yes
Main Chatham	74,572	Yes	Yes	Yes
Stewart	174,600	Yes	Yes	Replaced by Pacific rat (kiore)

Table compiled by Author

Tables 3.1 and 3.2 suggests that a similar process of mutual avoidance between rats and mice applies to both ships and islands.

Genetics as a Historical Tool

Until only a few decades ago, it was nearly impossible to know anything about the movements of vessels and people in the absence of archaeological remains and written records. Now, we have a powerful tool capable of revealing whole new libraries of information. Genetic studies of human populations, when linked to those of our unavoidable travelling companions, can lead to otherwise inaccessible conclusions about pre-documentary human history in wonderful detail.[51]

The DNA of all mammals carries a detailed record of their ancestry and distribution written in a molecular language that geneticists can now read with ease. Small rodents are ill-equipped to move far independently, but small enough to travel unnoticed among piles of baggage and supplies. They carry with them the coded information in their genes that can tell us much about where they and their human companions came from. The impact of DNA analysis on our understanding of the natural world is proving as significant as the impact of the printing press was on medieval scholarship, and for similar reasons.

Two independent DNA libraries are carried in all mammalian cells, inherited in significantly different ways. One, the mtDNA, is held in the mitochondria, the energy-generating units scattered through all the body cells. Its information is inherited unchanged through successive cell divisions, solely down the maternal line. Mitochondrial sequences (haplotypes) are grouped together in larger family units (clades). Family trees (phylogenies) constructed from mitochondrial data have given us invaluable information about the origins of many endemic species (Chap. 1) and the migrations across the Pacific of the ancestors of the Māori and their domestic animals (Chap. 2).

Important and interesting though mitochondrial data are, they give us only half the story, because they can convey information only down the female line. To get a complete picture, we have to know more about the paternal ancestry of our animals. For that, we need to venture into the more complicated matter of genomic analyses. Genomics is more than the study of the functions and inheritance of individual genes (simple genetics): rather, genomics seeks to map and understand the interactions among the entire DNA contents of a given animal.

Genomic analyses are expensive, but needed to examine the nuclear form of DNA, held on the chromosomes (nDNA), which record the more complicated story of bi-parental inheritance. When the germ cells are divided during sexual reproduction (by a different process from that of the simple division of body cells) and then reunited by fertilisation, information from both parents gets juggled, so it comes out differently in every generation. Both forms of genetic analysis have proved vital in untangling the stories of how many species came to New Zealand.

DNA is not itself the information: rather, it is directly analogous to the paper on which words are printed. Information conveyed by words is held in the *order* of the letters on the page: so, the three letters DOG do not mean the same if read backwards. Likewise, the information comprising a gene is held in the *order* of the four chemical bases grouped in three-letter 'words' (triads) along the DNA strand. That information can now be read in genetic language nearly as easily as any literate person can read a book, but it takes the right skills and equipment, and these have been available on the scale required only for the last few decades. Now, we can identify copying errors arising during cell division that inserts a letter in the wrong place (a mutation), that will either make the information incomprehensible, much like a typographical error in a printed document, or change it to mean something entirely different.

Such mutations can be fatal if they affect a vital function, but if not, they tend to accumulate down the generations. Independent lineages become distinctly different over time, but so long as they can still interbreed, they remain as no more than subspecies of a single biological species. For example, the house mouse is a single species with three recognisable subspecies, each with many separate haplotypes (strains), which can all still interbreed, although not all the hybrids are equally fertile. By contrast, the three species of rats that have arrived in New Zealand also include many haplotypes, but none are recognised as distinct subspecies. None of the three rat species can interbreed with each other. In fact, humans are more closely related to chimpanzees than these three rats are to each other.

Most animals do not randomly move across the entire range of their species—they are more likely to stay near their place of birth, and to mate with those they meet nearby. Barriers to dispersal such as rivers, mountains also restrict gene flow between populations. Hence, mutations accumulate differently in widely separated parts of a species range; these mutations can be used to identify the location where a particular animal's ancestors lived. The larger the range and the smaller the average dispersal distance of individuals, the greater the resolution of genetic data to reveal the origins of its ancestors. For small animals like mice, whose species is distributed across Europe and Asia, but which, as individuals, doesn't naturally move very far while at home, we can use genetic information found in a new invasion to determine the precise origin of its ancestors. The diversity of DNA within its population can even show the number of individuals that arrived. On the other hand, for a highly mobile species like an albatross, that move across the entire Southern Ocean, genetics can reveal very little about the origin of any particular bird.

The genetics of the Norway rat and the house mouse are among the best known of all mammals, because they are the animals most widely used in medical research labs. That means we can benefit from the advanced databases and genetic tools already developed for them, making our task of understanding their invasion histories easier for them than for any other invasive pests. But although genetic data alone can tell us a lot about the ancestry of the rats and mice living in New Zealand now, it leaves unanswered the greater, and historically more interesting, questions of how their ancestors could cross the oceans to reach such a remote corner of the world, or how to decide between conflicting explanations of where they came from. Fortunately, shipping records can often provide clear historical reasons for distinguishing between possible *post hoc* explanations. Combining genetic information and documentary records into a strong, consistent narrative can change our understanding of history. Outlining how to do that was the purpose of this chapter: the next three chapters explore the implications.

Summary

Understanding the history of three species of European commensal rodents in New Zealand has been greatly advanced by recent analyses of data from two independent but mutually supportive sources. First, the known movements of European ships have been well recorded in documentary records since the mid-eighteenth century. Shipping data defines some potential invasion routes as possible, and rules out others. Second, genetic analyses from the descendants of the original stowaway rodents often provide independent evidence confirming other records, or even suggesting otherwise unknown shipping activity. Between them, these two sources confirm a relatively clear and consistent description of how and when these three species reached New Zealand, and to some extent, where from. The movements of ships bringing traders, settlers and cargo throughout the nineteenth century provided multiple immigration opportunities for rodents from Europe and Australia. In addition, the sealers, whalers and traders sailing around the ports of Southeast Asia could bring rodents direct from India, the East Indies and China, depending on official shipping regulations. The details are clearest for house mice; two species of European rats travelled the same routes but are less informative because their source populations are more widely distributed and less genetically diverse.

Notes

1. King, M. (2007). The Penguin History of New Zealand Illustrated. Penguin: Auckland. pp. 115–116.
2. This chapter is based on King, C. M. (2016). How genetics, history and geography limit potential explanations of invasions by house mice *Mus musculus* in New Zealand. Biological Invasions, 18, 1533–1550, which gives a more complete list of page citations to primary sources.
3. Beattie, J. (2015). Plants, Animals and Environmental Transformation: Indian-New Zealand Biological and Landscape Connections, 1830s–1890s. In V. Damodaran, A. Winterbottom, & A. Lester (Eds.), The East India

Company and the Natural World (pp. 219–248). Basingstoke: Palgrave Macmillan.
4. Hainsworth, D. R. (1972). The Sydney traders: Simeon Lord and his contemporaries. Melbourne: Cassell Australia.
5. McNab, R. (Ed.) (1914). Historical Records of New Zealand, Vol 2. Wellington: Government Printer. P. 38.
6. Keneally, T. (2005). The Commonwealth of Thieves: The story of the founding of Australia. Australia: Random House.
7. McNab (1914). P. 108.
8. Hainsworth, D. R. (1972). P. 12.
9. Church, I. N. (2002). Opening the manifest on Otago's infant years: Shipping arrivals and departures, Otago Harbour and coast 1770–1860. Dunedin: Otago Heritage Books.
10. McNab (1914). P. xii.
11. McNab (1914). P. 127.
12. McNab (1908). P. 147.
13. McNab (1914). P.147.
14. Hainsworth (1972). Pp. 63–64.
15. McNab (1914). P. 167.
16. McNab (1909). Murihiku: A history of the South Island of New Zealand and the islands adjacent and lying to the south, from 1642 to 1835. Wellington: Whitcombe & Tombs Ltd., reprinted by Cambridge University Press, 2011, p. 138. This is the third and greatly expanded edition of McNab's classic work.
17. McNab (1908). Historical Records of New Zealand, Vol 1. Wellington: Government Printer. P. 272.
18. McNab (1908). Pp. 279–280.
19. Ross, J. O. C. (1987). William Stewart: Sealing captain, trader and speculator. Canberra: Roebuck Society. P. 59.
20. Gibson, J. R. (1992). Otter skins, Boston ships, and China goods: the maritime fur trade of the Northwest Coast, 1785–1841. Seattle: University of Washington Press. P. 149.
21. van Dyke, P. A. (2005). The Canton Trade: Life and enterprise on the China Coast, 1700–1845. Hong Kong: Hong Kong University Press. P. 58.
22. Hainsworth (1972). Pp. 205–206.
23. Hainsworth (1972) p. 72. A List of these shipments is provided by Stevens, M. I. (1965). Merchant Campbell 1769–1816: A study of colonial trade. Melbourne: Oxford University Press. P. 77.
24. Baker, A. I. M. (1994). Stowaway transport rates of house mice (*Mus domesticus*) and deermice (*Peromyscus maniculatus*). In W. S. Halverson & A. C. Crabb (Eds.), Proceedings of the 16th Vertebrate Pest Conference (pp. 106–112). Davis, California.

25. Veitch, C. R., & Bell, B. D. (1990). Eradication of introduced animals from the islands of New Zealand. In D. R. Towns, C. H. Daugherty, & I. A. E. Atkinson (Eds.), Ecological restoration of New Zealand islands (Vol. 2, pp. 137–146). Wellington, New Zealand: New Zealand Department of Conservation.
26. Russell, J. C., Towns, D. R., & Clout, M. N. (2008). Review of rat invasion biology: implications for island biosecurity. Science for Conservation, 286, 1–53.
27. McNab, R. (1909). P. 100.
28. The definitive version is Beaglehole, J. C. (Ed.) (1967). The Journals of Captain James Cook on his Voyages of Discovery. Vol. 3. The Voyage of the Resolution and Discovery, 1776–1780. Cambridge, UK: Cambridge University Press for the Hakluyt Society.
29. Gibson, J. R. (1992). P. 56.
30. Gibson, J. R. (1992). P. 26.
31. Ross (1987). P. 14.
32. van Dyke (2005). P. 58.
33. Salmond, A. (1997). Between worlds: Early exchanges between Maori and Europeans 1773–1815. Auckland: Viking. Pp. 299, 313.
34. Richards, R. (2010). Sealing in the Southern Oceans, 1788–1833. Wellington: Paremata Press.
35. Begg, A. C., & Begg, N. C. (1973). Port Preservation. Christchurch: Whitcombe & Tombs.
36. Begg, A. C., & Begg, N. C. (1966). Dusky Bay. Christchurch: Whitcombe & Tombs.
37. McNab (1909). P. 92.
38. Gibson (1992). P. 202.
39. Hainsworth (1972). P. 149.
40. Moon, P. (2012). A savage country: the untold story of New Zealand in the 1820s. Auckland: Penguin. Pp. 207–210.
41. Prickett, N. (2002). The archaeology of New Zealand shore whaling. Department of Conservation, National Historic Heritage Workshop, Wellington.
42. Hall-Jones, F. G. (1945). Historical Southland. Invercargill: Southland Historical Committee.
43. Salmon, J. H. M. (1963). A history of goldmining in New Zealand. Wellington: Government Printer.
44. Anon. (1864, 8 November). A gold-digger's notes on Otago. Otago Daily Times, 6.
45. McNab, R. (Ed.) (1908). Pp. 331–399.
46. Richards, R., & Chisholm, J. (1992). Bay of Islands shipping arrivals and departures 1803–1840. Wellington: Paremata Press.

47. Polack, S. J. (1838). New Zealand: Being a Narrative of Travels and Adventures during a Residence in that Country between the Years 1831 and 1837, Vol. 1. London: Richard Bentley. P. 252.
48. Church, I. N. (2002). Opening the manifest on Otago's infant years: Shipping arrivals and departures, Otago Harbour and coast 1770–1860. Dunedin: Otago Heritage Books.
49. Ling, J. K. (1999). Exploitation of fur seals and sea lions from Australian, New Zealand and adjacent subantarctic islands during the eighteenth, nineteenth and twentieth centuries. Australian Zoologist, 31(2), 323–350.
50. Atkinson, I. A. E. (1985). The spread of commensal species of *Rattus* to oceanic islands and their effects on island avifaunas. In P. J. Moors (Ed.), Conservation of Island Birds (pp. 35–81). Cambridge, UK: International Council for Bird Preservation, Technical Publication No. 3.
51. Jones, E. P., Skirnisson, K., McGovern, T. H., et al. (2012). Fellow travellers: a concordance of colonization patterns between mice and men in the North Atlantic region. BMC Evolutionary Biology, 12, 35–42.

CHAPTER 4

European Rats (Since 1769)

> Wherever you pitched your camp away in the wilderness, where never human foot before trod, there rats would be as abundant as near the settlers' homes ... they would come peering in at the door or under the curtain looking at you straight in the face with their earnest sharp gaze, and would only go when you shied something at them. Gillies (1878)[1]

Most new species deliberately brought to New Zealand by Europeans were relatively large animals, such as game for shooting and stock for farming. Certainly, none of the teeming hordes of small wild mammals native to the northern hemisphere—the shrews, voles, lemmings and wood mice of Eurasia—were ever considered. The commensal species of Old World rats and mice are only a select few of the more than 700 species in the murid family of rodents, but they are the ones with the greatest significance for New Zealand. Commensal rodents are experts at boarding all kinds of ocean-going vessels uninvited, and living independent lives hidden among the stores and cargo, often in huge numbers and small spaces.

The greatest rodent exponents of this hitch-hiking strategy are (1) the Norway or brown rat; (2) the ship or roof rat (another common name for the ship rat is 'black rat', but that is to be avoided because many of them

C. M. King, *Invasive Predators in New Zealand*,
Palgrave Studies in World Environmental History,
https://doi.org/10.1007/978-3-030-32138-3_4

are brown, so easily confused with the Norway rat); (3) the kiore (Pacific rat), and (4) the house mouse (Table 4.1). Small body size is sufficient to distinguish the kiore and the house mouse. Norway and ship rats are closer in size, but have very different facial features (Fig. 4.1). All these rodents have lived with humans for thousands of years, and they have been distributed around the world together with human companions for centuries.

Norway Rats

Norway rats originated in north-eastern China and Mongolia, and spread from there to reach almost every corner of the world inhabited by people. They appear in the historical records of Western and Northern Europe from the early eighteenth century, because their arrival was enabled by advances in urbanisation, market economics, global trade and shipping technology dating from about that time. British taxonomists, observing these new rats from about 1728,[2] gave them the specific name *norvegicus* ('Norway') rats, from the mistaken belief that they first reached Britain from Norway.

The story of the expansion of Norway rat populations around the world can be deduced from DNA analyses and fossil records. A large team led by Emily Puckett concluded that rats travelled from their centre of origin in the cooler parts of northern China and Mongolia, possibly via the strings of human settlements established along the Silk Road trade routes.[3] They spread south, east and west along five dispersal tracks identifiable from genetic analyses,[4] dividing as they went into 104 haplotypes grouped into ten clades. The far end of the fifth dispersal route expanded the range of Norway rats into Europe, and from there, across the world to the southern hemisphere.

The only two haplotypes common in New Zealand belong to clade number 10, which hitch-hiked around the world with European explorers, whalers, sealers and traders throughout the eighteenth and nineteenth centuries (Chap. 3). One, labelled by a team led by James Russell[5] as Norhap01, common in the UK, has occupied the whole of the North Island; the other, Norhap02, is reported only from China, plus the whole of the South Island and Stewart Island. Another form is found only around the port of Bluff (Norhap03). Consistent with early Otago and Bay of Islands shipping records showing how separate these two trading areas were,[6] these mutually exclusive distributions imply that the two main islands were colonised independently and from different sources (Chap. 3)—and so, for presumably the same reason, did mice

Table 4.1 Distinguishing marks of commensal rodents in New Zealand

	Kiore, Pacific rat	*Ship rat*	*Norway rat*	*House mouse*
	Rattus exulans	*Rattus rattus*	*Rattus norvegicus*	*Mus musculus*
Adult weight (g)	Normally 60–80, up to 180	Normally 120–160, up to 225	Normally 200–300, up to 500	Normally 15–20, up to 30
Max. head-body length (HBL) (mm)	180	225	250	115
Tail	Slightly shorter or longer than HBL Thin and uniformly dark all over	Much longer than HBL Uniformly dark all over	Clearly shorter than HBL Thick with pale underside	Slightly shorter or longer than HBL Uniformly grey-brown
Ears	15.5–20.5 mm; cover eyes when pulled forward Fine hairs do not extend beyond edge of ear	19.0–26.0 mm; otherwise as for kiore	14.0–22.0 mm; do not cover eyes when pulled forward Obvious hairs extend beyond edge of ear	12.0–15.0 mm
Fur on back	Brown	Grey-brown or black	Brown	Dull grey-brown
Fur on belly	White-tipped grey giving irregular colour	Generally uniform grey, white or creamy-white; rarely irregular colour	White-tipped grey giving irregular colour	Uniformly grey or white
Habits	Agile climber, digs small holes; nests mainly on ground; feeds on ground and trees; infrequent swimmer	Very agile and frequent climber; rarely burrows; nests mainly in trees and shrubs; infrequent swimmer	Burrows extensively; climbs much less frequently than other rats; strong swimmer; nests underground; very wary	Mainly ground-dwelling though capable climber; nests in small holes

King, C. M. (Ed.) (2005). The Handbook of New Zealand Mammals, 2nd edition. Melbourne: Oxford University Press. P. 160. Third edition in press, Eds C.M. King and D.M. Forsyth. CSIRO Publishing, 2020

Fig. 4.1 Portraits of a Norway rat (front) and a ship rat. (Photograph by Bruce Patty)

(Chap. 5). Norhap01 has produced three variations scattered around the North Island (Norhap04–06).

Both the two common mitochondrial haplotypes of Norway rats found on the North and South Islands have high proportions of West European ancestry, consistent with the assumption that they arrived on ships based in British ports, but their nuclear ancestries are different. Norhap01 has higher proportions of North European and Vancouver ancestry; it would have had plenty of chances to swim ashore from the many rat-infested whaling ships that cruised around the Pacific in the early 1800s, and called at the Bay of Islands to trade muskets for fresh water, pork and potatoes.[7] By contrast, Norhap02 has higher proportions of ancestry from Southeast Asia and western North America. Puckett's paper underlines the comparison with mice (which also have Asian connections in the same area) by adding that the Southeast Asian components of South Island Norway rats 'may be attributed to the seal skin trade with southern China by sealers from the USA'. Russell's team agree that 'The Asian link with Norhap02 might indicate a non-European origin' for Norway rats in the South Island.

Norway Rats on Sailing Ships

Norway rats are large, tough animals adapted to cold, wet conditions. They are very good at swimming in cool water and digging burrows, but they are too heavy to be good at climbing. On land, generations of rats have gnawed through solid wooden barn doors to get at stored grain. At sea, they thrived in the damp, unheated cargo holds of sailing ships, finding feed and shelter in inaccessible corners well out of danger from crew, fumigation or ships' cats.

Ships of the early nineteenth century were constructed of natural materials (timber for hull and masts, and flax for cordage and sails), all eminently gnawable. Wooden structures, sails and rigging, even the interior of the hull itself, were all defenceless against a rat's sharp incisors. Contemporary records frequently mention the damage done by stowaway rats, which were never named, but were probably all Norway rats.

Rodger's detailed description of the British Navy of the Georgian era (1714–1830) comments that:

> Rats were always present, and always hungry, [and, along with scorpions, centipedes and other pests which were regarded as] native to the wooden world. Whenever stores of any sort were missing, ... and had to be accounted for, the standard explanations were 'lost overboard' or 'eaten by rats', and there was virtually nothing which could not plausibly have gone one way or the other.[8]

Rats could easily destroy most of the easily accessible sacks of dry provisions on board. Meat packed in wooden casks was better protected, but that made it only slightly more challenging. In the days before any efficient means of preserving perishable food had been invented, the usual way to keep meat edible during a long voyage was pickled in salt. Rats could gnaw through the casks to get at the meat, and even if they did not eat it all, the leakage would allow the rest to spoil.

Rat-infested whaling ships, whose very timbers must have been redolent with delicious-smelling oil and strewn with scraps of blubber and meat, cruised around the coasts of New Zealand for decades between the 1780s and 1830s. Their generally filthy condition supported hordes of cockroaches, which in turn supplied abundant additional food for rats. Living quarters for the crew on one of the worst, the *Julia* on a whaling cruise in the south Pacific, were described by Herman Melville as bad enough already, without the additional misery that:

> we had not the undisputed possession of them. Myriads of cockroaches, and regiments of rats, disputed the place with us. ... it is almost impossible to get rid of them ... the vermin seem to take possession, the sailors being mere tenants by sufferance ... [On] the *Julia*, these creatures [the cockroaches]

> never had such free and easy times as they did in her crazy old hull; every chink and cranny swarmed with them; they did not live among you, but you among them … [The rats were] tame as Trenck's mouse, they stood in their holes peering at you like old grandfathers in a doorway. Often they darted in upon us at mealtimes, and nibbled our food.[9] [Baron von der Trenck was an imprisoned Austrian military officer who had a tame mouse as a companion.]

Fumigation was almost the only method of cleaning out the rats from the hold, and even that, as Melville noted, didn't work very well. It was also dangerous to the crew. The master of the *Lancaster*, an American whaling ship from New Bedford at anchor in Kororareka, accidentally caused the death by suffocation of one of his own crew while the ship was being fumigated for rats.[10]

The ships maintained and manned by the Royal Navy may have been kept somewhat cleaner, but even they could not escape their unwelcome passengers. A large part of the reason why not is that they routinely carried their fresh meat on the hoof. For the sake of avoiding months of dependence on salt pork, ships carried live cattle and sheep, pigs and goats, hens and geese in great numbers. The noise and mess they made in the already crowded accommodation below decks was disruptive, but accepted as part of everyday life for officers and men on board. Somehow, they managed to combine the disciplined efficiency of a man-of-war with large elements of the farmyard and travelling circus.[11] Rats were, of course, the inevitable beneficiaries of these chaotic arrangements.

After a long voyage far from any port, a ship's hull would accumulate a heavy load of weed and barnacles that could easily become a drastic brake on its speed. Then, the captain would look for a safe opportunity to beach the vessel in order to careen it. On James Cook's first visit to New Zealand, he took the chance to careen the *Endeavour* on the beach at a quiet anchorage in Queen Charlotte Sound. The crew laid the hull over on the larboard (port) side, scrubbed the bottom clean, then righted it and laid it over on the starboard side for scrubbing, before finally righting and refloating it. The whole procedure took three days, from 16 to 18 January 1770,[12] thereby providing the absolutely ideal conditions for rats to reach the land. No one thought of the possible consequences for native fauna, or preventing them from leaving the ship, even had that been possible—on the contrary, it is more likely that the crew would help them go. We do not have any written evidence as to whether the *Endeavour* had any rats on board, but Cook scholars agree that it would be very surprising if it had not.

During Cook's second visit to New Zealand, he brought the *Resolution* close in to shore in Dusky Sound in March 1773, and made use of a fallen tree trunk to provide a handy gangway for the crew and their animal companions to reach the land (Fig. 4.2). The *Resolution* was certainly 'much pestered' by rats, and therefore, also carried cats, so Cook was glad to see that 'the vermin could walk ashore across the bridge'.[13] He could hardly have made it easier for the first of the four-legged Europeans, Norway rats and cats, to reach the previously inaccessible Fiordland forests which his companions described so well. Cook could not at that time have guessed the massive scale of the consequences that could follow, but at least one member of his ship's company was apprehensive. Anders Sparrman was working in Cape Town when Cook arrived in 1772, and was invited to join the crew as an assistant naturalist to Johann and Georg Forster. In Dusky Sound, Sparrman noticed with regret the exodus of alien animals from the *Resolution*.[14]

Fig. 4.2 *HMS Resolution* in Pickersgill Harbour, Dusky Sound, March 1773. (Painting by William Hodges, National Maritime Museum)

In May 1773, Cook moved the *Resolution* north to meet its companion the *Adventure* in Queen Charlotte Sound, where Pickersgill's log records the 'Sailmakers repairing Sails which we find much damag'd by the Rats'.[15] Meanwhile, William Bayly, the astronomer aboard the *Adventure*, cleared a patch of bush for his observatory.[16] Sparrman commented on the abundance of rats in the disturbed area. He was 'mortified at the thought of being responsible for introducing such animals'.[17] Sparrman, an experienced naturalist, must surely have noticed the very large difference in body size between Norway rats and kiore (Table 4.1). That makes it more likely that the 'abundant' rats he observed in Queen Charlotte Sound were the descendants of those that could have escaped the *Endeavour* in January 1770, rather than the kiore staging one of their post seed fall irruptions (Chap. 2).

Norway Rats in New Zealand

Over the next 100 years, many sealing, whaling and trading ships followed Cook to the Fiordland and Otago coasts, thereby offering plenty of opportunities for rats to colonise the adjacent shores. Judging from more closely documented recent records of rats willingly swimming ashore from ships,[18] it seems much more likely than not that Norway rats got ashore, if not in Cook's time, then soon after. The closer inshore a ship was brought, the greater the chances of rats landing, especially from those vessels that gave rats easy access to land along mooring lines. One of them was HM cutter *Mermaid*, commanded by Captain J R Kent, on a visit to the southern South Island between 8 May and 23 August 1823. *Mermaid* was certainly infested with rats, which Kent's journal of 25 May reports had damaged a new sail. On 1 June, he followed Cook's chart to Pickersgill Harbour, where he moored the ship with a warp (a cable attached to an anchor) to the shore.[19] There are no written records of rats establishing in the South Island until the mid-nineteenth century, but that is not evidence of their absence. Rats were too common on ships to be worth commenting on by seamen, and there were few scientifically literate visitors or residents along that wild and uninhabited coast interested in writing about them.

The early explorers were soon followed by sealers, and then, in the more accessible parts of New Zealand, traders (Chap. 3). After the 1840s, wave after wave of European settlers arrived, accompanied by familiar animals from their homelands, each chosen to serve human purposes.

Each new ship was also potentially capable of disembarking yet more new arrivals from the inexhaustible supply of commensal rodents in their home ports. It says a lot for the original abundance of the native birds in the Bay of Islands that, even so late as 1823, Richard Cruise could report that:

> The plumage of the birds is rich, and their notes the sweetest we ever heard, particularly at daybreak, when their united harmony is called into action. The *Dromedary* lay so close to the shore at Wangarooa [Whangaroa], that each morning we were aroused by this most enchanting music.[20]

Later visitors often commented on the swarms of rats they saw, and on their apprehensions for the long-term effects of their arrival for the resident species—including, at first, the kiore. Ernst Dieffenbach was one of the earliest European observers of Māori society as it was in 1839–41, and he described the kiore as

> a frugiferous native rat, called Kiore Māori (indigenous rat) by the natives, which they distinguish from the English rat … which is introduced, and called Kiore Pakea (strange rat). On the former they fed very largely in former times; but it has now become so scarce, owing to the extermination carried on against it by the European rat, that I could never obtain one.[21]

Actually, the kiore was neither native nor frugivorous, although that would be an understandable opinion, from its length of residence, usual diet and cultural association, if not by indigenous origin. Dieffenbach wrote at a time when Norway rats were actively displacing kiore from the mainland, thereby causing a significant loss of resources to those (not all) Māori with a strong tradition of hunting them (Chap. 2). In response, Māori at first coveted the larger Norway rats as potentially better kai (food) than the disappearing kiore[22]—until they rejected them as unclean.

Dieffenbach's interpretation of this rat replacement process was similar to that of no less an authority than Charles Darwin himself, who had visited the Bay of Islands less than five years previously, on 21–30 December 1835. Darwin observed during his nine-day visit in the *Beagle*[23] that in the short space of only two years, Norway rats, which 'I was forced to own as my countrymen' had exterminated the kiore from the Bay of Islands. In his later and much more famous book, *The Origin of Species*, Darwin drew on his brief experience in New Zealand to support his argument against a fixed, divine order of individual creations. From the way that:

> the extraordinary manner in which European productions have recently spread over New Zealand, and have seized upon places that must have been previously occupied, we may believe, if all the animals and plants of Great Britain were set free in New Zealand … [most] would become thoroughly naturalized there, and would exterminate many of the natives.[24]

Darwin added that the same was widely assumed to apply to the Māori. He wrote that:

> It was melancholy at New Zealand to hear the fine energetic natives saying, that they knew the land was doomed to pass from their children.[25]

Darwin was right in many things, but not, thankfully, in that.

Throughout much of the nineteenth century, Norway rats continued to take over New Zealand in their millions, setting back human endeavours across both main islands. In 1840, a Sydney firm attempted to establish a rural farming settlement at Riccarton, now part of suburban Christchurch. They sent a manager, two farmhands and two teams of bullocks to plough and sow a 30-acre block. Within less than a year, they abandoned the effort after 'numberless rats attacked the garnered stores'.[26] Remains of mummified rats have been recovered from under-floor and ceiling spaces of historic buildings in the Bay of Islands, including the old school room in the Stone Store at Kerikeri, and in Pompallier House in Russell (Fig. 4.3).

Some European observers recounted their experiences in chilling detail. In 1877, the hordes of rats pestering the early settlers in Otago were described by R. Gillies, a Fellow of the Linnaean Society and member of the Council of the New Zealand Institute:

> Wherever you pitched your camp away in the wilderness, where never human foot before trod, there rats would be as abundant as near the settlers' homes … in the Taieri Plain near Otohiro in the year 1852 [I saw] rats running here and there in all directions from the horse's feet. When a new settler settled anywhere alone … the rats … stole everything from him … [when camping] we got so accustomed to the rats that we never felt inconvenienced by feeling them running … even on top of us as we lay in bed. So tame were they that when the candle was lit in the tent they would come peering in at the door or under the curtain looking at you straight in the face with their earnest sharp gaze, and would only go when you shied something at them.[27]

Fig. 4.3 Mummified rat preserved at Pompallier House, Russell, resting on a dishcloth stolen from the kitchen. The case label identifies it as a ship rat, but the large size, small ears and short thick tail are much more typical of Norway rats. (Photograph by Author)

Andreas Reischek, an Austrian explorer and collector who came to New Zealand for two years in 1877 and stayed for 12,[28] visited Chalky Sound in Fiordland, where he found the forest swarming with brown rats.

> At night they kept me awake with their noise, knocking things down from the walls, gnawing at my stores and digging holes round the hut. They dug up the potatoes in the garden and dragged them away ... they used to gnaw our books before our very eyes ... these rats are the great enemies of birds, and any bird living or breeding near the ground has but a small chance of existing. They play havoc alike with eggs and young, and even attack the parent birds.[29]

Eventually, towards the end of the nineteenth century in the South Island, astute explorers and surveyors like Charles Douglas began to notice an unexpected change in the rampant rat populations of the bush. While engaged in surveying work in the valley of the Waiho River in Westland in 1893, he reported that:

> The Norwegian rat ... which swarmed in the country at one time, is now becoming extinct from some cause or other, and the native and black rats are taking its place – two animals perhaps not quite so destructive as the grey gentleman.[30]

The 'two animals' mentioned by Douglas were, at that date, more likely to have been the brown and black forms of the ship rat (Table 4.1), and alas, the exchange of personnel was anything but good news for the birds.

Coastal areas frequently visited by European ships in the early days had a good chance of being invaded by Norway rats, which can swim up to 2 km in seawater[31]—their alternative name is 'water rat'. At one time, Norways were the most common rat on offshore islands, recorded from >60 islands of >5 ha, and so far have been eradicated from nearly 60% of them. Rats eradicated from various locations in the Bay of Islands all shared a common genetic profile, suggesting a lot of unaided dispersal by rats swimming between islands. Re-invasions of inshore islands previously cleared are so common as to be considered inevitable.

The results of the first arrival of rats for a previously mammal-free island's wildlife are always predictably disastrous, but the details are hard to work out because there is usually no information on what was there before the rats arrived. We do, however, have some idea of what happened on some of the islands with the longest records as sealing and whaling bases.

Kāpiti Island

Kāpiti (1965 ha), off the southwest coast of the North Island, became an important base for Ngati Toa chief Te Rauparaha in 1823, from which his raiding parties sailed their large war canoes up and down the coast, both north and south.[32] Kāpiti was also well placed to intercept the migration routes of whales, and during the height of the whaling era in the 1830s, up to 2000 people lived at the five shore whaling bases there, accompanied by Norway rats and kiore. When the whaling trade collapsed in or before 1840, the workers turned to farming, and much of the island's bush was cleared for pasture. The island was bought for a government wildlife sanctuary in 1897, and the livestock were removed, but restoration of the lost bird fauna had to wait another 100 years.

Rats destroyed great numbers of kākā nests within 1 m of the ground, and repeated attempts to reintroduce North Island saddlebacks failed while rats were still resident.[33] Now, three species of destructive pests have been eradicated; brushtail possums in 1996, and Norway rats and kiore in 1998.

Since then, Kāpiti has become one of New Zealand's most important refuges for threatened mainland birds, and for breeding seabirds. Kāpiti supports the largest remaining population of little spotted kiwi, translocated there before 1910. Other land bird species that survived the rats or have returned since the eradications, with or without human help, include the brown kiwi, red-fronted parakeet, North Island kākā, North Island robin, bellbird, takahē, North Island kōkākō, tomtit, fantail, morepork and weka. The recovery of so many populations of insectivorous birds is wonderful, although it has meant that native invertebrates haven't benefitted so much.[34]

Unlike many other island sanctuaries, Kāpiti can be visited by the public, so plays a very important part in conservation advocacy and education. That work is greatly helped by its long and well-known history, and the realistic understanding of the aims of restoration ecology by its current management team:

> We can't return to the past, [rather] Kāpiti Island has been 'reset' by human activity such as land clearance, mammal and weed introduction, mammal eradications and the arrival of other exotic species (e.g. blackbirds). We assume that, if threats are managed and appropriate species added, Kāpiti Island will continue to restore itself, driven by the natural processes (e.g. succession) currently underway.[35]

Campbell Island

Campbell Island (11,331 ha) lies in the subantarctic zone, baring its rocky teeth against the freezing southerly gales all year round. Hardly a good place for a human settlement, one might think, and yet, it has a long history. Norway rats got there soon after the island was discovered in 1810 by a sealing expedition in the brig *Perseverance*, whose captain named it after the brig's owner, Robert Campbell of Sydney (Chap. 3). Sealers and whalers soon wiped out the marine mammals, which used to throng the island's beaches and surrounding waters. They were also probably responsible for bringing in Norway rats and cats, and giving them abundant easy pickings at the two whaling stations that operated on the island while the going was good. Sheep farming started in 1895, but was challenging and often unprofitable, except, perhaps, for the rats. When the last lease-holder left in 1931, the remaining sheep were abandoned, until the last of their feral descendants were removed in 1992. There were coastal defence and meteorological stations there from 1941 to 1995, often staffed by trained naturalists, but since then, there have been no permanent inhabitants.

By the time the first scientific expedition arrived on Campbell Island in 1840, all the land birds had gone. Only in retrospect has it been possible to deduce that the original avifauna, other than seabirds, included three local endemics, the Campbell Island snipe, pipit and teal, plus the New Zealand merganser, New Zealand falcon, an unidentified rail, a plover, one or more parakeets, a tomtit, and a bellbird.[36] The rats ruled the main island for more than 160 years.

At a conference in 2000, the Department of Conservation announced what sounded like a wildly ambitious plan to clear Campbell Island of Norway rats. Campbell is two orders of magnitude larger than any island that had been cleared before, but the idea was promoted as feasible on an island with no human inhabitants or mammals of any value, provided it was done in winter, when all the nesting seabirds would be away. But—a very large but—it depended on faultless, military-grade organisation of transport, personnel and supplies separated from any backup by 700 km of the heaving southern ocean, and a sufficient run of weather good enough for flying helicopters with satellite navigation equipment to ensure no gaps in bait distribution. With the best will in the world, many in the conference's audience were not convinced.

All the greater credit to the team, then, when in the winter of 2001, four helicopters spread 120 tonnes of toxic brodifacoum pellets so thoroughly over such a large area that the job was done in four weeks. A few years later, several Campbell Island snipe, which had survived by camping out in exile on tiny Jaquemart Island, just off the southern coast of main Campbell, re-colonised the main island after the rats had gone.[37] By 2011, surveys confirmed dramatic increases in numbers of a flightless wētā; the unaided reappearance of the Campbell Island pipit, grey-backed storm petrels and white-chinned petrels; and the Campbell Island teal has been successfully reintroduced.[38]

Ship Rats

Like their larger cousin, the ship rat had eastern origins, starting from India and spreading across Eurasia from there. The commensal habit evolved in this species several times independently in different countries, which helps to account for their extensive geographical variation. Genetic analyses reveal a complicated history of six major lineages, whose distributions reflect known human movements documented from archaeology and written records since ancient times.[39] One of them, Lineage 1, reached

England from the east much earlier than did Norway rats, by AD 400 at least—as always, assisted by expanding trade routes. Surely, it cannot be simple coincidence that:

> Within a few years of the [Roman] occupation of Egypt [after the Battle of Actium in 30 BC], 120 Roman boats were sailing for India each year from the port of Mysos Hormos on the Red Sea. Commercial exchange with India did not open up so much as explode ... coin finds dating to the reigns of Augustus and his successors [are abundant on] the west coast of India.[40]

As first to invade Britain, for centuries, ship rats were the common and only commensal rats of the countryside and the cities. (Popular legends hold that the Black Death (bubonic plague) was brought to Europe in 1347 by black rats accompanying Crusaders returning from the East. No doubt Eastern rats did accompany them, but they were not the first to arrive.) Ship rats are smaller in body size than Norways, with disproportionately large ears (Fig. 4.1). Their slender physique and long, semi-prehensile tails give them tremendous climbing agility, so they spend a lot of time in trees (Fig. 4.4). They have long been common in the roofs of houses and of dockyard warehouses at which ships bound for ports all round the world were loaded with cargo—giving them many opportunities to stow away.

Fig. 4.4 Ship rat raiding a bird's nest, well above ground. (Photograph by David Mudge)

The ship rats that got to New Zealand after the mid-nineteenth century are all descendants of Lineage 1 arriving from Europe on trading and immigrant ships. The karyotype (chromosome number) of Lineage 1 is ($2n = 38$). The independent Lineage 2 ($2n = 42$) is represented in Southeast Asia, including the regions of Canton, Batavia and other ports of historic significance. Some rats of $2n = 42$ have been found on Korean vessels in New Zealand waters, but not on the adjacent land.

Russell's team identified 17 haplotypes, of which only two are widespread. Rathap01 occupies the northern North Island and all of the South Island except the southernmost region, plus Big South Cape Island and Port Jackson (Australia). Rathap02 occupies all the rest of the North Island, the southernmost South Island and Chatham Island. Rathap04 has Great Barrier Island to itself, while Rathap07 dominates Stewart Island. The township of Bluff supports unique forms of both European rat species (Rathap03 and Norhap03). All the other 12 haplotypes of ship rats turn up in singles, two or three on a single island each. These distributions imply at least four separate invasions by ship rats of the North, South, Great Barrier and Stewart Islands, plus some local variations.[41]

So far as can be deduced from historical observations, ship rats did not spread throughout the North Island until the 1860s, and in the South Island, not until the 1890s.[42] The genetic data confirm that they came not just once, but as multiple separate invasions, and some of the founder populations are still detectable. For example, ship rats can be found in three different body colours, all independently heritable but completely interfertile, and loosely correlated with haplotype. Their variable proportions in different parts of New Zealand probably represent separate arrivals. The 'rattus' phenotype (with a black back, and grey belly) is especially common in Westland and Canterbury, but scarce or absent in Waikato and Wellington, where the commonest form is the 'frugivorous' phenotype (brown back, white belly). The 'alexandrinus' form (brown back, grey belly) is dominant on Stewart Island and Fiordland, but unknown, for instance, in Manawatu and Taranaki.[43]

With such a head start in Britain, how come ship rats were so much later than Norway rats in colonising New Zealand? Even more puzzling is the question of why, having arrived up to 100 years after Norway rats had already established exclusive possession of most of New Zealand's wild places, are ship rats now the dominant and often the only rats to be found in virtually any contemporary forest?[44] Conversely, why are ship rats, resident and widespread throughout the British mainland for centuries before the Norway rats arrived, now confined to dockyards and, until eradicated, a few Scottish offshore islands? These reciprocal replacement stories in Britain compared with New Zealand are all the more surprising because they involve the same two species of European commensal rats invading two island archipelagoes of comparable size and temperate climate but in opposite hemispheres. How can we explain two island invasions by the same species but with different outcomes?

To solve this odd conundrum, we need to consider these two slightly different questions separately: first, why did ship rats take so long to reach New Zealand? and second, why did they so swiftly replace Norways as the common rat in the New Zealand bush?

Ship Versus Norway Rats on Ships

The first question is easier to answer than the second. For up to 150 years after 1700, Norway rats were the common rat in the dockyard stores and warehouses of ports throughout Britain and continental Europe. So the first European sailing ships making long voyages to the South Pacific, including to New Zealand after the 1770s, were made at a time when Norway rats were by far the most likely rat to be carried along the main routes of trade and exploration from Europe around the world. But at some time around 1830–1850, although Norway rats still lived in cities and dockside sewers and basements, Ian Atkinson's analysis shows that ship rats replaced Norways as the common shipboard rat.[45] From then on, ship rats spread around the globe as shipping routes diversified and trade volumes increased. The reason ship rats did not reach New Zealand sooner may simply have been that they could not get their passage any earlier.

There is no way now to explain this unexpected switch in rat passengers, but Ian Atkinson has made a suggestion that does make *post hoc* sense.[46] The hold and storage spaces on leaky wooden sailing ships were very often wet and cold. Norway rats could tolerate such conditions well, as Melville's descriptions illustrate (pp. 75–76). The slender bodies and tropical origin of ship rats would make them more sensitive to cold, and less likely to survive the trip. But on the steam ships that came into service from the mid-nineteenth century, the cargo spaces surrounded by iron plating were more likely to be dry and warmed by the engine, hence favouring ship rats.

Sailing ships were not immediately displaced from their accustomed trading routes around the world; many continued working well into the

1900s. Even though ship rats were gone from the British countryside and from city streets by the mid-nineteenth century or long before, they were still able to retain their foothold in the rafters of dockyard warehouses, as they still do,[47] and thereby, they could retain the niche of onboard residents. The agile rats aboard the emigrant ship *Mary Ann* in 1859, vividly described by Robert Booth, could only have been ship rats:

> I woke up during the first night at sea, with an indescribable feeling that I was being crawled over by some loathsome things. … I started up in my bunk, when, as I did so, it appeared that a large family of rats had been holding high carnival upon me and my possessions; fully a dozen must have been in bed with me … The captain said there were doubtless hundreds of rats on board, and seemed to regard the fact with complacency rather than otherwise. Sailors consider that the presence of rats is a guarantee of the seaworthiness of the ship, and they will never voluntarily take passage in a vessel that is not sound … it was not unusual of an evening to see these friendly rodents taking an airing on the ropes and rigging, and upon the handrails around the poop deck, and while so diverting themselves, I have endeavoured to shake them overboard, but always in vain; they were thoroughbred sailors, knew exactly when and where to jump, and flopping on the deck at my feet would disappear, with a twist of their tails, amidships.[48]

Competition Between Norway and Ship Rats

Two species that both depend on a necessary resource in short supply can often decide who gets the most simply by one beating up the other, in which case, it is usually the larger and/or most aggressive one that wins. This is straightforward interference competition, but it depends on direct personal confrontations—ranging from convincing threats to actual physical violence. The smaller species is likely to opt for safer foraging grounds, or at least to stay hidden until the larger one moves on.

Norway rats are larger and more aggressive than ship rats, so it is hard to imagine the long-resident British ship rats winning many direct confrontations with the newly arrived Norways. Captive studies (done years ago when regulations governing animal experiments were slack) confirmed that resident ship rats were less aggressive towards a Norway rat intruding into their colony than Norways were to an intruding ship rat. When two established colonies were allowed to invade each other, Norways took over all the nest boxes belonging to ship rats, which retreated from stressful confrontations into the roof; many died there.[49] In the British countryside, ship rats could not always escape such encounters by retreating to the treetops, which were inhospitably cold in winter and already occupied by a larger, specialist arboreal rodent, the red squirrel. The only likely retreats in the sheds, barns and farmyards were rapidly invaded and surrounded by Norway rats. So aggressive interference competition seems to be an entirely reasonable explanation for why Norways replaced ship rats in Britain. But in that case, how to we explain the opposite effect in New Zealand?

Norway rats, once massively abundant everywhere in New Zealand, now survive only in habitats where their large size, aggressive natures and skills in burrowing and swimming give them an edge—swamps and wetlands, stream and river banks, rubbish dumps and sewers—and on islands not occupied by ship rats. Norway rats are still occasionally seen swimming to and from ships moored near islands.[50] But they are no longer the commonest rat in mainland forests.

Part of the explanation could be that although both countries have temperate climates, the warmer end of New Zealand (34–46°S) is closer to the equator than is that of Britain (50–58°N: the antipodes of the North Island is in Spain, not the UK, and that of Scotland is in the subantarctic), so New Zealand's position might favour ship rats. Yet, ship rats have replaced Norways on the hill country of the South Island, which is a lot higher and colder than anywhere in the ship rats' tropical native range. They once thrived on subantarctic Macquarie Island, where Norways were absent (Chap. 6). So, there must be more to it than climate alone.

The quick reactions and fearless agility of ship rats give them skill in climbing trees (Fig. 4.4 and cover), even to the ends of very small branches, greatly exceeding that of Norway rats. It is not that Norway rats cannot climb, they can: it is that their extra weight makes climbing straight up much more of an effort, and small branches more liable to bend or break under them. These differences were neatly demonstrated by an elegant experiment conducted by a master's student, Stacey Foster, using captive rats in an enclosure. She showed that both species of rats could climb, but ship rats were much faster both going up and going down head first. Both could run across thin ropes, but the fastest average transit time for Norway rats was slower than the slowest average time for ship rats. Ship rats spent more time above ground, climbed up and down more often, and jumped between footholds without falling. In a real 2 m tree in a pot, ship rats climbed rapidly up and down, frolicked fearlessly among the topmost leaves and ventured out along an unsupported, slightly bending branch and back again. Norway rats climbed slowly, sometimes losing their grip and sliding ignominiously down the trunk again, tail-first. When, with effort, they made it to the top, they retreated carefully, and the unsupported branch bent down hard and often decanted them onto the floor.[51]

For Norway rats living in New Zealand forests in the mid-late 1770s, their lack of climbing expertise was not a problem. Despite centuries of occupation by kiore, the forest floor was still a larder bountifully stocked with ground-dwelling animals and plants, and easily harvested. A huge variety of defenceless endemic species (Chap. 1) that had evolved in secure isolation free of four-footed predators—including invertebrates (giant snails and worms, large flightless beetles and weevils), lizards, birds and native short-tailed bats—foraged and nested on the ground. Kiore could not compete for these resources with Norway rats, which simply added the kiore to their menu. Norway rats soon had it all to themselves—and that fuelled a population explosion of rats that lasted for nearly a hundred years.

When ship rats arrived, Norway rats were faced with a new and quite different sort of competition, in which the outcome of the confrontation between two species both depending on a necessary resource in short supply is decided by who is most efficient in monopolising that resource. The two parties don't have to meet face to face, so physical inferiority is not a disadvantage so long as the smaller species can avoid direct encounters and find the food first. This is exploitation competition, and ship rats are masters at it.

By the mid-1860s onwards, the easiest prey available on the ground were scarcer than they had been, while the best of what remained was out of reach, up in the canopy. Foster's climbing experiments suggested that ship rats could access those arboreal resources with ease, bounding fearlessly through the branches and nimbly dodging out of the way of the Norway rats when on the ground. But would the difference in the foraging habits of the two species be enough to explain the ship rats' unexpected victory? A follow-up study explored the implications of Foster's work.[52]

Most food resources available to wild animals are patchy in time and space—fruiting trees and nectar-filled flowers for birds, clutches of eggs and colonies of rodents for rats, fresh cow-pats for dung flies. So in theory, most foraging animals face the problem of how long to stay in one patch before moving on to the next. The dilemma becomes most difficult when the time and energy it takes to find or travel to the next patch is substantial relative to the amount of food remaining in the present patch, and the amount of food available in other patches is unknown. The food available in one patch will diminish the longer the searcher stays in it, whereas there might be more available elsewhere. How to decide how long to stay? In theory, we would expect that heavier rats would incur greater travel costs in moving between food patches, and that they might well prefer present gain to future possibilities. For them, the best plan is to make sure they clean out as much food as possible from each patch before making the effort of looking for another one. Is this true? It took months of tedious second-by-second analyses of 144 hours of video tapes to come up with the answers.

Foster constructed an artificial forest of wooden stands of three different heights. Each stand had a platform at the top supporting a plastic tray filled with sand, in which were buried 50 sunflower seeds. The number of seeds left by morning was taken as an indirect measure of foraging effort, indicating when a rat considered that a tray was no longer worth searching even though it had all night to go on doing so, and there was no other food available.

The rats started each evening by systematically visiting all the seed trays in turn. The ship rats climbed the highest (2 m) stand three times more often than did Norways, but stayed there only briefly. Ship rats checked each tray quickly, then spent the rest of their time racing up and down the stands checking for undiscovered trays, or scanning the walls for a way out to a more profitable foraging area. By contrast, once a Norway rat had reached a seed tray, it spent longer digging than did the ship rats, cleaning out all the trays one after another. Then, they climbed down, and spent the rest of the night on the floor. Ship rats behaved as 'skimmers'—they took only the easiest seeds from many different trays, a good option for an agile animal searching for widely scattered resources. By contrast, Norway rats behaved as 'crumb-pickers'—they found almost all the seeds from fewer trays, but then, they had less chance of finding new ones.

This experiment helped to refine the question of how Norway and ship rats compete for food in a forest. It is not that Norway rats cannot reach arboreal resources—such as, especially, birds' nests, a textbook example of a patchy resource. It is that making the effort to climb more than a few trees in order to slowly and thoroughly search for scattered patches of infrequently renewed food above ground is too inefficient and expensive to be a profitable foraging strategy for them. They once had no trouble finding the previously abundant food on the untouched forest floor; now, they can get most benefit by aggressive competition for the rights of intensive local exploitation of rich and constantly replenished local resources available in sewers and at rubbish dumps and seabird colonies.

By contrast, the ship rats we observed were much less tolerant of the accelerating costs of finding the last few seeds in each tray. In a forest environment, the ship rats' rapid, fleeting visits and superficial harvesting at minimal cost in energy and risk of falling is ideal for finding widely scattered, patchy and temporary resources among the treetops (seeds, fruit, insects, roosting birds or bats, and especially, birds' eggs and chicks).

Ship Rats in New Zealand

Historians are very familiar with the frustrations of finding that two events, almost certainly known from other evidence to have had independent causes, cannot be disentangled because they took place at the same time. So while it is a safe bet that much of the rat damage to native fauna mentioned in early descriptions really was due to the activities of Norway rats, it is much more difficult to distinguish the first effects of ship rat arrival on land from that of stoats, which arrived at roughly the same time as did ship rats. In other words, the damages newly added by ship rats, on top of that already done by the other two species of rats—which had already had had successive possession of the mainland forests and of many offshore islands for nearly 600 years between them—are now very difficult to distinguish from each other, or from that of stoats. Worse, deciding what damage has been done depends on knowing what was there to be damaged, and there are few unequivocal before-and-after observations for pristine places newly invaded by ship rats.

However, there are a couple of clues that might help to identify the additional changes brought in by ship rats. One turns on the different foraging strategies of Norway and ship rats. Because Norway rats cause most damage on the ground, native fauna that can live mostly in the canopy could stay safe. The carnage extended to the treetops only when the agile ship rats arrived, because they are capable of swarming up the slender lianes and supplejacks giving access to the canopy as easily as up a staircase. What happened next is illustrated by an intensive study that surveyed 106 fantail nests, and at the same time, monitored the abundance of ship rats with chew cards (giving a simple index of rat density from tooth marks). Fea and Hartley found that the probability of nest survival dropped to <50% when the rat index reached 25%. When the rat index got to 45%, only 20% of nests were likely to survive. Fantails are still common, thanks to their resilient breeding strategy, their habit of building nests perched on the ends of thin branches (see cover illustration) and their ability to offset some risks against others.[53]

The arrival of ship rats in the mid-1860s changed everything for people too. In the central North Island, long occupied by Norway rats, traditional Māori hunters could still harvest birds from Pureora Forest in abundance as they had for centuries, sufficient to supply an important feast in 1856, but they were puzzled and alarmed to see that bird numbers never recovered.[54] No one at the time made any connection with the arrival of ship rats in that area around the 1850s–60s.[55]

Other and much more convincing cases are those of islands that have never been reached by stoats or any other rodents whose effects could mask the impact of ship rats. By far, the worst and most famous example is that of the invasion of Big South Cape Island (1040 ha), off the west coast of Stewart Island.[56]

Big South Cape Island (Taukihepa)

Until an astonishingly late date (the mid-1960s), Big South Cape Island had managed to remain free of any rats. Earlier visits by ornithologists (Herbert Guthrie-Smith, Edgar Stead, Robert Falla and Robert Wilson) had recorded how different the island was from islands invaded by rats, and emphasised the importance of protecting it.

Big South Cape Island is Māori land, still covered in forest, and home to immense colonies of sooty shearwaters, hugely abundant seabirds that nest in their millions all around the southern coasts. The large, fat chicks of the shearwaters sheltering deep in the nesting burrows are known as muttonbirds. The island and its birds have always been fiercely protected by a limited number of indigenous South Island families holding the traditional rights to the annual harvest of muttonbirds.[57]

The island's long rat-free status had made it a precious sanctuary, not only for the muttonbirds, but also for six endemic species/subspecies of New Zealand birds that had long since vanished from the mainland. Two of these six refugees, the Stewart Island snipe and Stead's bush wren, were specialised endemics for which the island was the *only* remaining refuge. Two were Stewart Island subspecies of robins and fernbirds, which still survive on Stewart Island. Then, there were dozens of noisy South Island saddlebacks, bellbirds, both red- and yellow-crowned parakeets, and the last known colony of greater short-tailed bats, one of only three living species of endemic bats.

Big South Cape was regularly visited by families of local Māori during the muttonbird harvest season. After a single rat reached the island in about 1955, the danger of rats scampering ashore along mooring lines was known and discussed with the muttonbirding families, and in 1961, some basic biosecurity measures were agreed upon.[58] The muttonbirders were well aware of the effects any rat invasion could have on their harvest, so they undertook to be careful. But fishing boats needing shelter from the stormy southern seas were also allowed to use three permanent mooring sites attached to the rocks with ropes.[59] Rats were seen on the island in May 1963, though not then in great numbers, and they were dismissed as just as insignificant as the 1955 one had been. But this time, things were very different. By the time the muttonbirders arrived for the start of the next harvesting season in February 1964, the first few rats had had time to produce several generations of well-fed offspring, and all the huts and stores were trashed by rat damage. In March 1964, the muttonbirders reported their losses to the then New Zealand Wildlife Service (NZWS), and their great concern for the island birds.

Senior conservation managers did not appreciate the need for urgent action, so they did not give permission for wildlife officers to mount a rescue expedition until August–September 1964. By then, it was almost too late. The north end of the island, the centre of the invasion, was eerily silent. Robins and fernbirds were already gone; the numbers of saddlebacks and bellbirds were devastated and the survivors restricted to the south end. The bellbirds and parakeets later recovered somewhat, but the saddlebacks disappeared from the island. So did the bats, and the forest floor was swept clean of seeds and seedlings.

Wildlife officers led by Brian Bell, Dick Veitch and Don Merton set about capturing as many surviving members of the rarest species as possible. In dreadful winter conditions, they collected 36 saddlebacks, 3 snipe and 9 wrens, kept them in an aviary and hand-fed them until they could be removed to safety on rat-free islands nearby. All the saddlebacks made the trip, and have since thrived in safe havens and recovered their numbers, but none of the snipe or wrens survived. Added to the robins, fernbirds and greater short-tailed bats, that makes five of the six endemic species or subspecies lost in a few months. They have joined the many other New Zealand birds, whose decline or demise can be attributed largely or solely to the arrival of ship rats: the New Zealand pigeon, three of the four species of parakeets, yellowhead, brown creeper, and North Island kōkākō, all listed in the latest (2010) roll call of tragedies reviewed by John Innes and colleagues.[60] The muttonbirds survived.

The question of where the rats came from has never been quite resolved. A genetic analysis of 152 rats from the South Island, Stewart Island, Big

South Cape and its surrounding islands by Judith Robbins and her team identified six mtDNA haplotypes. Of these, 63 carried haplotype 7 dominant on nearby Stewart Island, and 49 carried haplotype 1, found worldwide and throughout southern New Zealand, except the Southland coast.[61] Only haplotype 1 has ever been found on Big South Cape itself (including in those rats collected at the time of the invasion). So the invading rats could not have simply swum the 1.5 km from Stewart Island, where haplotype 1 is absent, or boarded boats at Bluff or Riverton, the main ports on the South Island coast used by muttonbirders, where rats of haplotypes 3 and 2 are dominant. So neither of the two most obvious and nearest sources seems definitely to have provided the haplotype 1 rats that reached Big South Cape, which means that they probably did not come with the muttonbirders, and it is unfair to blame them. Rather, they seem to have been unintentionally assisted by strangers, perhaps on non-local fishing boats whose crews were not familiar with the muttonbirders' precautions. It is not yet possible to say exactly where they came from, except that it was from some distance way.

The Big South Cape invasion was, by far, the greatest single short-term catastrophe to hit New Zealand wildlife in modern times, but it also had a positive outcome: it marked a critical turning point in official attitudes towards invasive rats. One might ask why this revelation took so long, since the idea was not new. Rats (most often ship rats) are the principal agent responsible for more than half of the 70% of island bird extinctions attributed to predation elsewhere in the world.[62] But among senior conservation managers of the 1960s whose opinions were influenced by reading the international literature, especially the works of Paul Errington,[63] there was a persistent idea that predators and prey normally reach a mutual balance, after which predators only take the surplus, 'doomed' members of a population, thereby acting as a valuable sanitary service, rather than a destroyer of whole populations. That idea is often correct for established predator-prey relationships in the continental mainland habitats where Errington worked, but is dangerously misleading when applied to island faunas that had evolved in predator-free isolation. Thankfully, the difference is appreciated now, helped along by all-too-vivid descriptions of these and many similar devastating losses of species from other islands listed in gruesome detail by (among others) William Stolzenburg.[64] Big South Cape is unusual only because it was one of a handful of islands whose original faunas had been described before the rats got to them.

The ship rats were eradicated from Big South Cape and three smaller neighbouring islands by a well-organised community effort in 2006. After declaration of rat-free status in 2009, the saddlebacks were returned in 2011 (38 birds) and more were added in 2012 (42 birds).[65] The Stewart Island robin and fernbirds re-colonised naturally. Smaller ground-nesting and burrowing birds, lizards and large invertebrates have also benefitted, but not as much, due mainly to the continued presence of weka—a predator that is native to New Zealand but introduced by the Māori to Big South Cape as a traditional food (p. 187) in the early 1900s. The snipe and the wren are totally extinct.

Rats Versus Mustelids

Until very recently, the universal assumption has been that the worst predators ever to invade New Zealand were the mustelids (stoats, weasels and ferrets), and that therefore conservation actions must be focused on them. That is certainly true for a particular subset of *species* especially vulnerable to them, especially in high-elevation beech forests (Chaps. 8, 9 and 10). But mustelids are not nearly abundant enough to account for the millions of *individual* common bush birds destroyed throughout New Zealand's lowland forests every six-month breeding season. That back-of-an-envelope calculation was made after the outcry caused by Greenpeace's report that more than 20,000 seabirds were killed in the Bay of Plenty over the summer of 2010–2011, after the oil spill from the wreck of the *Rena*. That was a terrible and very public event; by comparison, the unseen, nocturnal activities of rats and possums account for many more bush birds (mostly eggs and nestlings) every night.[66]

The only positive outcome of all these sad stories is that, at long last, ship rats are now recognised as by far the most widespread, agile, voracious and massively abundant predators of bush birds in what remains of the lowland forests. In such habitats, rats have small home ranges (<10 ha) and visit most parts of them every night; stoats have large home ranges (50–150 ha) and visit parts of them infrequently. Both are good climbers, but climbing and searching a large home range takes longer than for a small one, so the probability seems high that the first predator to find any given bird's nest will be a rat rather than a stoat. This important piece of information has caused a helpful update to traditional predator-control assumptions, just in time to benefit the Predator Free New Zealand programme (Chap. 12).

Summary

European ships carry two species of commensal rats, the Norway rat *Rattus norvegicus* and the ship rat *R. rattus*. Both originated in Asia, travelled to Western Europe with traders, and thence around the world on European shipping. Norway rats arrived in New Zealand first, possibly as early as 1769, and spread rapidly throughout the main and many offshore islands. For decades, they were hugely abundant, adding yet more damage to that already done by Polynesian rats. The native fauna on the ground was seriously depleted, but arboreal species remained relatively safe. Ship rats by contrast are agile climbers, so when they arrived in the mid-to-late nineteenth century, they replaced Norway rats as the most abundant rat in forests. The effects of rats on native fauna are clearly illustrated by well-documented data on their invasions of islands previously surveyed by ornithologists, from which the likely course of events on the mainland can be guessed. The technology of removing rats from uninhabited islands out of reach of re-invaders is now routine. The benefits of doing so are inspiring dreams of doing the same on the inhabited mainland, where rats are vastly more abundant and damaging than any mustelid.

Notes

1. Gillies, R. (1878). Notes on some changes in the fauna of Otago. Transactions and Proceedings of the New Zealand Institute, 10, 306–324.
2. Yalden, D. (1999). The History of British Mammals. London: Academic Press.
3. Frankopan, P. (2015). The Silk Roads: A new history of the world. London: Bloomsbury.
4. Puckett, E. E., Park, J., Combs, M., Blum, M. J., Bryant, J. E., Caccone, A., & et al. (2016). Global population divergence and admixture of the brown rat *(Rattus norvegicus*). Proceedings of the Royal Society B, Biological Sciences, 283 (1841), 20161762.
5. Russell, J. C., Robins, J. H., & Fewster, R. M. (2019). Phylogeography of invasive rats in New Zealand. Frontiers in Ecology and Evolution 7, 48. doi: https://doi.org/10.3389/fevo.2019.00048
6. Church, I. N. (2002). Opening the manifest on Otago's infant years: Shipping arrivals and departures, Otago Harbour and coast 1770–1860. Dunedin: Otago Heritage Books.
7. Wolfe, R. (2005). Kororareka: hell-hole of the Pacific. Auckland: Penguin.

8. Rodger, N. A. M. (1986). The Wooden World: An anatomy of the Georgian Navy. London: Collins. P. 53.
9. Melville, H. (1847). Omoo: A Narrative of Adventures in the South Seas. London: John Murray. Pp. 38–39.
10. Wolfe, R. (2005). P. 185.
11. Rodger (1986). Pp. 54–55.
12. Beaglehole, J. C. (Ed.) (1955). The Journals of Captain James Cook on his Voyages of Discovery: Volume I: The Voyage of the *Endeavour* 1768–1771. Cambridge: Cambridge University Press for the Hakluyt Society. Pp. 235–236.
13. Beaglehole, J. C. (Ed.) (1961). The Journals of Captain James Cook on his Voyages of Discovery. Vol. 2. The Voyage of the *Resolution* and *Adventure* 1772–1775. Cambridge, UK: Cambridge University Press for the Hakluyt Society.
14. Andrews, J. R. H. (1986). The Southern Ark: Zoological Discovery in New Zealand, 1769–1900. Auckland: Century Hutchinson NZ. Pp. 27–28.
15. McNab, R. (Ed.) (1914). Historical Records of New Zealand Vol. 2. Wellington: Government Printer. P. 190.
16. McNab, R. (Ed.) (1914). P. 206.
17. Andrews (1986).
18. Russell, J. C., Towns, D. R., & Clout, M. N. (2008). Review of rat invasion biology: implications for island biosecurity. Science for Conservation, 286, 1–53.
19. Begg, A. C., & Begg, N. C. (1973). Port Preservation. Christchurch: Whitcombe & Tombs. Pp. 310–339.
20. Cruise, R. A. (1823). Journal of a Ten Months' Residence in New Zealand [1820]. Christchurch: Republished 1957 by Bagnall, A. G. (Ed.), Pegasus Press. P. 288.
21. Dieffenbach, E. (1843). Travels in New Zealand, Volume 2. London: John Murray. Pp. 184–185.
22. Hochstetter, F. von (1867). New Zealand, its physical geography, geology and natural history. Stuttgart: J. G. Cotta. Footnote on p. 195.
23. Darwin, C. (1845). Journal of Researches into the Natural History and Geology of the Countries Visited during the Voyage Round the World of HMS Beagle under the Command of Captain Fitzroy RN. London: John Murray (Reprinted from 1860 edition by Folio Society, London, 2003). Pp. 426–427.
24. Darwin, C. (1859). The Origin of Species. London: Reprinted 2006 by The Folio Society. P. 267.
25. Darwin (1845). P. 434.

26. Thomson, G. M. (1922). The Naturalisation of Animals & Plants in New Zealand. Cambridge: Cambridge University Press. P. 79.
27. Gillies, R. (1878).
28. King, M. (1981). The Collector: A biography of Andreas Reischek. Auckland: Hodder and Stoughton.
29. Reischek, A. (1930). Yesterdays in Maoriland. London (translated by H.E.L. Priday): Jonathan Cape. P. 251.
30. Douglas, C. E. (1894). On the Westland Alps. Appendices to the Journal of the New Zealand House of Representatives C1, 30, 71–75.
31. Russell, J. C., Towns, D. R., & Clout, M. N. (2008). P. 34.
32. King, M. (2003). The Penguin History of New Zealand. Auckland: Penguin Books.
33. Lovegrove, T. G. (1996). Island releases of saddlebacks *Philesturnus carunculatus* in New Zealand. Biological Conservation, 77, 151–157.
34. Sinclair, L., McCartney, J., Godfrey, J., Pledger, S., Wakelin, M., & Sherley, G. (2005). How did invertebrates respond to eradication of rats from Kapiti Island, New Zealand? New Zealand Journal of Zoology, 34, 293–315.
35. Brown, K., Rolfe, J., Adams, L., de Lange, P., & Green, C. (2016). Kapiti Island ecological restoration strategy. Wellington: Department of Conservation Technical Report.
36. Barker, D., Carroll, J. W. A., Edmonds, H. K., Fraser, J. R., & Miskelly, C. M. (2005). Discovery of a previously unknown *Coenocorypha* snipe in the Campbell Island group, New Zealand subantarctic. Notornis, 52, 143–149.
37. Miskelly, C. M., & Fraser, J. R. (2006). Campbell Island snipe (*Coenocorypha* undescribed sp.) recolonise subantarctic Campbell Island following rat eradication. Notornis, 53(4), 353–359.
38. McClelland, P. J. (2011). Campbell Island – pushing the boundaries of rat eradications. In C. R. Veitch, M. N. Clout, & D. R. Towns (Eds.), Island Invasives: Eradication and Management (pp. 204–207). Gland, Switzerland: IUCN.
39. Aplin, K. P., Suzuki, H., Chinen, A. A., Chesser, R. T., ten Have, J., Donnellan, S. C., . . . Cooper, A. (2011). Multiple geographic origins of commensalism and complex dispersal history of black rats. PLOS One, 6, e26357.
40. Frankopan (2015). P. 16.
41. Russell, J. C., Robins, J. H., Fewster, R. M. (2019).
42. Atkinson, I. A. E. (1973). Spread of the ship rat (*Rattus r. rattus* L) in New Zealand. Journal of the Royal Society of New Zealand, 3, 457–472.
43. Innes, J. G. (2005). Ship rat. In C. M. King (Ed.), The Handbook of New Zealand Mammals, 2nd edition. (pp. 187–203). Melbourne: Oxford

University Press. Third edition in press, Eds C.M. King and D.M. Forsyth. CSIRO Publishing, 2020.

44. Innes, J. G., King, C. M., Flux, M., & Kimberley, M. O. (2001). Population biology of the ship rat and Norway rat in Pureora Forest Park, 1983–87. New Zealand Journal of Zoology, 28, 57–78.
45. Atkinson, I. A. E. (1985). The spread of commensal species of *Rattus* to oceanic islands and their effects on island avifaunas. In P. J. Moors (Ed.), Conservation of Island Birds (pp. 35–81). Cambridge, UK: International Council for Bird Preservation, Technical Publication No. 3. See the updated version of Atkinson's figure 5, extended to 2008, by Russell et al. 2008.
46. Atkinson (1985).
47. Harris, S., & Yalden, D. W. (Eds.). (2008). Mammals of the British Isles: Handbook, 4th edition. Southampton: The Mammal Society. P. 157.
48. Booth, R. B. (1912). Five Years in New Zealand (1859 to 1864). London: J. G. Hammond & Co. Pp. 4–5.
49. Barnett, S. A., & Spencer, M. M. (1951). Feeding, social behaviour and interspecific competition in wild rats. Behaviour, 3, 229–242.
50. Harper, G. A., Dickinson, K. J., & Seddon, P. J. (2005). Habitat use by three rat species (*Rattus* spp.) on Stewart Island/Rakiura, New Zealand. New Zealand Journal of Ecology, 29, 251–260.
51. Foster, S., King, C., Patty, B., & Miller, S. (2011). Tree-climbing capabilities of Norway and ship rats. New Zealand Journal of Zoology, 38, 285–296.
52. King, C. M., Foster, S., & Miller, S. (2011). Invasive European rats in Britain and New Zealand: same species, different outcomes. Journal of Zoology, 285, 172–179.
53. Fea, N., & Hartley, S. (2018). The balancing act of nest survival: survival of a small endemic bird in the face of ship rat predation and other risk factors. Avian Conservation and Ecology, 13(2), 11.
54. Waitangi Tribunal. (1993). The Pouakani Report. Waitangi Tribunal, Wellington. S. 15.2.
55. King, C. M., Gaukrodger, J. D., & Ritchie, N. A. (2015). The Drama of Conservation: A history of Pureora Forest, New Zealand. Berlin and Hamilton: Springer Verlag and New Zealand Department of Conservation. P. 64.
56. Bell, E. A., Bell, B. D., & Merton, D. V. (2016). The legacy of Big South Cape: Rat irruption to rat eradication. New Zealand Journal of Ecology, 40, 212–218.
57. Moller, H., Kitson, J. C., & Downs, T. M. (2009). Knowing by doing: learning for sustainable muttonbird harvesting. New Zealand Journal of Zoology, 36, 243–258.

58. Bell et al. (2016).
59. Robins, J. H., Miller, S. D., Russell, J. C., Harper, G. A., & Fewster, R. M. (2016). Where did the rats of Big South Cape Island come from? New Zealand Journal of Ecology, 40, 229–234.
60. Innes, J., Kelly, D., Overton, J., & Gillies, C. (2010). Predation and other factors currently limiting New Zealand forest birds. New Zealand Journal of Ecology, 34(1), 86–114.
61. Robins et al. (2016).
62. King, W. B. (1973). Conservation status of birds of central Pacific islands. Wilson Bulletin, 85, 89–103.
63. Errington, P. L. (1946). Predation and vertebrate populations. Quarterly Review of Biology, 21, 144–177.
64. Stolzenburg, W. (2011). Rat Island: Predators in Paradise and the world's greatest wildlife rescue. New York: Bloomsbury.
65. Bell, E. A., Bell, B. D., & Merton, D. V. (2016). The legacy of Big South Cape: rat irruption to rat eradication. New Zealand Journal of Ecology, 40, 212–218.
66. Hill, G. (2012). Why I enjoyed the Rena disaster. Forest & Bird, 346 (November), 49, citing J.G. Innes.

CHAPTER 5

House Mice (Since About 1792)

> To whom the natives are indebted for the rat is not known, but they assign to the missionaries the credit of introducing mice. Jameson (1842)[1]

Of all the successful commensal animals that have accompanied extensive human traffic around the world, none are as skilled as are house mice at hiding away in the tiniest spaces, and travelling unseen among even small cargoes and possessions. These little hitch-hikers have been carried frequently but unwittingly by traders across continental Eurasia for thousands of years,[2] and more recently, overseas, even to the remotest oceanic islands. New Zealand provides one of rather few relatively clear demonstrations of how these processes have unfolded since the late eighteenth century.

Like the ship rat (Chap. 4), the original home of the house mouse was somewhere in Asia, probably northern India, about 350,000 years ago. Over time, the species separated into three distinct subspecies, each widespread within their separate homelands, labelled *domesticus* in Western Europe from Spain north to Denmark, *musculus* in northern and eastern Eurasia from Scandinavia to Russia and northern China, and *castaneus* across Southern Asia from India eastwards to Southeast China.[3] The three are genetically distinct but still capable of interbreeding (though generally with reduced fertility), which is why they are not regarded as fully separate species. For simplicity, we can call them simply western, northern and Asian mice.

C. M. King, *Invasive Predators in New Zealand*,
Palgrave Studies in World Environmental History,
https://doi.org/10.1007/978-3-030-32138-3_5

Fig. 5.1 A wild house mouse in the beech forests of Fiordland. It is impossible to tell, just by looking at it, which of the three subspecies is represented by its maternal genes. (Photograph by Author)

Because the three separate lineages of mice are indistinguishable in general appearance (Fig. 5.1), but were carried to New Zealand along different routes (Chap. 3), we can use both genetic and shipping information to construct a fair picture of where the ancestors of the mice now living in New Zealand came from. In other countries, genetic analyses of house mice have provided rich and varied, and sometimes, very surprising, information about human activities invisible to documentary history.[4]

The propensity for mice to hide among cargo on boats, and to escape to land and establish new populations wherever these boats moored, has led to the most complex story of ancestry, origins and intermingling of any introduced New Zealand mammal. Mice in New Zealand have 33 known mitochondrial haplotypes, representing three different subspecies, and eight different divergent clades (Table 5.1). For comparison, stoats and weasels came from one location, the UK—primarily Lincolnshire (Chap. 10)—and they have five and four known mitochondrial haplotypes, respectively. There were two separate invasions of Norwegian rats from different sources, with six known mitochondrial haplotypes (Chap. 4). Ship rats, another great stowaway like mice, also came from the UK, probably via Australia, with 17 known mitochondrial haplotypes (Chap. 4).

The ancestors of New Zealand mice range across Europe and Asia, and their diverse descendants have mixed with each other, before they arrived in other ports, on board ships, and since they arrived on our shores. More so than any other mammal in New Zealand, genetic information has been critical to inform our understanding of this complicated story.

Mice arrived much later in New Zealand than did kiore, which are equally skilful and inconspicuous stowaways. The reason that mice could not have arrived with the ancestors of the Māori was simply because there were no house mice in the Pacific Islands in pre-European times. No prehistoric mouse subfossils have been found there, and not simply because they are small and might be overlooked, because kiore (not much larger than a mouse) are common in subfossil deposits around the Pacific islands.[5] The native laughing owls of New Zealand readily added the kiore to their diet after the fourteenth century, but no mice were found in the pellets of any native predatory birds until after European colonisation began.[6]

Therefore, mice could only have arrived in New Zealand after the entry of European/American shipping in the SW Pacific, facilitated by the establishment after 1788 of Port Jackson (modern Sydney) as a port of call and the first important trading hub of the SW Pacific (Chap. 3). Over time since then, hundreds of thousands of human colonists have settled in Australia and New Zealand, and the great majority of them came from Europe. It is a fair bet that both the Western and the Northern European subspecies would have arrived with them, since shiploads of goods have accompanied immigrants from countries occupied by either subspecies. Because there were no native mice already in possession of the vast New Zealand forests, the newly arriving house mice soon escaped domesticity, spread widely and settled down to take the place of the absent northern-hemisphere wood mice (Fig. 5.1).

In 1999, master's student Paul Jamieson was the first to ask questions about which of the two European subspecies of mice came with the first settlers reaching New Zealand after 1840. Paul collected wild mice from all around the country, and analysed their body measurements, skull features and coat colours as well as their genetic lineages. Paul's results were linked with those of a broader programme examining the origins of mice in Britain, coordinated by Jeremy Searle.[7] Of all their combined results, those identifying the distributions of the separate haplotypes of mitochondrial DNA were the most informative.

Mice stowing away with the early nineteenth-century traders and colonists bound for the South Pacific from Britain and the Continent would certainly have belonged to the West European subspecies, and probably also those that got on board with supplies loaded at the Cape of Good Hope or in Port Jackson. Those from Ireland, Scandinavia or Russia were more likely to bring mice belonging to the northern subspecies. So, as expected, Paul's results provided ample grounds to conclude that New Zealand had been colonised by both the European subspecies (multiple times, and overwhelmingly from the West). But when Searle and his colleagues looked further into Paul's samples, they made an unexpected discovery: the Asian subspecies was also represented, by mice which must have come from somewhere in Southeast Asia, most likely China (Chap. 3).[8] So far as anyone knew, there had been no significant trading to or colonisation of New Zealand from that direction at the time.

This discovery was very exciting, but, as is usual in preliminary surveys, there were some large gaps in the geographic coverage. To fill them in, a later and more extended survey incorporated further collections into a final set of 502 mice from 74 sites all around New Zealand. Mice were contributed by two more master's students, Tanya Chubb and Helen McCormick, plus Elaine Murphy from the Department of Conservation (DOC) and other colleagues from other institutions and friends around the country. These new samples produced enough DNA data to enable a detailed mitochondrial phylogeny for mice to be constructed by Alana Alexander, which confirmed Searle's conclusions and added many more local details.[9]

The problem was that, for reasons explained in Chap. 3, mitochondrial genes describe only half the story, since they are inherited only down the maternal line. Fortunately, the original mouse collections had been kept in storage in the hope of getting the chance to look at the nuclear DNA, supplemented where necessary with additional material. It took many more long and expensive sessions in the genetics lab to unravel the more complicated story of bi-parental inheritance.

A new project led by Andrew Veale chose a selection of the same animals for which we already had data from the mitochondria, plus some additional NZ sites and one from Sydney, and put 182 samples through a sophisticated new tool, the GigaMUGA array. This formidable equipment, the third generation of the Mouse Universal Genotyping Array, is designed to search for diagnostic markers identifying each subspecies at short intervals along all 21 mouse chromosomes. This procedure allowed the assembly of dataset orders of a magnitude larger than any previous one, and hence, the genetics of New Zealand house mice are now better understood than any other introduced mouse population in the world. The results confirmed the main geographic variation in ancestry patterns across the country already outlined by the mitochondrial data alone, but added more complicated details.[10]

History of West European Mice in New Zealand

West European mice (*M.m. domesticus*) are divided into six mitochondrial clades labelled A to F, each dominant in a different part of Europe, although with a lot of overlap and mixing (Table 5.1). The great majority of mice in England belong to Clade E; those in northern Scotland and Ireland more often belong to Clade F.[11] Both are, as expected, very widespread in New Zealand, represented by 14 of the 23 haplotypes we found in 279 mice. They confirm the multiple different mouse invasions accompanying the colonisation of New Zealand by waves of British and Irish settlers.

Mice were not observed in the Bay of Islands in 1819,[12] but they were there within 20 years. The farmer-naturalist Herbert Guthrie-Smith reported that:

> Archdeacon Samuel Williams informed me that he had no recollection of mice in New Zealand until the 'thirties. About that date they were noticed at the Bay of Islands Mission Stations. Vessels were then beginning to reach the colony, laden or partly laden with cargoes of a kind that for the first time offered shelter, harbourage and breeding accommodation for the small creature … only in the intricacies of a miscellaneous cargo could the mouse have obtained during the long voyage, shelter and safety from his voracious kinsman the rat.[13]

Archdeacon Williams' opinion was confirmed by a contemporary source: 'To whom the natives are indebted for the rat is not known, but they assign to the missionaries the credit of introducing mice'.[14] Indeed, one elderly vicar with a strong life-long interest in missionary history assured me that he once read an unpublished diary recording how mice ran out of a consignment of goods being unpacked by missionaries in Pahia.

Further south in the North Island, another missionary, Richard Taylor,[15] reported seeing a mouse for the first time ever, at his house near Whanganui, in about 1844. The surveyor John Rochfort[16] found a house 'swarming with mice' at Moutoa (46 km from the mouth of the Manawatu River) in 1852.

Table 5.1 Original and New Zealand distributions of the major lineages of house mice in New Zealand

Subspecies	*Mitochondrial DNA*				*Nuclear DNA*
	Clade	*Haplotypes*	*Countries occupied by ancestral lineage*	*Distribution in New Zealand*	*% non-domesticus*
M.m. domesticus	A	domNZ.6, domNZ.7, mac.domNZ.1	Greece to Turkey, Iran	Saddle I., Taranaki, Canterbury, Pitt I.***	Zero throughout NZ except as below
West European mouse	B	mac.domNZ.2, domNZ.12, domNZ.14 domNZ.23–26	Portugal, to Greece, Turkey	Napier, Cape Kidnappers, Mahia Peninsula, Waikawa Island, Waipukurau*	
	C	domNZ.8, domNZ.19	France, Germany Italy, Spain, Portugal	Antipodes I., Ruapuke I.**	
	D	domNZ.11	Central Europe to Spain	Nelson, Rotoiti*	
	E	domNZ.1–5, 15, 16, 18, 20,21,22	UK, central and northern Europe	Dominant throughout mainland north of Canterbury, absent from far south, isolated on Macquarie, Auckland Is	
	F	domNZ.9–10, 13, 17	UK, central and northern Europe	Frequent on mainland north of Canterbury	
M.m castaneus South-east Asian mouse	HG2	casNZ.1	SE Asia, India to China	Dannevirke, Eketahuna, Wellington; dominant south of Canterbury; absent in northern N. I., northern S. I.	Northland 3% Taranaki 1% Wellington 2%
		casNZ.2		Chatham I.; absent from mainland	
		casNZ.3–6		Wanaka, Waianakarua, Grebe Valley, Dunedin	

(*continued*)

Table 5.1 (continued)

Subspecies	*Mitochondrial DNA*				*Nuclear DNA*
	Clade	*Haplotypes*	*Countries occupied by ancestral lineage*	*Distribution in New Zealand*	*% non-domesticus*
M.m. musculus North European mouse	?	musNZ.1	Germany, Austria (very similar to other haplotypes in the westernmost part of *M. musculus musculus* range)	Wellington	Nelson 7% Hurunui 5% Franz Josef 1% Wellington 0.02%

King C, Alexander A, et al. (2016)
Entries marked * represent the seven independent invasions of non-British/Irish stock

Why did it take so long for Clade E and F mice, once established in the North Island, to spread inland along with the settlers and their goods? Partly, perhaps, because few of the new ports needed to receive European settlers (Whangarei, Whakatane, Napier, New Plymouth, Whanganui), or the railways, and the roads necessary for heavy carts to reach inland, laden with mouse-infested goods, were not developed until after 1840. All these ports that have been sampled are occupied by western mice.

Guthrie-Smith added that:

> Although twice during my occupancy of Tutira [1882–1940] irruptions of mice have overrun the station, it is unlikely the place has been directly stocked from the north ... the local origin of Tutira mice was almost certainly the port of Napier. Thence there may have been a migration sufficiently sustained to reach the station. On the other hand, mice may have been directly packed up on horseback amongst station stores. Certainly from the homestead they have been carried over every part of the run in grass-seed bags; I have myself sown the half-smothered little creatures from sacks of grass and clover seed.[17]

We also found a few mice belonging to each one of all the other four western mitochondrial clades A, B, C and D (<3 individuals each) (Table 5.1). These represent a total of at least seven independent arrivals of colonists from other Western (non-British) countries, belonging to clades found nowhere else in New Zealand.

Clade A is the most diverse of these, found in four separate locations, together providing evidence of three independent invasions. Two are in the North Island (in the Bay of Islands and in Taranaki), and two in the South (in Canterbury and on Pitt Island). The mice on Pitt Island are an especially interesting lot because they belong to domNZ.7, a mitochondrial lineage found nowhere else in our samples except around the Canterbury port of Timaru, the home base of the shipping company that carries stock and goods to and from Pitt Island. So it is not hard to explain that link, although we cannot tell whether this lineage of mice first became distinct on the island and sent colonists to the mainland or vice versa.

Clade B is represented by three haplotypes found only in Hawke's Bay, presumably the diversified offspring of a single invasion. Likewise, Clade D is represented by a single haplotype not found anywhere else, but dominant in a chunk of the northern South Island from the port of Nelson inland.

The most complex examples of single independent invasions are provided by the mice of Clade C that have become established only on two offshore islands not reached by any other mice. Offshore islands are especially challenging for mice, because although these little passengers are small enough to escape visual detection during the loading of ships, and to survive a long voyage by hiding among the stores and in confined spaces too narrow for rats, they are less likely than are rats to swim ashore. So their opportunities for reaching an island depended mainly on intact packages of stores being landed, or at least cast ashore undamaged. That is less likely in places not being colonised by organised groups of settlers arriving with all their goods and chattels, but is, nevertheless, how mice of Clade C got ashore on Ruapuke and Antipodes Islands.

Among the new trading ventures permitted by the change in shipping regulations after 1821, even before sealing declined (Chap. 3), was an expedition by the Sydney-based *Elizabeth Henrietta* carrying 'European manufactures' to trade with the Māori for dressed flax. Increasing numbers of Port Jackson-based commercial firms were sending ships to explore the southern coasts for this valuable product, such as the expedition to Bluff harbour by the brig *Perseverance* in 1813, described by the Sydney-based rope-maker Robert Williams,[18] and the cutter *Mermaid*, whose survey in the winter of 1823 is recorded in her captain's journal (Chap. 4).

The *Elizabeth Henrietta* was stranded on Ruapuke Island, in Foveaux Strait, between Stewart Island and the southern coast of Otago, in 1824, and attempts to lighten the ship required landing all stores, trade goods and 10 tons of flax, valued at £1000.[19] After the ship had re-floated, retrieved its property and sailed away, the islanders found it had left a population of mice behind. The Māori residents of Ruapuke Island did not know what the strange creatures were and reportedly referred to them as 'henriettas' after the ship they came from.

These were the first mice known from documented records to have reached southern New Zealand, but they were completely different in origin from the mice that independently colonised Fiordland and Southland, only 15 km away. On the contrary, their descendants belong to a haplotype of western mice (domNZ.19) not recorded anywhere else in New Zealand, but with relatives in Iberia and on Antipodes Island.

Antipodes Island[20] supported hugely productive fur seal colonies, and at the height of the sealing trade, many sealing gangs landed with their piles of provisions.[21] Curiously enough, it seems unlikely that any mice came with them. Mice were not found on the island until a hundred years later in 1909, a year after the *Président Felix Fauré*, a French four-masted barque, was wrecked on rocks in Anchorage Bay, on the north side of the island, in 1908. All 22 men and a dog on board made it ashore, and survived for two months before being rescued. They must have retrieved some mouse-infested stores from the wreck, because mice collected from this island all belong to domNZ.8, belonging to western Clade C—a lineage originating in southern Europe, including France, which has so far not been found in mainland New Zealand or in Sydney.

History of Asian Mice in New Zealand

The origins of the six haplotypes of Asian mtDNA (*M.m. castaneus*) we have found so far in 218 mice, all belong to the clade HG2, which is widespread in Southeast Asia,[22] so are much harder to distinguish. Five of these were found on the main islands, and a quite different one, only on main Chatham Island. They represent at least two totally independent invasions by different mitochondrial lineages, each not found in the other location.

Mitochondrial haplotype casNZ.1 is completely dominant over very wide areas of the southern South Island. Mitochondrial data alone suggested at first that this could have been the first haplotype to arrive and establish a bridgehead on the Otago coast, from where colonising mice of pure Asian origin could stage a rapid and unopposed spread inland from the first entry point—a common pattern on islands, including Australia—until they met western mice spreading south from Canterbury.

On the Chatham Islands group, 840 km east of Christchurch, mice live on both main Chatham Island and on Pitt Island, only 25 km apart and yet have completely different ancestry. The mitochondrial DNA of mice on main Chatham is inherited from the Asian lineage casNZ.2, a completely different one from that found anywhere in New Zealand or Australia. Nearby Pitt Island has mice that are west European in origin, showing that they have an invasion history different from that of the main Chatham Island.

Historic data show that there were plenty of opportunities for Asian mice to reach Sydney from India or Canton. One appealing idea is that Indian/Batavian food supplies and cattle imported to Sydney by, among others, Robert Campbell (Chap. 3) could be a potential explanation for how Asian mice could have reached Sydney, and thence New Zealand. But it has until now foundered on the inconvenient fact that so far, there are no traces of Asian mitochondrial DNA in Australian mice.[23]

The Australian samples available so far are small, and were not taken from Sydney's historic sites such as Campbell's Wharf (Fig. 3.2), where Campbell's ships unloaded cargoes from India throughout the early 1800s. We know from our intensive surveys in New Zealand that huge variations in ancestry can be found in samples of mice taken from very limited areas. In one single calving shed, we found nine mice representing four different haplotypes of western mice of two different clades, coexisting with seven mice of Asian ancestry.

Campbell's imports of cattle must have been accompanied by stocks of feed for them, and, perhaps, Indian mice. But both of the haplotypes of the Asian subspecies found in New Zealand are closer to known Fukien or Taiwan haplotypes, rather than to Indian ones.[24] There is a lot of noise in the mitochondrial data, and the evidence is far from conclusive, but given these results, the safer bet at the moment is that Asian mice reached New Zealand from China rather than India. So also did Norway rats, potentially on the same ships or at least by the same shipping routes (Chap. 4).

So, we need to undertake more intensive and targeted sampling of mice from the Rocks, the historic port area of Sydney, before we conclude that no Asian maternal lineage was ever there, at least at first. Mummified mouse remains are on display at the Hyde Park Barracks erected to accommodate newly-arrived convicts. One of them, the oldest securely dated mouse remains so far known from Sydney, was found in a matchbox dated to sometime between 1862 and 1886,[25] but has not been haplotyped. New collections of rodent bones from excavations of historic buildings at the Rocks are now being processed.[26]

The other possibility is that Asian mice could have reached New Zealand, not via Port Jackson, but directly from China. That would mean that any Asian signatures still detectable in New Zealand mice must have come by some more direct route. At a time when Port Jackson could barely feed its own settlement, let alone offer surplus provisions for export, it must have been very difficult for Sydney-based sealing masters to find the large quantities of supplies that had to be landed with New Zealand shore sealing gangs in the 1790s–1810s. Provisions could be sourced, or at least supplemented, with back cargo from Asian ports. While it lasted, the sealing trade might have been expected to offer plenty of chances for Asian mice to be carried to the south coast with provisions direct from Canton (Chap. 3).

By the 1810s, the export of New Zealand sealskins to Canton was so uneconomic that the Sydney-based traders abandoned the attempt. None of the mice on any of the southern islands where sealing was established after 1810 have any trace of Asian ancestry, suggesting that the opportunity for Asian mice to reach New Zealand in provisions from Southeast Asia via Sydney had already passed. Mice on Macquarie Island and Auckland Island belong to the same Clade E western lineages common on mainland New Zealand and in Sydney, although the nuclear DNA of the Auckland Island mice suggests a distant ancestral connection with mice from the USA. American sealing and whaling ships certainly visited this and all other subantarctic islands during the boom years. But there is no evidence that Asian mice reached any of the subantarctic islands worked intensively by Sydney-based sealers.

Evidence of Interbreeding in the South

Mitochondrial data are relatively easy to extract and analyse and are very widely used to deduce colonisation histories like the ones summarised earlier. The nuclear DNA are needed to fill in the missing parts of the story hitherto told from the mother's point of view only. But nuclear data are more difficult to handle, so less often employed.

We found good reason to doubt our initial interpretation of mouse history based on mtDNA alone. The mitochondrial (maternal) DNA found on both the southern South Island and on main Chatham was almost entirely derived from Asian ancestors, but the nuclear DNA showed that the remaining input came almost entirely (99–100%) from western lineages. Put simply, all of these mice's mother's mother's mother's … mothers were Asian, but all of their remaining detectable ancestry was from Western Europe. That means that these two independent populations of New Zealand mice provide unusually dramatic examples of genetic discordance following incomplete hybridisation. That is, the libraries of mitochondrial and nuclear DNA data of these mice did not agree.

These mice are not true hybrids, because there is no nuclear admixture; rather, their nuclear DNA is derived purely from Western European ancestors, while their body cells still carry mitochondrial 'ghost' DNA derived from a previous encounter with Asian ancestors. Geneticists call this 'mitochondrial lineage capture', and New Zealand is remarkable in having seen it happen twice independently—on both the South and main Chatham Islands. The same process has been observed in other mammals, but rarely so completely as in these mouse populations. It is evidence of a greater reproductive success by western males compared with Asian males during interbreeding between mice of different lineages. These observations also show that there are genetic incompatibilities between these two subspecies, so that some combinations of their genes cause reduced fertility or death. In that case, evolution will then re-segregate their genomes, like water and oil separating, because mice with mixed ancestry risk surviving less often. So, did the two subspecies arrive separately and interbreed after landing, or were they hybridised already?

Our best clue as to what really happened comes from a detailed study of mitochondrial lineage capture in the southern South Island mice with discordant genomes.[27] One possibility is that a population of Asian mice, originally established in Fiordland from sealing vessels before 1810, was invaded by West European mice arriving on trading and whaling ships from Sydney touching the south coast during the 1820s–30s (Chap. 3). If so, the proportion of hybrids would decrease northwards as the invading genetic markers spread inland. The geographic details show no smooth increase in genetic diversity northwards, so ruling out the first interpretation.

The alternative suggestion is that there was no large established population of purebred Asian mice already widespread on the South Island invaded by a later-arriving group of western mice. It seems more likely that the two groups met on the boat, or in a previous port, and arrived and spread as 'hybrids'. If the first arrivals were inbred 'hybrid' mice already carrying this discordant genome, markers from both subspecies would have been able to spread together from the coasts of Fiordland and Otago north into previously mouse-free territory. So the interpretation of these results best fitting the geographic details is that a small number of 'hybrid' individuals arrived in the southern South Island when it was still mouse-free and established a population that then spread north. All of these 'hybrid' mice are genetically very similar to each other (this population has the lowest genetic diversity found anywhere among mainland mice) and they are far less morphologically variable than are other New Zealand mice.

Confirmation of this alternative idea can be found in the eastern South Island, where there is a clear contact zone showing the meeting point of mice of different origins, one group spreading northwards from the Otago coast, and the other spreading southwards from Canterbury (Fig. 5.2). The contact zone is only about 50 km wide, and runs from just north of Temuka to around Waimate, both in South Canterbury. Helen McCormick found that all 46 mice she examined from north of the zone to Cook Strait were purebred westerners, and all but one of 172 from south of it were Asian/western 'hybrids'.

Contemporary reports commented on the unwelcome invasion of mice in the southern South Island. Gillies[28] was quite certain that there were no mice in Otago in 1852 (he arrived in 1848), but within two years, they were noticed in Dunedin. Taylor White arrived in Canterbury in 1855, and when he went out rat-catching with dogs on his farm some 60 km inland, there was 'not a mouse to be seen'. Then, hordes of mice appeared in the tussock and in every corner of the house; they were everywhere.[29] Shortly afterwards, he reported that the gold diggings in the Nelson district already had both rats and mice. In the late 1850s, he witnessed the arrival of mice on another farm in Canterbury, and then, he saw the same thing happening again in Southland a few years after he settled there in 1861.

White put these observations down to mass migrations of mice dispersing into Otago, without apparent human help, before the mid-1860s. But there would have been plenty of time since the 1830s for mice to spread north from the shore whaling villages along the Southland coast with the whalers-turned-farmers.

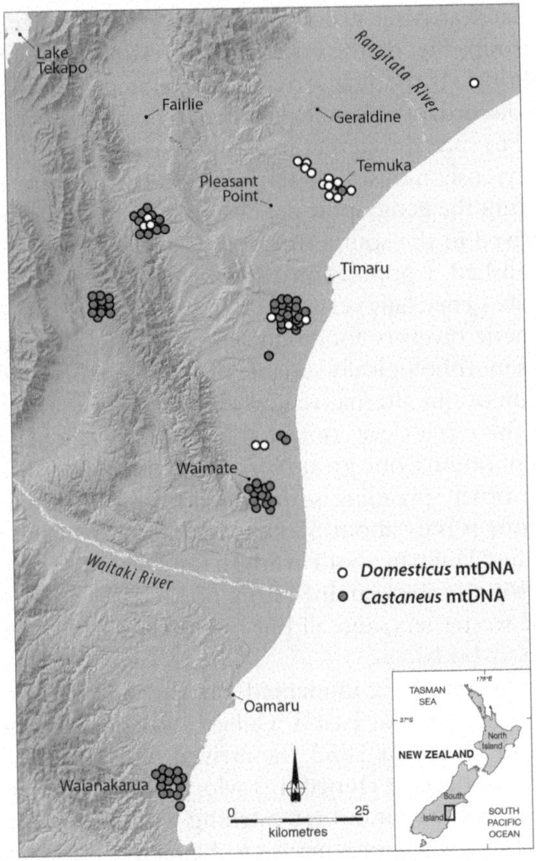

Fig. 5.2 Map of the contact zone between independently colonising lineages of house mice in the South Island. (Redrawn by Max Oulton from McCormick et al. (2014). Location of a contact zone between *Mus musculus domesticus* and *M. m. domesticus* with *M. m. castaneus* mtDNA in southern New Zealand. Mammalian Biology, 79, 297–305)

An article in the London *Spectator* of January 1860, describing the changes in the Australian landscape since European settlement, commented that the English settler's

> sense of home is rendered complete by the common residence of the house rat and mouse. He has the retrospective felicity sometimes of finding them help him to eat his cheese and bacon ... [they have] followed him as his ordinary coadjutors.[30]

The same problems, and ultimately from the same source, soon became widespread in New Zealand.

Discussions of how Asian rodents got to New Zealand inevitably raise questions about the Chinese gold miners who sailed to Otago in droves after 1866. It is a reasonable question, but the short answer is that any mice accompanying the miners came too late: anywhere they went, they met populations of mice and rats, some with Asian ancestry, already resident in the pastoral landscape established from the 1850s onwards (Chap. 3).

For example, on the Shotover River (a goldfield in inland Otago), one digger wrote in July 1864:

> We have been over-run with rats and mice lately; these plagues swarm, I believe in all parts of New Zealand. I caught a great many in traps of my own invention. ... I could not keep meat, flour, candles, soap, or anything at all eatable. My head, as I lay in bed, was a favourite springing-place for the rats who wished to get upon the table.[31]

Evidence of Interbreeding Further North

On the North Island and in Nelson, we found a rather different form of genetic discordance. All 107 mice we examined north of Hamilton carried maternal mtDNA belonging to the western subspecies, dominated by a single haplotype domNZ.4, found in 98 of them. An identical lineage is also dominant in Australia[32] and in Britain.[33] From mitochondrial data alone, this conclusion amounted to a cut-and-dried case of straighforward, predictable hitch-hiking with traders and settlers straight from Europe via the Bay of Islands, which then spread widely throughout the northern North Island. But unexpectedly, the nuclear DNA analyses detected occasional traces of Asian paternal input, even in mice carrying wholly European mitochondrial (maternal) DNA. These nuclear genes show that the ancestors of some of these apparently all-western mice also carried traces of previous encounters with Asian mice.

The proportion of non-western paternity in the North Island mice is tiny. Up to 3% of nuclear DNA in Northland, and about 1% in Taranaki, came from the eastern subspecies, even though we never found any Asian mitochondrial genes in either place. In the Wellington region, mtDNA of all three subspecies is present, representing multiple European lineages of Clade E west European mice living in and around the city and extending up State Highway 2 towards Hawke's Bay, plus one haplotype each of the other two subspecies. The predator-proof enclosure in the Wellington suburb of Karori is free of rats and therefore well-populated with mice. All the Karori mice we sampled carried either Asian or north European maternal mtDNA, but only 2% of their paternal ancestry came from Asian mice, plus a trace (0.02%) from northern Europe. All the rest (97%) represented western ancestors (Table 5.1).

In the Nelson area, only West European mice are represented in all mitochondrial DNA found there so far, but three separate populations carried evidence of northern paternal ancestry, declining southwards from 7% to 8% on the Nelson coast and Lake Rotoiti, 5% at Hurunui and 1% at Franz Josef, Westland.

The interesting question is, how could all these mixtures have happened? More interestingly, which subspecies got here first, and invaded the other?

The conditions best favouring these outcomes are found when the source of the mitochondrial input is rare, the combined breeding population is small, and hybrid sterility ensures that cross-breeding does not work equally well in both directions. All three conditions could be met when a lot of European mice met a few Asian mice for the first time—for example, en route to New Zealand on ships that had picked up both subspecies on voyages to Canton, or from goods brought from there on European ships loading in Sydney or a previous port, or immediately after two purebred propagules (colonising groups) both arrived on land.

One scenario can be ruled out: European males could invade an established population whose females were all of Asian origin, but not vice versa. From other information, we know that hybrid sterility prevents males of the Asian subspecies from producing fertile offspring from a female of the European subspecies, but European males have no such problem, so the 'hybrids' are biased against the Asian lineage. So we cannot explain the traces of Asian paternity in the northern North

Island by supposing an invasion by Asian mice; no new Asian males could have invaded an established population of European females and bred successfully enough to leave viable offspring with traces of Asian paternal input.

So those mice we collected in the Bay of Islands and Taranaki that carried traces of Asian nuclear ancestry could be the descendants of colonists that were the distant offspring of parents that had met on more equal terms elsewhere. There could have been many visits by ships that had previously picked up supplies, and mice, from ports in Southeast Asia, before calling briefly at Kororareka or other coastal villages in the early days. Whalers were among the earliest and unregulated visitors to the North Island, in search of fresh water, pork and potatoes in exchange for muskets. Any 'hybrid' mice they landed have been almost entirely replaced by wave after wave of pure-bred westerners. But the story is complicated and indisputable data are few. No doubt there will be more surprises down the track.

History of North European Mice in New Zealand

The northern subspecies (*M.m. musculus*) is represented in the northern South Island by some nuclear DNA, and in the Wellington region by a lot of mitochondrial DNA plus some limited nuclear ancestry. There are several reasons to suspect that these mice came from central Europe. First, the mitochondrial haplotypes are identical with some found in Germany and Austria. Second, the Nelson area in the northern South Island was colonised primarily by German settlers during the nineteenth century.[34] The northern South Island was also the only area in New Zealand where we found mice carrying Clade D1 mtDNA, common in Germany, strengthening the implication that these early German immigrants brought mice with them from home.

The proportions of northern nuclear ancestry detected in the Nelson region is similar to that found in Europe, in some mice sampled near the hybrid zone between the western and northern subspecies that crosses the continent from Denmark south to Greece. It is not yet known whether pure northern house mice ever arrived in New Zealand, or if individuals from the European hybrid zone were the founders of this population.

Mice as Conservation Pests

Like rabbits (Chap. 7), house mice are not usually regarded as predators. But in New Zealand, wild house mice behave quite differently from the way they do as commensals in their home countries, and they are significant to conservation in New Zealand for two reasons. One, they are, in fact, significant predators in their own right, at least of invertebrates; and two, they are important prey for other predators, especially cats and mustelids.

The effects of mouse predation are easiest to see on islands that have no other introduced mammals. For example, mice lived on Antipodes Island from 1908 (P. 111) until they were eradicated in 2016. James Russell's survey found fewer large invertebrates there compared with nearby mouse-free islands,[35] and now, they presumably will have the chance to recover. Mice probably arrived on Mana Island (217 ha), off the North Island coast, in 1886, when it was cleared for farming, with associated losses of native fauna. Farming was abandoned in the 1980s, and the mice were eradicated in 1989, to the immediate benefit of two lizards, the McGregor's skink, the Wellington green gecko and the Cook Strait giant weta.[36] The New Zealand shore plover was reintroduced, had started breeding and was doing well until a single Norway rat arrived in a bale of hay, and wiped most of the new population out.[37]

Mass seeding events in native forests produce periodic irruptions of mice and ship rats, comparable with those of the voles and lemmings of the northern hemisphere. Stoats and weasels evolved in that environment, so are well adapted to take advantage of such feast-or-famine conditions. Hence, every mast year has catastrophic consequences for conservation. After a seed fall, stoats happily concentrate on rodents, their favourite food, and produce extra-large cohorts of young.[38] Wild house mice are simply too small and too few to distract the stoats' attention from inflicting collateral damage on native birds or lizards, so a larger than usual number of stoats each eating the usual number of birds per head adds up to a serious challenge for beech forest birds during the summer after a mast season. Those most at risk are the ones that nest in tree holes with only one entrance. Stoats are slim, agile climbers and can easily access the nest holes, and when they do, the sitting female has nowhere to go.[39]

Eradication of rats from islands is now relatively routine, but mice are more difficult targets. Removal of rats is nearly always followed by increased numbers of mice.[40] Successful clearing of rats out of a fenced sanctuary will predictably produce swarms of mice, whose scent will attract more weasels and stoats to patrol the fence looking for a way in. Fourteen weasels have succeeded at Maungatautari, and a stoat at Orokonui.[41] The volunteers in charge of monitoring operations at sanctuaries and pest-free islands can never relax their guard.

Mice as Historians

Historic patterns of international trading and migration have largely prevented European and Asian mice from meeting elsewhere in the world (Chap. 3). But the picture that has emerged shows that New Zealand has hosted a unique geographical encounter between West European and Asian mice, two independently evolved subspecies that have remained separate for millennia, plus a scattering of north European mice in a few places. Explaining how, when and from where these mice reached New Zealand, the last major landmass occupied by humans, has turned into an intriguing biogeographical puzzle. These data tell us something new about the history of New Zealand mice, and of pre-colonial trading activities in the southern Pacific.

Historical records of shipping routes, political decisions, market prices, trading patterns and immigration help to explain the earliest invasions by mice native to Western Europe and established in the developing port of Sydney after 1788. Apparently, one-off arrivals have included mice from northern and eastern Europe in Wellington; mice of a European lineage unknown elsewhere in New Zealand on Antipodes Island after a shipwreck in 1908; and a related one, also unknown elsewhere, on Ruapuke Island with stores in 1824.

The first mice to reach New Zealand from Asian sources are most likely to have come, several times independently, as hybrids already carrying genetic markers from both Asian and European ancestors, signalling previous encounters outside New Zealand. An independent invasion of a completely different lineage of Asian origin, not found on the main islands, has colonised main Chatham Island. By piecing together the evidence from both shipping records and genetic analyses, we have found that what originally seemed like a simple story has turned into a remarkably interesting saga (Figure 5.3).

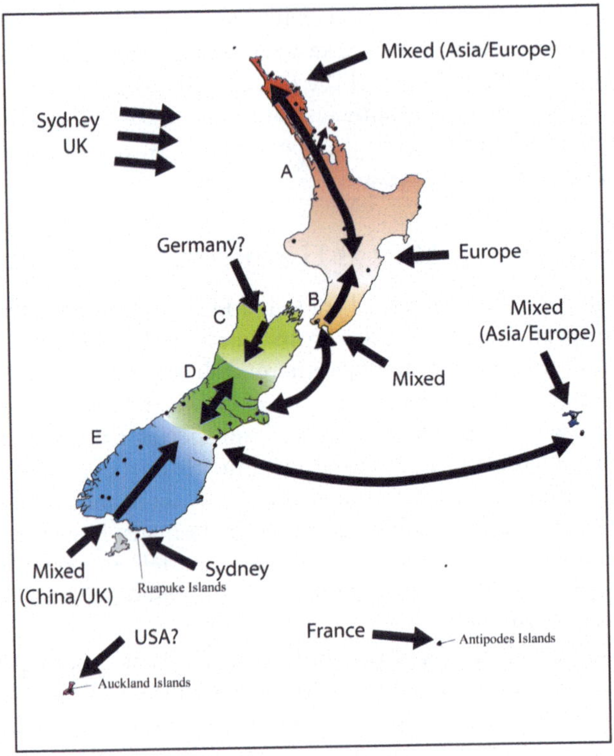

Fig. 5.3 Broad history of invasions by house mice into and within New Zealand. (Map by Andrew Veale, based on Veale, A. J., Russell, J. C., & King, C. M. (2018). The genomic ancestry, landscape genetics and invasion history of introduced mice in New Zealand. Royal Society Open Science, 5, 170879)

Summary

The house mouse *Mus musculus* comprises three distinct subspecies. In New Zealand, it has 33 distinct haplotypes from 8 clades of separate geographical origins. The distribution of so many distinguishable lineages of mice around the main and subantarctic islands throws new light on settlement history. The northern and southern ends of the country were colonised separately from different sources. Living mice in Otago, main Chatham Island and patchily in the North Island have discordant genomes, clear evidence of incomplete hybridisation on ships or in ports outside

New Zealand. They show in unexpected detail how, and roughly when, west European *Mus musculus domesticus* encountered an independently evolved subspecies of house mice, the Southeast Asian *M. m. castaneus*. These 'hybrids' were the first mice to arrive in the far south of the South Island, probably with sealers, and have spread north to meet *M.m. domesticus* at a clear contact zone. Most of the rest of the country was colonised multiple times by western mice from British clades E and F. Clades A-D represent seven independent invasions of non-British stock, including on two offshore islands after shipping accidents. The northern subspecies *M.m. musculus* is represented in only two locations.

Notes

1. Jameson, R. G. (1842). New Zealand, South Australia and New South Wales: a record of recent travel in these colonies, with especial reference to emigration and the advantageous employment of labour and capital. London: Smith, Elder & Co. P. 335.
2. Berry, R. J., & Scriven, P. N. (2005). The house mouse: a model and motor for evolutionary understanding. Biological Journal of the Linnaean Society, 84, 335–347.
3. Bonhomme, F., & Searle, J. B. (2012). House mouse phylogeography. In M. Macholán, S. J. E. Baird, P. Munclinger, & J. Pialek (Eds.), Evolution of the house mouse (pp. 278–296). Cambridge: Cambridge University Press.
4. Jones, E. P., Skirnisson, K., McGovern, T. H., Gilbert, M. T., Willerslev, E., & Searle, J. B. (2012). Fellow travellers: a concordance of colonization patterns between mice and men in the North Atlantic region. BMC Evolutionary Biology, 12, 35–42.
5. Steadman, D. W. (2006). Extinction and biogeography of tropical Pacific birds. Chicago: University of Chicago Press.
6. Worthy, T. H., & Holdaway, R. N. (2002). The lost world of the moa. Bloomington: Indiana University Press.
7. Searle, J. B., Jones, C. S., Gűndűz, I., Scascitelli, M., et al. (2009). Of mice and (Viking?) men: Phylogeography of British and Irish house mice. Proceedings of the Royal Society B, 276, 201–207.
8. Searle, J. B., Jamieson, P. M., Gunduz, I., Stevens, M. I., Jones, E. P., Gemmill, E. C., & King, C. M. (2009). The diverse origins of New Zealand house mice. Proceedings of the Royal Society B, 276, 209–217.
9. King, C. M., Alexander, A., Chubb, T. L. A., et al. (2016). What can the geographic distribution of mtDNA haplotypes tell us about the invasion of New Zealand by house mice *Mus musculus*? Biological Invasions, 18, 1551–1565.

10. Veale, A. J., Russell, J. C., & King, C. M. (2018). The genomic ancestry, landscape genetics and invasion history of introduced mice in New Zealand. Royal Society Open Science, 5, 170879.
11. Jones, E. P., Jóhannesdóttir, F., Gündüz, I., Richards, M. B., & Searle, J. B. (2011). The expansion of the house mouse into north-western Europe. Journal of Zoology, 283, 257–268.
12. McNab, R. (Ed.) (1908). Historical Records of New Zealand, Vol 1. Wellington: Government Printer. P. 543.
13. Guthrie-Smith, H. (1969). Tutira. The Story of a New Zealand Sheep Station 4th edition. Wellington: A. H. & A. W. Reed. Pp. 333–334.
14. Jameson, R. G. (1842). New Zealand, South Australia and New South Wales: A record of recent travel in these colonies, with especial reference to emigration and the advantageous employment of labour and capital. London: Smith, Elder & Co. P. 336.
15. Taylor, R. (1855). Te ika a Maui, or, New Zealand and its inhabitants. London: Wertheim and MacIntosh. P. 213.
16. Rochfort, J. (1853). The adventures of a surveyor in New Zealand. London: David Bogue. P. 36.
17. Guthrie-Smith (1969). P. 334.
18. McNab, R. (1909). Murihiku: A history of the South Island of New Zealand and the islands adjacent and lying to the south, from 1642 to 1835. Wellington: Whitcombe & Tombs Ltd. 1909, reprinted by Cambridge University Press 2011. Pp. 263–268.
19. McNab, R. (Ed.) (1908). Pp. 655–656.
20. Russell, J. C. (2012). Spatio-temporal patterns of introduced mice and invertebrates on Antipodes Island. Polar Biology, 35, 1187–1195.
21. Ross, J. O. C. (1987). William Stewart: Sealing captain, trader and speculator. Canberra: Roebuck Society. P. 44.
22. Rajabi-Maham, H., Orth, A., Siahsarvic, R., Boursot, P., Darvish, J., & Bonhomme, F. (2012). The south-eastern house mouse *Mus musculus castaneus* (Rodentia: Muridae) is a polytypic subspecies. Biological Journal of the Linnean Society, 107, 295–306.
23. Gabriel, S. I., Stevens, M. I., da Luz Mathias, M., & Searle, J. B. (2011). Of Mice and 'Convicts': Origin of the Australian house mouse, *Mus musculus*. PLOS One, 6(12), 1–6.
24. Rajabi-Maham, H., Orth, A., Siahsarvie, R., Boursot, P., Darvish, J., & Bonhomme, F. (2012).
25. Davies, P., & Garvey, J. (2013). Early zooarchaeological evidence for *Mus musculus* in Australia. International Journal of Osteoarchaeology, 23, 106–111.
26. A. Veale, C. King et al., in progress. Preliminary results have so far found no trace of *M.m. castaneus*.

27. Veale, A. J., Russell, J. C., & King, C. M. (2018).
28. Gillies, R. (1878). Notes on some changes in the fauna of Otago. Transactions of the New Zealand Institute, 10, 306–322.
29. White, T. (1890). On rats and mice. Transactions and Proceedings of the New Zealand Institute, 23, 194–201.
30. Anon. (1860, 21 January). Man's work upon the globe. Otago Witness [reprinted from Spectator].
31. Anon. (1864, 8 November). A gold-digger's notes on Otago. Otago Daily Times.
32. Gabriel, S. I., et al. (2011).
33. Searle, J. B., Jones, C. S., Gűndűz, I., et al. (2009).
34. King, M. (2003). The Penguin History of New Zealand. Auckland: Penguin Books.
35. Russell, J. C. (2012). Spatio-temporal patterns of introduced mice and invertebrates on Antipodes Island. Polar Biology, 38, 1187–1195.
36. Newman, D. G. (1994). Effects of a mouse, *Mus musculus*, eradication programme and habitat change on lizard populations of Mana Island, New Zealand, with special reference to McGregor's skink, *Cyclodina macgregori*. NZ Journal of Zoology, 21, 443–456.
37. Dowding, J. E., & O'Connor, S. M. (2013). Reducing the risk of extinction of a globally threatened shorebird: Translocations of the shore plover (*Thinornis novaeseelandiae*), 1990–2012. Notornis, 60, 70–84.
38. King, C. M., White, P. C. L., Purdey, D., & Lawrence, B. (2003). Matching productivity to resource availability in a small predator, the stoat (*Mustela erminea*). Canadian Journal of Zoology, 81, 662–669.
39. Elliott, G. P. (1996). Productivity and mortality of mohua (*Mohoua ochrocephala*). NZ Journal of Zoology, 23(3), 229–237.
40. Bridgman, L. J., Innes, J., Gillies, C., Fitzgerald, N., Rohan, M., & King, C. M. (2018). Interactions between ship rats and house mice at Pureora Forest Park. New Zealand Journal of Zoology, 45, 238–256. Wilson, D. J., Innes, J. G., Fitzgerald, N. B., Bartlam, S., Watts, C., & Smale, M. C. (2018). Population dynamics of house mice without mammalian competitors and predators. New Zealand Journal of Ecology, 42, 192:203.
41. Innes, J., Fitzgerald, N., Binny, R., Byrom, A., Pech, R., Watts, C., Gillies, C., Maitland, M., Campbell-Hunt, C., Burns, B. (2019). New Zealand ecosanctuaries: types, attributes and outcomes. Journal of the Royal Society of New Zealand. https://doi.org/10.1080/03036758.03032019.01620297

CHAPTER 6

Cats (1769–1900)

> Ages of domesticity have failed to eradicate the native wildness of the cat: opportunity serving, puss soon relapses into feline barbarism, and forsakes the hearth to prowl the forest. Hursthouse (1857)[1]

Cats are usually classed as domesticated animals, but in rural areas at least, many of them are borderline commensals that retain their independence to variable extents. Neither is it always clear whether cats exploit their humans for shelter and food, or humans exploit cats for companionship and pest control. Either way, cats, humans and pest rodents have lived together for thousands of years,[2] and they have travelled around the world together for centuries.

Because it was always so difficult to protect ships from being boarded by rats, most ships also carried cats, which were supposed to keep rat numbers down. This was often a forlorn hope, although they dutifully killed rats and mice when opportunity offered. One midshipman on Cook's ship the *Resolution* had trained his cat to bring him its catches, which he cooked and shared with it, roasted and peppered.[3]

Shipboard cats could not possibly keep control of the numbers of rats hidden in cargo spaces too small for cats to enter, and full of unlimited supplies of food to sustain rats under cover indefinitely. Hence, the cats'

C. M. King, *Invasive Predators in New Zealand*,
Palgrave Studies in World Environmental History,
https://doi.org/10.1007/978-3-030-32138-3_6

main effect was simply to keep the rats out of sight. Cats had the same opportunities to explore the world as did rats, so had better chances than did wild carnivores such as mustelids of reaching remote islands populated by naïve endemic fauna. Cats were unlikely to swim ashore independently, but very quick to take advantage of any new dry-shod hunting opportunity.

Cats on the Mainland

The first and best documented example is that of one of the cats on the *Resolution* in Dusky Sound, which was moored 'head and stern, so near the shore as to reach it [via] a large tree, whose end or top reached our gunwale'.[4] Georg Forster noticed the ship's cat, who

> had no sooner perceived so excellent an opportunity of obtaining delicious meals, than she regularly took a walk in the woods every morning, and made great havoc among the little birds, that were not aware of such an insidious enemy.[5]

In the west and south of the South Island at that time, forests were thick and human settlements few. As the shipping traffic around New Zealand increased in the years following Cook's visits (Chap. 3), many more cats could have abandoned shipboard life in favour of the delights of hunting the unwary forest birds, or perhaps were left stranded if they were not on board when the ship sailed. Those that survived abandonment became truly feral, and their descendants are still there.

More cats again were brought in by the gold miners from 1865 onwards, and some of these got left behind in the sudden panic after a new rush was announced.[6] Some others found a new commercial value among miners plagued with rats. During the gold rushes on the Shotover River, one man wrote in July 1864 that he had

> bought a kitten for seven and sixpence, of a man who was going away … The general price here for cats is now from thirty to fifty shillings. It was still more. A man passed through this place with a horse-load of them in boxes, which he brought from Dunedin a few weeks ago.[7]

As the miners dispersed, the numbers of feral cats increased. The alpine explorers A.P. Harper[8] and Charles Douglas[9] (Fig. 6.1) both commented in the 1890s that wild cats must kill many birds, because the tracks of cats

Fig. 6.1 Harper and Douglas at the Cook River, April 1894. (Canterbury Museum, Day Book, p. 577. J.J. Kinsey Collection 1940.193.12)

were becoming almost common in the unexplored forests clothing the high country valleys of Westland, right up to the snowline.

Cats reached the settled regions of the northern North Island in the earliest days of European ship visits, where the Māori had a rather different view of them. One of the best descriptions is provided by Joel Polack, a trader who lived in the Bay of Islands from 1832 to 1837, built a large house and trading base in Kororareka and published detailed observations of his life there:

> The puhihi, the pronunciation for pussey cat, has been introduced within the last twenty-five years [since about 1813]. This animal is looked upon with much affection by these kind-hearted people, for the delicacy of its flesh when cooked, and its skin; it is accounted as very nutritious food. A large-sized cat, belonging to a European, is not likely to preserve its situation for a length of time, as it is almost sure to be stolen by the natives, who pay but little attention to that code of morality, *meum et teum* [mine and yours].[10]

Likewise, an independent report by Richard Cruise recounts an incident in which one of the natives 'brought us a present of a cat, which he must have intended for a repast, as he seemed a good deal surprised that we did not eat it'.[11] Cruise added that the cat must have come from the shipping at the Bay of Islands or the Coromandel.

These accounts tell us a couple of important things about how the Māori made sense of the new animals invading their previously settled lives. First, unfamiliar observations can at first be understood only in relation to what is already known. Māori were impressed by the large size of Norway rats compared with their familiar little kiore, and at first, coveted them as new and more generous supplies of meat (Chap. 4). The valued the first chickens as sources of feathers for making traditional cloaks, easier to collect than those of wild birds (Chap. 2). Likewise, they saw the first cats as equivalents to kuri as valuable sources, both of meat and of skins for fine quality clothing.

Second, the Māori held tribal property in common, and were unfamiliar with European ideas of the difference between *mine* and *yours.* More significant, any violations of tapu (laws of sanctity) were seen as serious threats to the whole tribe, so had to be reported and dealt with. Polack added:

> Many a poor cat have I thus lost, through these informers, pretending to have found puss in a tapued [sacred] place, whereas the only fault of grimalkin consisted in having a handsome outside, which answered for dress mats, and was in good condition inside, which equally answered for the oven.[12]

European explorers often commented on the activities of cats they saw in the bush from the 1840s onwards:

> The cat probably introduced itself from some early ship. Like every other alien animal introduced into this teeming country, it has taken full possession of the soil; and, like the dog, has multiplied beyond due limits. … When the New Zealand settler moves deeper into the Bush, the cat usually accompanies [him]: the first bush-born cats are always shy, the second are more so, and will prowl about the near woods; here, they meet neighbour cats similarly inclined to forest life; and, thus, in three generations, a cat race is produced which is so far wild as to shun the house, to have its lair about the farm buildings and to subsist itself on birds, lizards and rats; and which would show small respect to any pullet straying beyond the sanctuary of the homestead.[13]

> The common cat is at present [1844] found in a wild state in the forests of New Zealand, and is another cause of the extermination of indigenous birds.[14]
>
> On 13 November [1843], north of Taumarunui, about 90 km from the sea and even further from pakeha settlement, the guides [accompanying Richard Taylor and George Selwyn] killed a wild cat. Its stomach contained the intact bodies of 30 lizards.[15]

Feral cats are still widespread in North Island bush, where rabbits are few except on the overgrown margins of broad gravel river beds. Their main food resources now are provided by rats (Fig. 6.2), mice, lots of invertebrates, and young rabbits when available.[16]

Fig. 6.2 A feral cat interrupted in the middle of eating a ship rat in the Orongorongo Valley, Wellington, bearing an eartag, which marks it as part of a study by B.M. Fitzgerald. (Fitzgerald, B. M., & Karl, B. J. (1979). Foods of feral house cats (*Felis catus* L.) in forest of the Orongorongo Valley, Wellington. NZ Journal of Zoology, 6, 107–126. Photograph by Dave Ward)

Cats in cities often bring some proportion of their catches indoors to present to their owners, but certainly not all of them.[17] This habit offers an easy method of monitoring the effects of cat predation in urban areas, although the catches recorded depend largely on the age of the cat, the hunting habitat available and the identification skill of their owners. Allowing for that variation, the results usually show that the main prey of urban cats are rats, mice and common small birds, nearly all introduced.

In most places, the significance of cat predation is unknown, but one study in the city of Dunedin was able to calculate the annual rate of cat-captures for three native bird species (bellbird, fantail and silvereye) and three introduced (house sparrow, song thrush and blackbird).[18] Yolanda van Heezik's team running the Dunedin study estimated that cats removed fantails, song thrushes and bellbirds faster than the local population could replace them, so these species remained present in the city only by immigration of new birds from outside. An alternative view is presented by John Flux, who recorded kills by fewer cats for much longer (two, for 17 and 10 years) on a hillside above Lower Hutt. He agreed that most cat kills are rodents and introduced small birds. But he concluded that the *net* effect of these cats' activities, even allowing for individual variation, was beneficial to native birds, suggesting that any proposed control of domestic cats would benefit mainly rats, mice and rabbits.[19] Rats are also common in cities, and also kill birds, so targeting cats alone to protect birds, or shutting them indoors at night, would not necessarily help; in this as everywhere, nothing in pest control is simple (Chap. 12).

Cats and Rabbits

In the mid-nineteenth century, as the vast grasslands of the dry eastern sides of both main islands began to teem with rabbits (Chap. 7), the distribution of cats was deliberately assisted by runholders buying up pet cats in towns and transporting them many kilometres inland for release on rabbit-infested pastures. For example:

> Messrs Beetham, of Brancepeth Station, Wairarapa, [have devised] a unique method of exterminating the rabbit nuisance on their property. … Mr Levin, town agent for Messrs Beetham, is prepared to give 2s 6d per head for any number of cats brought to his warehouse. … Up to the present 700 cats have been presented for sale.[20]

Charles Tripp, the long-time resident of the Orari Gorge run, inland from Geraldine, bought up cats from Timaru—often tame ones, which soon gave up the hard work of hunting rabbits, and resorted to the homestead to be fed by his housekeeper. He had to cart them back to the city, avoiding their angry owners.[21]

In 1883, William Acton Adams, an MP and prominent South Island landholder, began an active programme buying up cats in Christchurch and transporting them in 'several crates' by wagon for liberation on Tarndale, a large sheep run later incorporated into the vast Molesworth estate in Marlborough. In March 1885, another 200 were brought in, and six months later, Acton Adams asked the Amuri County Council to upgrade the Acheron road so that he could run a frequent four-horse express with cats. He got no support for his suggestion that Tarndale would be a good site for experiments on imported rabbit diseases, so continued to buy cats in Christchurch. He would carry up to 50 cats in his buggy when driving home from a visit to town.

> One wonders now how many pets belonging to maiden ladies in Christchurch were collected by boys for sale to Acton Adams; also how many of the cats died an early death on Tarndale before they had a chance to become acclimatized.[22]

Many other vivid descriptions of confused and traumatised cats being carted inland by the crateful could be repeated.[23] But the South Island runholders discovered, the hard way, that they had expected too much.

> The cat is a good hunter, but … although many cats have been turned out they have not increased as it was hoped they would do.[24]

The runholders could not explain why their confidence in cats as the natural enemy of rabbits was misplaced. Why, when surrounded by tasty prey, did not cats breed up to the expected huge numbers? They could not guess, because the mysteries of predator-prey dynamics were still shrouded in the future (Chap. 12). Worse, the runholders did not anticipate the long-term consequences of their naïve assumptions. Those cats that did become truly feral certainly killed many rabbits when rabbits were abundant, but without affecting the long-term numbers of such a fast-breeding and familiar prey. Then, when rabbits became scarce, cats would make do on lizards, or the eggs and chicks of ground-nesting birds. One five-year study using infra-red cameras on the open gravel river beds of the Waitaki Valley found that feral cats were responsible for nearly half (43%) of 'lethal events' recorded at nests of dotterels, stilts and terns.[25]

Cats on Offshore Islands

By a painful contrast with the minimal effects of cats on rabbits, cats reaching slow-breeding native birds, both on the mainland and especially on hitherto untouched islands, were immediate personifications of disaster. The best evidence for these doleful events comes from islands which had been visited by ornithologists whilst still in their pristine state. Such islands still represented the variety and abundance of wildlife on the pre-human mainland, which means that the toll on the birdlife on that island can be actually counted, and by implication, can illustrate the previous catastrophic events that followed the arrival of the first humans.

The list of proven local extinctions of endemic fauna attributable to cats, although shorter than the equivalent list for rats, far exceeds that for stoats, weasels or ferrets, largely because cats have often been deliberately or accidentally carried by people to offshore islands that no mustelids ever reached. Islands invaded only by cats, but not by rats or any other predators, have provided startling evidence of what cats can do when given the chance.

Herekopare

Herekopare is a small (28 ha) island off the northeast coast of Stewart Island (the side opposite to Big South Cape, see Chap. 4). There were no cats or any other mammals present when Herbert Guthrie-Smith observed the rich bird life there in 1911. He saw vast breeding populations of seabirds—diving petrels, broad-billed prions, fairy prions, sooty shearwaters—along with land birds, including red- and yellow-crowned parakeets, robins, Stewart Island snipe, fernbirds, brown creepers, tomtits and banded rails. Muttonbirds (the chicks of the sooty shearwater) grow large in their burrows, and are a valued traditional source of food for Māori who visit the islands every breeding season to harvest them. Alas, muttonbirders reported that cats were accidentally introduced in 1924.[26] Intermittent attempts to get rid of the cats started in the early 1940s. Whilst mainland scientists debated what to do, muttonbirders continued their efforts; they killed six cats by mid-1970, but they never got the last one.

Finally, an organised operation started in June 1970. Of the 33 cats left alive at that date, 30 were caught in traps, and two that evaded traps were shot after being cornered by dogs. The dogs then thoroughly searched the island until July and found the last cat dead of natural causes, and no further sign of live ones. They also found the decayed bodies of the six cats that muttonbirders had killed a few months earlier.

The effects of the cats on the birdlife of Herekopare are well documented by staff of the New Zealand Wildlife Service, who could compare bird species lists for 1932, 1940s and 1968–70 with Guthrie-Smith's for 1911. Six species of land birds, which mostly feed on the ground (yellow-crowned parakeet, fernbird, robin, brown creeper, banded rail and snipe) were completely exterminated. Red-crowned parakeets and tomtits were lost from the island, but later returned; weka still survive. Enormous nesting colonies of diving petrels and broad-billed prions were exterminated; fairy prions and sooty shearwaters persisted, and are now presumably recovering their numbers. But Herekopare is privately owned, and there is no public access or official management mandated by the Department of Conservation.

Stephens Island

Stephens Island (150 ha, bounded on all side by steep cliffs and thunderous waves) catches the full force of the gales that roar through Cook Strait, so was never an attractive place for human settlement in Māori times. Two important reviews published together in 2004 have greatly clarified the often-oversimplified story of what happened after 1892. By integrating a series of original unpublished reports, David Medway,[27] and Ross Galbreath and Derek Brown[28] were able to tell a fuller and much more interesting tale than the one that has been repeated for decades.

By 1891, little Stephens Island was, among very few other inshore islands, still rodent-free and supporting its original land bird fauna of 25 native species, including representatives of three endemic families—all by then scarce or gone from the mainland. The most famous of these was Lyall's wren, a tiny flightless avian mouse that once thrived throughout the pre-human mainland, but was, by then, surviving only on Stephen's Island. Other irreplaceable and vulnerable endemics that shared the island with the wrens included many of the same familiar list of vulnerable bird species whose other island refuges had already been invaded—South Island saddleback, piopio (native thrush), robin, banded rail, rifleman—some of them in huge abundance.

Unfortunately, that same windswept position that had once protected the island from visits by kiore or people, also made it an important site for a lighthouse to guard the dangerous shipping lane at the entrance to Cook Strait, and that guaranteed the end of the world for its most sensitive birds. The top of the island was still covered in thick forest at the time construction began on the lighthouse in 1892. It was finished in January 1894, and the first three families of light keepers moved in.

Lonely lighthouse keepers or their children were understandably keen to bring cats as companions, even where there are no rodents to control. The first cats must have been taken to Stephens Island by one or more of the families at the same time as, or at least within a few months of, the beginning of the lighthouse's operations, because during the winter of 1894, one of the keepers' domestic cats began to bring wrens to the house. David Lyall, an assistant keeper interested in natural history, retrieved the unusual birds and skinned them. Ironically, that particular cat may not have been owned by David Lyell's family, because there is a report that he obtained one specimen from 'the owner of the cat that had brought in the others'.[29] The key point was that Lyall wanted to know what it was, so the next time the government steamer called on its regular supply run, he sent a skin to Walter Buller, New Zealand's leading ornithologist, who, with great excitement, immediately recognised it as a new species.

While Buller was preparing a paper describing it, a rival bird collector and dealer, Henry Travers, persuaded Lyall to sell him nine more skins collected in late-winter/spring 1894. Travers—recognising a golden commercial opportunity when he saw one—tried to sell them at high prices in England, even before Buller's paper established their identity. Travers then visited the island with three assistants in search of more wrens in February 1895, but found none. Lyall concluded the wren was already extinct by November 1894. In fact, a few more specimens did turn up later, but the assumption of extinction helped to keep up the price of skins. Still, by December 1898, Travers was sure that the wren, piopio and the saddlebacks were all gone, and few of the robins and other birds that used to throng the bush were still hanging on.

Lyall has been unfairly blamed for aiding the extinction of the wren, but, contrary to the popular story, the last wrens were not finished off by a single domestic cat, but by the cats that had escaped the keepers' families, turned feral and multiplied in the forest that then still stood on the rest of the island. Altogether, only 16–18 specimens can now be tracked down worldwide, other than the subfossil bones often found in cave deposits and owl middens on the mainland. Only two were ever seen alive.

The last of the wrens and the rest of the abundant birdlife of the bush succumbed to feral cats so quickly that 'the demise of the native land bird fauna of Stephens Island was complete by 1903'.[30] The light keepers needed to clear space for their buildings, and for pasture for their sheep and cattle, which inevitably opened up the forest interior to wind throw and salt sting. By 1905, the principal light keeper reported that the island's forest was fast disappearing, and some commentators have made the point that felling the bush could have been enough on its own to have caused the loss of the birds. But that process was not sufficiently advanced in time to account for the *immediate* loss of the forest birds; only the cats could have done that.

Today, the cats have gone (eradicated in 1925), and Stephens Island is an internationally important location for conservation of what is left of its original fauna. Pride of place is taken by the tuatara, the ancient reptile once found throughout the mainland but now confined to offshore sanctuaries like Stephens. Other rarities surviving there are Hamilton's frog, plus several rare invertebrate species, and thousands of skinks and nesting seabirds. They all have to battle against the gales, but now at least they are safe from cats.

Stewart Island

The third largest island of New Zealand (Stewart, 174,600 ha) was frequently used by sealers, whalers and traders in early days, and by fishermen ever since, as shelter from storms. Cats live there along with all three of the rat species (kiore, Norway and ship rats), but without mice or any mustelids—the only island in the New Zealand archipelago with that peculiar combination of predators.

It is not certain when cats first arrived on Stewart Island, but they were certainly there when the last, largest and only functional breeding population of the endemic kākāpō, an iconic flightless parrot, was discovered in 1977. Field staff were elated to find that the island supported the first female kākāpō seen in a hundred years. They offered the only hope of rescuing the species, since only a few old males were left alone in Fiordland.

By early 1982, 29 out of more than 40 known kākāpō on Stewart Island were banded, and the population appeared still to number a healthy 50–150 birds despite the attentions of the cats.[31] Intensive cat control was launched, and was apparently helping, but even as the birds were being found by trained dogs, and many birds were being removed to safety elsewhere, the remaining cats were finding and removing the surviving kākāpō almost as fast. Just in time, all the 61 remaining living birds were transferred to cat- and rat-free islands by 1997. In this as in so many other cases, the number of small offshore islands available as refuges has made the difference between survival and extinction for many of New Zealand's ancient species. Kākāpō now have the benefit of an intensive breeding management programme on predator-free Codfish Island (Whenua Hou)., and their future looks good.[32]

Little Barrier Island

Little Barrier Island (2817 ha) was one of the earliest islands set aside for government wildlife sanctuaries at the end of the nineteenth century, and is uninhabited except for conservation staff. Its story illustrates two important issues: first, that nature is generally on the side of the pests, and second, that things get much more complicated when cats are not the only significant cause of a problem.

Kiore had been on Little Barrier since Polynesian times (Chap. 2), and cats since 1867. It was not at all clear which of the two was most directly responsible for the catastrophic decline in numbers of birds, lizards and tuatara on the island before organised surveys began in 1977. It seemed a good idea to start with the cats, but introducing cat flu, which had had some success in reducing cat numbers on Marion Island in the Indian Ocean (Chap. 12), didn't work at all, because the Little Barrier Island cats were descended from mainland ancestors, which were already immune to it.[33] So the cats were removed by conventional means, in a massive effort that succeeded only in 1980.

Over three years starting in 1977, 149 cats were trapped in 80,000 trap-nights, and 25,500 toxic baits removed unknown numbers more, yet the expected benefit to bush birds was undetectable. Unexpectedly, the breeding success of Cook's petrels actually declined. They recovered only after further operations also removed a large population of kiore suddenly free of suppression by cats but equally interested in raiding petrel nests.[34]

This operation not only took many years to complete, but also illustrated the need to consider the entire interacting community to be protected. The results suggested that, contrary to all expectations, the cats affected seabird nesting success less than did the kiore. In conservation, things are not always what they seem.

Macquarie Island

Macquarie Island (12,785 ha) is a bleak Australian subantarctic outpost, an important island for breeding seabirds and fur seals, but invaded by four species of pest mammals. Rabbits, landed in the late 1870s by sealers as a reliable source of food, caused great damage to the vegetation; ship rats and mice followed by accident; within a century, cats, brought in to control rodents and rabbits, were killing about 60,000 seabirds a year. From 1979, the introduced disease myxomatosis diminished rabbit numbers but failed to eradicate them, and soil erosion, cliff collapses and vegetation damage continued to damage the seabirds and their nests.

To protect the birds, the entire population of about 2500 cats was removed, the last individuals in 2000, whereupon rabbit numbers surged. Finally, in 2007, the Federal Government funded a massive A$24 million poison baiting programme, combined with a new disease, rabbit haemorrhagic disease (RHD; Chap. 12) and hunters, followed up by specialist dog teams. The last 13 rabbits were removed in April 2012, and two years later, the island was declared pest-free.[35] Again, the secret of success was adaptive management: when one strategy fails, go on to the next, or a combination, or try something new: just never give up!

All these sad stories have raised serious concerns about what can be done to undo the mistakes of the past. Fortunately, conservation and restoration of endemic species on islands are advanced and serious businesses, better developed in New Zealand than almost anywhere else in the world. As Winston Churchill pointed out, 'Success consists of going from failure to failure without loss of enthusiasm'.

Summary

Cats were frequently carried on European ships, supposedly to control rats, so both had similar opportunities to reach New Zealand from 1769 onwards. On the North Island, the first cats were often killed by Māori for their meat and skins, but by the 1840s, cats were accompanying European settlers into the bush. On the South Island, cats were uncommon until the 1870s, when they were often transported inland by pastoralists to control rabbits, and by gold-diggers plagued by rats and mice. Feral populations of cats established from these sources are now common throughout all the inhabited islands. Cats also had multiple opportunities to reach at least 35 offshore islands, by accident or as deliberate introductions.

Their exterminations (local or total) of vulnerable native birds and lizards, previously protected by their isolation, have been well documented on Stephens, Herekopare, Little Barrier, Stewart and Macquarie Islands. Eradication techniques targeting both cats and rodents, preferably together, have now cleared all these islands and many others. Restoration of island habitats, and reintroduction programmes are observing the encouraging recovery of surviving endemic fauna on cat-free islands.

Notes

1. Hursthouse, C. (1857). New Zealand, or Zealandia, the Britain of the south (Vol. 1). London: Edward Stanford.
2. Ottoni, C., Van Neer, W., Peters, J., Spassov, N., Bălăşescu, A., Chahoud, J., … Geigl, E. (2017). The palaeogenetics of cat dispersal in the ancient world. Nature Ecology & Evolution, 1 (0139).
3. Sparrman, A. ([1953]). A Voyage Round the World with Captain James Cook in H.M.V. *Resolution*. London: Robert Hale Ltd.
4. Beaglehole, J. C. (Ed.) (1961). The Journals of Captain James Cook on his Voyages of Discovery. Vol. 2. The Voyage of the *Resolution* and *Adventure* 1772–1775. Cambridge, UK: Cambridge University Press for the Hakluyt Society. P. 62.
5. Forster, J. G. A. (1777). A Voyage Round the World in His Britannic Majesty's Sloop *Resolution* (Vol. 1). London: B. White. P. 128.
6. Salmon, J. H. M. (1963). A history of goldmining in New Zealand. Wellington: Government Printer.
7. Anon. (1864). A gold-digger's notes on Otago. Otago Daily Times, 8 November.
8. Harper, A. P. (1896). Pioneer work in the Alps of New Zealand. London: T. Fisher & Unwin. P. 43.
9. Douglas, C. E. (1894). On the Westland Alps. Appendices to the Journal of the NZ House of Representatives C1, 30, 71–75.
10. Polack, S. J. (1838). New Zealand: Being a Narrative of Travels and Adventures during a Residence in that Country between the Years 1831 and 1837 (Vol. 1). London: Richard Bentley. P. 314.
11. Cruise, R. A. (1823). Journal of a Ten Months' Residence in New Zealand [1820]. Christchurch: Republished 1957 by Bagnall, A. G. (Ed.), Pegasus Press. Pp. 202, 316.
12. Polack (1838) Vol. 2. P. 60.
13. Hursthouse, C. (1857). New Zealand, or Zealandia, the Britain of the south (Vol. 1). London: Edward Stanford. P. 127.

14. Angas, G. F. (1866). Polynesia. London: Society for Promoting Christian Knowledge. P. 55.
15. Young, D. (2004). Our islands, our selves. Dunedin: University of Otago Press. P. 58.
16. Fitzgerald, B. M., & Karl, B. J. (1979). Foods of feral house cats (*Felis catus* L.) in forest of the Orongorongo Valley, Wellington. NZ Journal of Zoology, 6, 107–126.
17. Loyd, K. A. T., Hernandez, S. M., Carroll, J. P., Abernathy, K. J., & Marshall, G. J. (2013). Quantifying free-roaming domestic cat predation using animal-borne video cameras. Biological Conservation, 160 (Supplement C), 183–189. Gillies, C., & Clout, M. (2003). The prey of domestic cats (*Felis catus*) in two suburbs of Auckland City, New Zealand. Journal of Zoology, 259(3), 309–315.
18. van Heezik, Y., Smyth, A., Adams, A., & Gordon, J. (2010). Do domestic cats impose an unsustainable harvest on urban bird populations? Biological Conservation, 143, 121–130.
19. Flux, J. E. C. (2017). Comparison of predation by two suburban cats in New Zealand. European Journal of Ecology, 3, 85–90, doi: https://doi.org/10.1515/eje-2017-0009
20. Cats as Rabbiters. (1881). Marlborough Express, 27 October.
21. Parry, G. (1964). N.M.A. The Story of the First 100 Years. The National Mortgage and Agency Company of New Zealand Ltd 1864–1964. Dunedin: The National Mortgage and Agency Company of New Zealand Ltd.
22. McCaskill, L. W. (1969). Molesworth. Wellington: A.H. and A.W. Reed. P. 150.
23. Burdon, R. M. (1938). High Country: The evolution of a New Zealand sheep farm. Auckland: Whitcombe and Tombs.
24. Anon. (1884). The rabbit plague. Marlborough Express, 27 February.
25. Sanders, M. D., & Maloney, R. F. (2002). Causes of mortality at nests of ground-nesting birds in the Upper Waitaki Basin, South Island, New Zealand: a 5-year video study. Biological Conservation, 106, 225–236.
26. Fitzgerald, B. M., & Veitch, C. R. (1985). The cats of Herekopare Island, New Zealand; their history, ecology and effects on birdlife. NZ Journal of Zoology, 12, 319–330.
27. Medway, D. G. (2004). The land bird fauna of Stephens Island, New Zealand in the early 1890s, and the cause of its demise. Notornis, 51, 201–211.
28. Galbreath, R., & Brown, D. (2004). The tale of the lighthouse-keeper's cat: Discovery and extinction of the Stephens Island wren (*Traversia lyalli*). Notornis, 51, 193–200.
29. Galbreath & Brown (2004).
30. Medway (2004).

31. Karl, B. J., & Best, H. A. (1982). Feral cats on Stewart Island; their foods, and their effects on kakapo. NZ Journal of Zoology, 9, 287–294.
32. Elliott, G. P., Eason, D. K., Jansen, P. W., Merton, D. V., Harper, G. A., & Moorhouse, R. J. (2006). Productivity of kakapo (*Strigops habroptilus*) on offshore island refuges. Notornis, 53, 138–142.
33. Veitch, C. R. (2001). The eradication of feral cats (*Felis catus*) from Little Barrier Island, New Zealand. NZ Journal of Zoology, 28, 1–12.
34. Rayner, M. J., Hauber, M. E., Imber, M. J., Stamp, R. K., & Clout, M. N. (2007). Spatial heterogeneity of mesopredator release within an oceanic island system. Proceedings of the National Academy of Sciences USA, 104, 20862–20865.
35. Terauds, A., Doube, J., McKinlay, J., & Springer, K. (2014). Using long-term population trends of in invasive herbivore to quantify the impact of management actions in the sub-Antarctic. Polar Biology, 37, 833–843.

PART III

The Years of Deliberate Introductions

CHAPTER 7

Rabbits (1860–1900)

> There is always a well-known solution to every human problem – neat, plausible and wrong. H.L. Mencken (1920)[1]

Rabbits are not predators, but they play a key part in our story because, like mice, they provide the essential resources to maintain large populations of other introduced animals that are. It could be argued that, without the plague of wild rabbits that swept the country in the nineteenth century, New Zealand might even have escaped the deliberate introduction of predators supposed to be able to control it.

At this distance in time from the crisis, it is hard for modern readers to appreciate why the colonists were so taken aback by this sudden, unexpected turn of events. But at least, to the extent that the original records still survive or can be deduced, we can now look at the issue from both sides, in historical/social and in biological contexts. The colonists' own words best capture the flavour of the times, and often read like messages from another planet.[2]

Establishment of Pastoral Runs

Shore whaling was one of the first of the short-lived extractive industries available in early mainland New Zealand (Chap. 3). As most whaling stations worked themselves out of business in the 1830s,[3] the people they

C. M. King, *Invasive Predators in New Zealand*,
Palgrave Studies in World Environmental History,
https://doi.org/10.1007/978-3-030-32138-3_7

formerly employed sought opportunities to work on the newly established farms on the south Otago shores. Then, as soon as the Treaty of Waitangi (1840) established New Zealand as a British colony, an increasing tide of European settlers arrived, determined to impose familiar farming methods on the new land. From the 1850s onwards, and especially after the Murihiku Purchase of 1853 (Chap. 3), the first and largest pastoral enterprises spread rapidly across the extensive fire-induced open country of the southern and eastern South Island (Fig. 1.1). In the eastern North Island, similar developments appeared in Hawke's Bay and the Wairarapa.[4]

The result was a large-scale takeover of huge areas of apparently unused native dryland vegetation for pastoral runs.[5] Where possible, the pastoralists also added imported grass better able to support great numbers of imported domestic livestock (cattle, sheep and horses).

Other animals also benefitted from these changes to the landscape in sheep country, especially rabbits. At least at first, small farmers in the wetter parts of the country were much less affected by rabbits than were the absentee lessees of large pastoral runs. This difference eventually led to much social unrest and bitter disagreements over proposals to introduce 'natural enemies' to tackle rabbit damage to the wool industry (Chap. 8).

Many Europeans of the nineteenth century (especially those keen to establish sport hunting) also seized the opportunity to populate what they thought of as an empty landscape by liberating game mammals and birds (deer, chamois, Himalayan tahr, wallabies, hares, wild rabbits, pheasants, quail and partridges, but not foxes; Chap. 11), so that the colonists could enjoy free access to the shooting sport that was restricted to wealthy landowners back home. Wild rabbits were added to the domestic rabbits that were already present because they offered better sport. To the extent that these enthusiasts thought about the New Zealand native fauna (endemic birds, lizards, insects) at all—many never did—they considered them better replaced by superior European forms.[6]

One of the keenest proponents of the acclimatisation of game species into the new colony was Francis Dyer Rich, owner of Bushey Park near Palmerston. (The alternative spelling 'Bushy' also appears in original sources.) Rich was a keen supporter of deliberate importations and acclimatisation of species potentially useful as sporting targets. He imported pheasants, deer and hares in order to make his estate into a hunter's paradise, mainly for wealthy fellow members of the Dunedin Club and other local runholders and businessmen.[7] But he objected to the introduction of wild rabbits, which he foresaw, even in 1867, would create a 'great evil'. Australian experience was already offering plenty of warning signs of what sort of evil could be in store.

The history of the arrival and impact of rabbits in New Zealand pastoral country has become one of the best-known examples of well-intentioned acclimatisation gone wrong.[8] Enterprising settlers first quite reasonably hoped that wild rabbits would be a prolific source of fresh meat and shooting sport, but their previous experience gave them no hint that rabbit populations might behave differently in response to environmental changes in New Zealand from the way they did in Britain at the time. They were surprised to see that, from the early-mid-1870s, burrowing and pasture damage by rabbits was beginning to cause serious reductions in sheep numbers, wool clip and lambing percentages associated with malnutrition of the breeding ewes. The consequent decline in the value of pastoral land set back by decades the development of pastoralism, the dominant source of income of the young colony.

Spread of Rabbits

Domestic rabbits were a normal article of trade at whaling shore stations, so were shipped across from Sydney along with other supplies for the whalers working on the south coast of the South Island in the 1820s–1830s (Chap. 3). There were reports of sightings of rabbits in that area from the 1840s and 1850s, although only in small numbers. It is not clear exactly when the docile domestic strains gave way to real wild rabbits, but by the early 1860s it became very obvious that the newcomers (still the same species) were quite different animals.

The European wild rabbit is a remarkably successful colonising species. Live domestic rabbits are easily carried in small hutches on ships, and in the days before refrigeration they were a vital source of fresh meat for sailors on long voyages. Rabbits were distributed on islands around the world to provide food for castaways, and from the earliest days of the European diaspora, they accompanied pioneer farmers travelling to colonise new territories overseas.[9] Rabbits prefer a dry climate and soft sandy soils easy to burrow in; thick, wet forests are completely alien territory to them. That is why, except during the very peak of their initial mass invasion, the 'rabbit problem' in New Zealand was largely confined to the grassy low-rainfall eastern sides of both main islands.

As colonisation proceeded, wild rabbits were deliberately transported and liberated by settlers in several other parts of New Zealand, for sport as well as for meat. One story about a carefully planned introduction of rabbits to Southland in 1863 describes the occasion when four rabbits, brought to Bluff on the immigrant ship *Helenslea* from Scotland, were liberated near Invercargill:

> There was a liberation ceremony with speeches and toasts drunk in champagne. But … an old Highlander named Mitchell … stood on a sandhill and cursed the rabbits in Gaelic, saying they would ruin the country. No one paid much attention at the time … but many of those present must have lived to see the day [when Mitchell was proved right] … and remember his words.[10]

The rabbits that arrived by the *Helenslea* were not the first brought to Southland, but by the 1870s, they were widely believed to have been the origin of what became an unstoppable grey army of rabbits moving northwards into the hitherto highly profitable sheep country of inland Southland and Otago. These were the areas first settled by pastoralists from the 1850s onwards, who grazed large flocks of sheep on runs (blocks of Crown, that is, government-owned, grassland) leased from the government under licence.

According to the first official survey of what was called, with remarkable understatement, 'The Rabbit Nuisance', wild rabbits first appeared on the finely grassed sandhills between Invercargill and Riverton 'about twelve years since [1864]'. Concentrated grazing by wild rabbits turned what had been a rolling sward into barren, shifting dunes. From there, rabbits spread inland along the rivers, assisted by farmers, runholders and gold miners who trapped and carried with them a few rabbits to supplement their often meagre diet.

The speed of the invasion, and its consequences, astonished everyone. Cuthbert Cowan had been a leading runholder of the Hokonui district of Southland since 1858 and member of the national House of Representatives (MHR). His property at Okaiterua on the Oreti River was one of the first to be invaded by rabbits. In 1872, he saw the first rabbits arriving on his 29,000 acre farm, moving north up the river from the coast and west from the Hokonui Hills. They increased in numbers so fast that, between March and July 1876, he killed 26,000 rabbits, by the efforts of men with spades, dogs and one gun, but his lamb crop still fell by 20%.

Further west, in the Waiau Valley, Robert Ferguson Cuthbertson lived with his older brother John Robert Cuthbertson on the Otahu run. They emigrated to New Zealand from Scotland in 1860–1861. Both brothers were deeply concerned about rabbits, especially as it was their misfortune to have liberated rabbits in the district, first securing assurances from their neighbours that the animals would be protected until they were firmly established! By 1867, rabbits were undermining the profits of their farm more and more, until by 1876 their pastoral investments were ruined.

Both Cuthbertson brothers were active in public affairs, Robert as a member of the Southland Provincial Council in 1864 and 1866, and John as MHR 1873–1875. When he lost his seat in 1875, John went to live in Invercargill and became a partner in the rural supply firm Macrorie and Cuthbertson (Chap. 9). Robert continued to work the run until the rabbits finally beat him in 1883.[11]

In the Marlborough province, a pair of silver-grey rabbits were presented to the Keenes, owners of Swyncombe station outside Kaikoura, in or just before 1861 by a Mr. Harwood, newly arrived from England.[12] Silver-greys were regarded as a superior form of rabbits, and their offspring were 'jealously guarded until they had multiplied enough to be liberated'.

Swyncombe was the first Marlborough station to be affected, and suffered the most damage, but until that doleful consequence became clear, other local landholders enthusiastically set up warrens of their own and entertained shooting parties. These rabbits too reached pest levels in the early-mid-1870s. In 1875, at least 120,000 rabbit skins were shipped from Kaikoura, of which 80,000 came from Swyncombe.

Swyncombe had carried 14,000 merino sheep on its 3850 ha in 1869, but the Keenes did not have the capital or the knowledge to combat the rabbits; by the time they left their ruined property in 1882, the flock was down to 6300. Nor were they the only ones to suffer: more than a million acres [404,685 ha] of sheep pasture in the Marlborough province were damaged by rabbits within <15 years.[13]

So there were two separate centres of rabbit invasion at opposite ends of the South Island: Marlborough/Kaikoura in the north, and Southland/Otago in the south. Most of Canterbury was still free of rabbits in late 1883, but the Canterbury runholders feared invasion from both directions.[14] In the North Island, the first populations of rabbits to reach very large numbers were confined to the dry areas of Wairarapa and Hawke's Bay.[15]

First Responses

Settler farmers first began to debate what to do about the increasing numbers of rabbits in Southland in the early 1870s. About 30 of them attended a meeting in 1874 to discuss what might be done. Robert Cuthbertson said that on his recent tour through the province, he was struck with the universal prevalence of rabbits over the entire district. He strongly recommended the introduction of weasels. The meeting voted in favour of the suggestion.

By 1875, reports from pastoral lessees of the Crown describing rapidly increasing pasture damage over the previous two years prompted the Southland Provincial Council to institute a Commission of Enquiry to investigate the extent of the 'rabbit nuisance' in the district and to consider means of dealing with it. The commissioners at first doubted that such a sudden increase in rabbits could have been possible in the short time frame reported, and suspected that the case might be somewhat exaggerated.

Members of the Commission travelled widely around the affected districts, interviewing witnesses, and consulting both those affected by it locally and the relevant authorities in New South Wales and Tasmania. They confirmed that that rabbits had indeed spread to the whole of Southland 'within the short period of two years'. They compiled figures documenting the reduction of the Southland wool clip by 700–800 bales compared with the previous year and losses of lambs averaging up to 20% across the province per season, both attributed to malnutrition of ewes following rabbit damage to pastures. They listed the difficulties of exterminating rabbits caused by their extraordinary fecundity and the absence of natural enemies. One man quoted by the *Otago Witness* said that, from his own experience, 'the plague of rabbits has become so formidable, that it is now a struggle for existence between "fur v. wool"'.[16]

The commissioners called for a remedy which would be 'immediate, compulsory and universal', such as the introduction of 'natural enemies' of the rabbit which, if too small to injure lambs, would relieve the country of a serious economic threat to pastoral agriculture. They urged the Colonial Secretary in Wellington to deem the matter of sufficient public importance to make it the subject of legislation. A few months later, the Rabbit Nuisance Committee of the House of Representatives recommended that the government make 'a grant of money for the purpose of introducing weasels, as a natural check, into the country, [which] would (if the object were attained) be of very great service'. Given that condition, the committee's reservations are clear, so perhaps they cannot be held entirely to blame for the consequences.

Landholders were already exploring every possible means of killing rabbits on their own properties. On smaller farms in the settled lowlands, the efficient use of traps, guns, working ferrets, dogs and poisoned bait or gas often worked well. Matters were quite different for the runholders running tens of thousands of sheep on large leased estates in the back country. Rabbits breeding in uncounted hordes on the unoccupied Crown lands, railway embankments and Māori lands quickly reinvaded any cleared ground. So runholders all around the country strongly supported the commissioners' conclusions.

The Idea of 'Natural Enemies'

Modern readers accustomed to contemporary ecological principles tend not to appreciate why the European migrants were so astonished by this turn of events. In Europe at the time, human conceptions of the relationships between people and nature were very different from now. From the time of the ancient Greeks, the universal assumption was that life on Earth has both balance and divine purpose.[17] 'Balance' was variously defined as a stable state of nature, steadily maintained by the interactions between natural communities and their environment. This reassuring stability was seen as the result of the long-term coexistence of native species all keeping each other in check. On this view, disturbances to this mutual harmony, such as a short-term outbreak of caterpillars, would be soon corrected by increased attention from their natural enemies, caterpillar-eating birds.

In Europe, rabbits and their predators (stoats, weasels, polecats and foxes, crows and raptors) have co-evolved a set of mutually adaptive population dynamics and flexible behaviour that allow long-term coexistence at landscape scale. So the obvious answer to the question of why rabbits suddenly became so much more abundant in New Zealand than in Europe must be that rabbits in New Zealand had been imported without their natural enemies. New Zealand native predators such as weka and harrier hawks did kill many rabbits, but did not count as 'natural enemies' under

Fig. 7.1 Ferret killing a rabbit. (Photograph by Patrick Garvey)

this definition, because, as part of the 'inferior' native fauna, they were perceived to be inadequate for the task.

To the afflicted runholders, the logic was straightforward. Natural enemies kill rabbits (Fig. 7.1); too many rabbits meant not enough natural enemies; this seemed to be a problem with a simple and obvious solution. Rabbits free of any restraints had disturbed the balance of nature, so an organised, large-scale importation and release of natural enemies would soon restore it. The colonists reasoned that, since killing individuals reduced their numbers, killing enough of them would get rid of the problem,[18] much as English gamekeepers had long-established methods of managing threats to game birds on shooting estates.[19] No one at the time understood why the evolution, size and diversity of pastoral holdings in New Zealand made the problem much more complicated than that.[20]

The assumption that natural enemies can control the numbers of their prey was then universally believed, and under certain conditions it can be true. It is quite possible when the natural enemies have reproductive rates roughly matching those of the prey, and when the prey population is normally regulated from the top-down, for example, by mortality, rather than from the bottom-up, for example, by food and productivity (Chap. 12). Neither of these conditions applied to rabbits and mustelids in the nineteenth century, but no one knew that. The problem is that human actions are largely determined by what we believe about the way nature works, so the consequences will be serious if what we believe is wrong. As philosopher

Mary Midgley points out, we in our day must not be too smug in criticising mistakes made in the past: past errors differ from current ones only in being easier to see.[21]

Of course, not everyone agreed with the runholders' diagnosis. So, understandably, arguments in nineteenth-century New Zealand over the wisdom of importing 'natural enemies' to attack rabbits soon developed into a serious political issue, focused on repeated and strongly worded protests from ornithologists. The debate was stimulated more by concerns about the potential damage to acclimatised game birds than by the potential threat to native birds. In 1876, prominent politician Sir George Grey attempted to prohibit the importation of foxes, polecats, stoats or weasels. He proposed a Noxious Animals Introduction Prevention Bill that would have made such imports illegal. Grey's Bill was passed by the House of Representatives with a large majority, but the upper House, the Legislative Council, refused to consider it. Many members of the Council were wealthy wool barons at risk of losing serious amounts of money if the rabbit pest could not be controlled. Actually, Grey's effort was too late to prevent the first private imports of mustelids (Chaps. 7 and 8), and not needed for foxes, which had been prohibited in 1867 (Chap. 11).

Southern newspapers applauded the Council's refusal to ban the imports, since they saw the natural enemies of the rabbit as more significant as defenders of the Colony's commercial income from wool and crops than as threats to domestic poultry and game birds (native birds were ignored). So, despite the vigorous disputes, the imports remained legal. Just to make sure, when McMeckan Blackwood, a firm of stock agents in Melbourne, received orders to export two dozen ferrets to Bluff in 1878, they took the precaution of asking for permission from the Colonial Secretary in Wellington. They received the reply: 'There is no law against it, and ferrets are not mentioned in Section 2d of the Protection of Animals Act 1873'.[22]

The Scale of the Damage

It is easy to see why the arguments over the search for effective solutions generated so much heat and why the idea of introducing the rabbits' 'natural enemies' was among the most hotly contested from the start. The dangers of compounding the problem by creating a real-life cascade of spider-and-fly-type interactions were clearly recognised, but counted for little with those whose economic survival was at stake.

In search of ways to protect both the national income and their own property, pastoralists and politicians saw this option as entirely rational. When so many livelihoods and fortunes were at stake, it seemed simply necessary to discount the fact that the enemies of rabbits are also enemies of birds. A common opinion among the landholders most at risk was that, although they were well aware of the hazard for birds, the economic realities shouted louder than could the powerless minority who valued the native biota more highly than the wool industry.

One runholder, Capt. Thomas Fraser, told Parliament that he 'was very fond of birds; but if it came to a question whether he would have birds or sheep, he would certainly vote in favour of the sheep'. (Almost the same arguments (conservation vs economic profit) were repeated a hundred years later during fierce debates about clear-felling native forests for timber.)[23] So the decision was not made in honest ignorance, as had the idea of bringing in wild rabbits in the first place, but against vehement and well-informed objections that predicted all too accurately the risk of irreversible damage to New Zealand's endemic fauna.[24] Historical documents preserve, often in startling detail, how and why the crucial decisions were made during this unfolding crisis.

Of the more than a thousand pastoral runs established by the late 1870s, all but 26 were in the South Island. Three quarters of the 1.3 million acres (526,091 ha) of grazing land occupied in 1882, and two-thirds of the runs, lay in Otago and Canterbury. These two provinces also supplied more than half of the national wool clip for export. Hence, the runholders in those areas, especially those in Otago, had a disproportionately large influence on the Colonial Secretary's policy concerning rabbits. Many of them were members of the Dunedin Club, the local equivalent of the best gentlemen's clubs in London, with the same connections with the levers of power.

One influential runholder badly affected by the crisis was the Hon. Robert Campbell (1843–1889), a member of a wealthy family-run mercantile business with branches in India and Australia. His grandfather Robert Campbell Junior (1789–1851) was the nephew of Robert Campbell of Sydney (1769–1846), a prominent trader responsible for shipping thousands of cattle from India to stock the earliest farms of New South Wales, probably accompanied by rats and mice (Chaps. 3, 4 and 5). Two generations later, the Hon. Robert came to New Zealand from Buscot Park, Berkshire, in 1859, where his father, Robert Tertius Campbell (1811–1887), and his eldest brother, William Henry Campbell, still lived.

The Hon. Robert was a member of the New Zealand Legislative Council from 1870 to 1889, and used his position to speak strongly on rabbits.

Native predators (harrier hawks and flightless weka; Fig. 8.4) killed many rabbits at first. In July 1877, Thomas Middleton, manager of Robert Campbell's Benmore Station in the Waitaki Valley, northern Otago (Fig. 8.3), had written to his employer in October 1880: 'I feel sure the rabbits are done for here … the weka are killing them fast'. But by May 1881, it was becoming clear that leaving the task entirely to them was clearly not going to be enough. Middleton changed his mind when he saw that rabbits were becoming a serious concern on the Benmore estate.[25] He begged Campbell to do something urgently:

> Now, if ever, is the time to prevent them becoming a nuisance … I can't help dreading the fate of Southland for the Waitaki unless we stir ourselves and combat the vermin before it is in their power to do us harm.[26]

Alas, serious harm was already being done. The unstoppable grey blanket was steadily rolling on into Otago from Southland, and by that same year of 1881, the rental income to the government from the half-million acres of runs worst affected by rabbits (202,342 ha) had dropped from £2288 a year to £619; the financial cost to the whole colony in lost exports had escalated to at least £500,000 a year. On one of the Hon. Robert Campbell's runs, Galloway (68,000 ha), rabbit control in 1880 cost him £3000 (=$486,193 now) for a return of £1500; at least 26,000 ha of the run was incapable of feeding sheep in winter. Another of his runs, Burwood, which used to carry 80,000 sheep, could by then support only 24,000. Part of the loss was due to heavy snow, but after it melted rabbits prevented any recovery of the pasture. The Hon. Robert gave up on seven of his runs in 1881; four were re-let at a lower rent, and three were simply abandoned; the national flock fell from 13 million in 1878 to 11.5 million in 1879.

The runholders' view of the future was well expressed in this letter to the *Marlborough Express*:

> It is certain that manual labour cannot prevent the rabbits increasing and spreading over the country, although many thousands are destroyed annually; unless stoats, weasels, and ferrets are introduced into the country in large numbers, very little wool will be expected from New Zealand ten years hence.[27]

The point was serious, considering that wool was then earning 30–40% of the Colony's income. Unsurprisingly, government officials accepted the argument that a measure calculated to deal more effectively with the nuisance was urgently required, but they still hesitated.

The Rabbit Nuisance Act 1881

Eventually, the government could no longer ignore the outcry from formerly wealthy runholders facing ruin from rabbit damage. Their influence had been the main reason for the failure of Grey's 1876 attempt to prohibit the import of the natural enemy of the rabbits (p. 153). Now, as rabbits swarmed and pastoral profits dived, they insisted on more decisive action.

The existing Rabbit Nuisance Act of 1876 had provided only for a system of district rabbit trustees with powers to raise funds from landowners to help pay for the work of dealing with rabbits (excluding unoccupied Crown and Native lands). This did not work well, largely because it could not prevent landholders who disagreed with it from simply walking away.

The July 1881 report of the Rabbit Nuisance Committee to Parliament cited detailed statements from 15 witnesses. They specifically wanted to see all landholders required to bear their proportionate share of the cost of destroying rabbits, and provision made for protecting the natural enemies of the rabbit already in the colony.

Among the witnesses was George Bullen of Kaikoura. He presented a succinct argument in favour of relying on ferrets to solve the rabbit problem:

> I am glad to say the ferrets have done, and are still doing, a great work for us in checking the rabbits; and I do believe the ferrets will be the salvation of the country. … My firm belief is, the rabbit-catchers [trappers] do more harm than good, they kill everything that kills a rabbit. … We turned out ferrets about eight years ago, and have been turning out … about 200 a year, which we shall continue as long as we see any rabbits around. This country … is considerably over 100,000 acres, and I believe there are many hundreds of ferrets on it; we leave it entirely to them to do the work, keeping one man to breed and act as keeper over the ferrets.[28]

The experience of Bullen, among others, convinced the colonial government to take official action. First, it toughened up the existing anti-rabbit legislation and passed the Rabbit Nuisance Act 1881. Second, it divided the country into rabbit districts, excluding unoccupied Crown and Native lands. Each district had a resident rabbit inspector, all under the direction of a zealous new superintending inspector, Benjamin P. Bayly (1837–1909), appointed in November 1881. Bayly helped to draft new legislation which held individual landholders responsible for controlling rabbits on their own land. Defaulters could be fined up to £20 for failing to control rabbits, which seemed the crowning insult to many runholders already facing unavoidable financial losses due to rabbits. The new act not only supported the proposed introduction of new natural enemies of the rabbit, but also gave legal protection for all natural enemies already present. With minor alterations this act remained in force for many years.

Reactions to the New Act

Rear-guard protests continued in the House of Representatives, especially by those most vehemently opposed to the very idea of encouraging natural enemies as a means of overcoming the rabbit nuisance. The government ignored them. On the contrary, it continued, not only commissioning many imports itself, but also subsidising private enterprises doing the same.

Settlers' meetings debated the new Rabbit Act exhaustively and often angrily. For example, the *Marlborough Express* reported in great detail the proceedings of two successive meetings in Blenheim in October and November 1881. In the face of the pastoralists' growing fear that, if the advance of the rabbits was not stopped, they might as well give up sheep farming altogether, every possible method of meeting the act's requirements had its passionate advocates. For some, poisoning and fencing were the only solution, and for others, organised breeding and distribution of ferrets. Some relied on trapping to remove the rabbit pest, while others pointed out that rabbit trappers could not be expected to remove the source of their employment, and their traps also killed many natural enemies of the rabbit.

Even after it was already too late to prevent the decision, the deliberate importation of natural enemies continued to be resisted by some sections of the public, especially by ornithologists and game bird enthusiasts. The government was unmoved. But, plagued by repeated and increasingly hostile correspondence, it was eventually forced to justify its policy.

Bayly's correspondence, preserved by Archives New Zealand in Wellington, show how fervently he dismissed all objections against government policy. They must be coming, he said in 1886, from those who had evidently never been south or seen the damage caused by the rabbit pest … [and are unaware that] private individuals have and are introducing the natural enemy and turning these animals out within the settled districts … I am as strongly as ever a believer in the introduction of the natural enemy … to save the large tracts of pastoral country in Canterbury that are threatened with an influx of rabbits from the boundary of the Otago country … [where it is] impossible to deal with them in the ordinary manner, it is here especially that the natural enemy is required.[29]

Bayly was not the only one fending off criticism. The same file includes a solemn protest sent on 10 September 1886 by the Hamilton Borough Council to the Hon. John Ballance, minister of lands, and forwarded to Bayly for reply, which received a terse response:

I am directed by the Colonial Secretary's Office to inform you that, although there is some diversity of opinion on the subject, the great majority of persons having practical acquaintance with the rabbit pest are strongly in favour of the attempt being made to destroy the rabbit by means of their natural enemy, and in adopting this plan the Government are only giving effect to the express wish of the Legislature as contained in the Report of a Joint Committee of both Houses of Parliament.[30]

Justification of Bayly's inflexible attitude had been provided in 1883 by runholder Charles de Vere Teschemaker, who attempted to account for the full scale of the damage done to the economy of the colony by rabbits. The logic of his estimate is worth considering in full:

The public pastoral estate of Otago and Southland alone consists of some 8,000,000 acres [3,237,485 ha]. In 1873, this land was worth from 5s to 40s an acre, say an average of 15s, and would have been readily purchased at that price. Total value, £6,000,000. Today [29 September 1883], solely on account of the rabbit, the public estate of Otago and Southland has depreciated fully 50 per cent … our loss … so far as the capital value of the country is concerned, up to the present amounts to £3,000,000; [plus] an annual loss [of income from reductions in lambing percentages and wool clip] of £1,700,000.[31]

No one doubted how far the scale of the losses devastated the value of land (Fig. 7.2), but not everyone was convinced that natural enemies would be able to control rabbits at all, let alone as effectively as Bayly claimed.

Fig. 7.2 High country black humour. David Henshaw's character Jock explains, to a visiting surveyor, the effects of rabbit damage on leasehold values and the profits of sheep runholders. (Reproduced by permission from *Jock's Country*, David Bateman, Auckland (2018), p. 208, and Alastair Henshaw)

Richard Henry was an ardent conservationist, later famous for attempting to protect New Zealand's endemic flightless birds from mustelids by translocating them to Resolution Island.[32] He was well aware that rabbits can be willing to stand up to a predator; indeed, a female rabbit with kits to protect has been known to attack and kick a stoat halfway across a paddock. He watched a ferret failing to catch a rabbit put in an enclosure with it. Henry recommended that anyone thinking of purchasing ferrets for release on a property should first undertake a few simple experiments, which would enable the experimenter to estimate whether there was any point in Bayly's official policy of buying up young ferrets at 8s or 10s each, and turning them straight out of cages by the thousands. Henry's warning was ignored.

Opponents continued to claim that ferrets would prove to be a cure worse than the disease, but any such prediction could be countered by reassurances from keen but misinformed supporters of the programme, such as farmer C.J. Tully. He told the Joint Committee on Sheep and Rabbits in 1884 that ferrets were so easy to catch that there was no need to fear that they will get too numerous.

The Acclimatisation Societies were sharply divided. Alexander Bathgate may well have been correct that 'the zeal of the earlier acclimatisers was greater than their knowledge', but probably not that 'Due care and consideration would have prevented the introduction of several undesirable immigrants which now, like the poor, are always with us'. Contrary to popular assumption, the Societies were not entirely responsible for the regrettable decisions of the time, and 'even when they were, their goals generally reflected what people of the time wanted … The general view was that change was inevitable and could not be stopped'.[33]

The long-term result of this ongoing conflict of interests was, as the *Otago Witness* reported sorrowfully in December 1907, that 'the operations of acclimatisation societies have to a great extent been restricted to fish culture'—especially trout and salmon, whose impact on native freshwater ecosystems paralleled that of rabbits on pastures, although that was not known at the time. By the early twentieth century, recognition of the need to protect the native species was rapidly increasing, but still, for some, concern about wasted investment in introduced gamebirds was still stronger.

Sheep Versus Rabbits as Sources of Income

Not everyone agreed with the runholders' assumption that the New Zealand economy would be ruined by rabbits. Sir James Hector, one of the colony's leading scientists, pointed to new economic opportunities created by the increasing international demand for rabbit skins for fur gloves and felts, and for canned or frozen rabbit meat. He predicted that much of the country could be more profitably occupied by rabbits than by sheep. Hector's opinion led to the obvious question: Could income from rabbits compensate for loss of income from wool? One possible answer can be indirectly indexed from the official records of national income from exports of rabbits compared with wool.

Exports of wool, one of the colony's principal products at that time, had increased steadily until 1876 as more runs were taken up and developed, but after 1877 exports of rabbit skins began an even faster rise. The export records clearly show that the rabbit invasion shifted a gear in 1877, and by the years 1878–1887, it was having a severely depressing effect on the colony's income from wool production. An official survey listed 99 pastoral runs abandoned in Otago alone between the years 1877 and 1884. More detailed records extracted by Peter Holland and Guil

Figgins from the farm diaries of two sheep stations in the Waitaki Valley (1874–1907 at Otematata, and 1891–1930 at Hakataramea) demonstrate the even-more dramatic impact of rabbits on wool production on the local scale.

For a while, the harvesting of rabbit meat provided an alternative source of income for distressed landholders. The owners of canning factories and (after the development of refrigerated shipping in 1894) dealers in frozen carcases paid them to cease using poison when rabbits were in their prime, and by the 1890s these inducements had indeed created a valuable buffer protecting rural incomes from the hardest times, but at a cost: it created a corps of professional rabbit farmers who conserved the stocks and resisted effective control. Far from adding to the rabbit control effort, trapping by professional rabbiters made things worse. On properties where rabbits were harvested for meat, rabbit numbers greatly increased during the off season. Government officers attempting to enforce the Rabbit Nuisance Act regarded such cosy arrangements as illegal.

New Zealand did not, of course, have this thorny conflict of interest on its own. The owner of a rabbit freezing works who had invested thousands of pounds into a new freezing works in Dubbo, inland New South Wales, estimated that the then current value of the rabbit harvesting industry was £1 million a year. He attempted to restrain the pastoralists advocating the introduction of a disease that might prove effective as a rabbit control, on the grounds that he had a right 'to protect my interests as a trader'.[34]

In 1949, A.H. Clark[35] asked whether income from rabbits could have compensated for losses from wool. The official statistics show that in 1873 wool generated an export income of £2.7 million, compared with only £1300 from rabbit skins. The profits of the wool barons rapidly increased to £3.7 million in 1877, when the export value of rabbits was £8600. From 1878 to 1888 inclusive, the value of wool exports stagnated between £2.9 million and £3.3 million, while the value of rabbit skins escalated to >£100,000 in three of those same years. Wool exports did not reach £4.0 million until 1889, and even in 1899 they earned only £4.3 million, but that was still a lot more than the £195,000 from rabbit skins and frozen meat. In Australia, the rabbit skin industry eventually became worth £3 million a year, but the losses due to rabbits amounted to £30 million a year.[36]

Kazimierz Wodzicki (p. 278), a thoughtful animal ecologist, also tried to answer the same question, using slightly different assumptions based on observation of animals rather than economics alone. He reckoned that in

1947 the economic value of one breeding ewe, producing about eight pounds weight of wool and one lamb a year, was about £1 a year. His calculation of stock equivalents was based on feeding trials showing that 15 rabbits ate as much as 1 sheep. If so, 50 million rabbits would displace 3.33 million sheep and thereby (after adding the costs of pasture damage and subtracting the income from rabbits exported) lead to a net loss of between £3.0 and 3.5 million a year.

Estimated either way, the compensatory income from rabbits was marginal at the gross national level, though it certainly did save some individual farmers from bankruptcy during the leanest years. More importantly, as Wodzicki pointed out, such naïve calculations do not account for the many other financial burdens imposed on runholders by a heavy rabbit infestation—loss of plant cover leading to irreversible soil erosion, pasture fouling, destruction and conversion to unpalatable sward, spread of weeds, and the £180,000 a year spent on killing rabbits.

Throughout the 1880s the need to find a solution to the rabbit problem just went on getting more pressing. In 1887, the *Oamaru Mail* cited a report by the Commissioner for Crown Lands in Canterbury, who found that the rabbit pest was

> developing into a difficulty quite as large and stubborn as any with which the colony is called upon to deal. Nobody seems to really know how to face it ... [the rabbits swarming just south of the border with Otago were eating the country bare, and were now] menacing Canterbury ... but whether poison, or ferrets, stoats, and weasles, or men and dogs, or all of them, should be employed is a matter on which experts only are capable of judging.[37]

On the contrary, the judgement had already been made by Superintending Inspector Bayly, and it was the wrong one. He had always been a strong supporter of the idea of importing mustelids to control rabbits, so he had applauded private importations. Immediately upon his appointment in 1881, which gave him authority to act on his convictions, he ignored all contrary advice and set to work on organising official purchases of ferrets for release on Crown lands.

The arguments for the attempt to control rabbits by wholesale releases of mustelids that so impressed the nineteenth-century colonial government are well preserved and worth careful attention, because they explain why events unfolded as they did, in the terms and language understood at the time. The interests and convictions of the players explain much of the story, and so the story itself cannot be fully understood without paying attention to them. In particular, they explain why the successful early strategy developed by George Bullen was used to justify the later indiscriminate releases by others, who were desperate enough to ignore the essential conditions Bullen had emphasised. Later experience amply confirmed that too few ferrets hunting individual rabbits year-round cannot reduce rabbit populations, mainly because ferrets could never remove rabbits faster than they could be replaced (Chap. 12).

Arnold Wall expressed the unequal odds of the human battle against rabbits, as seen from the rabbits' point of view:

> Where the sheep feed, there feed I, /Depleted lands behind me lie, /Of dogs and guns I take no heed, /I only breed and breed and breed.
>
> At traps and guns and dogs I smile, /I laugh at cats' and weasels' guile, / I frolic, nibble, frisk and feed, /But all the time I breed and breed.
>
> They send my skin to clever folks, /who turn me into sealskin cloaks, /A sorely hunted life I lead, /But still I breed and breed and breed.
>
> When foes attack I cannot bite, /I have no spirit for a fight, /By other methods I succeed, /I merely breed and breed and breed.[38]

The original poem, first published in 1922, has three more verses, the last recommending that nations aiming to attain global leadership should do the same.

Many critics at the time and since have blamed the early runholders for altering the natural environment of the southern pastoral grasslands in favour of rabbits, by overstocking, repeated burning and inadvertently creating refuge areas in railway embankments, sod walls and under rampant gorse hedges, and for the holding by absentee landlords of huge high country runs where labour-intensive rabbit control was impossible. In reassessing this argument, the environmental historian Robert Peden suggests that uninhabited island ecosystems could also be wrecked by rabbits; Peden and Holland found that burning did not destroy native vegetation wholesale; and Holland and Figgins emphasised that local variation in soil and climate strongly modified the impact of rabbit damage. Without denying these points, there remains ample evidence that repeated burning and overstocking does destroy native vegetation and change its character, and the species composition of vegetation, largely controlled by rainfall and soil type, is as important as total volume.

Arguments about the proper management of rabbits on leasehold land in the semi-arid, rabbit-prone high country have continued to the present. Fortunately, in most of the rest of the country where damper habitats are less favourable to rabbits, predation by cats and ferrets is now counted as among the reasons rabbits are no longer the pest they once were. But eradication was impossible, and at present it still is.

Summary

Wild European rabbits brought to New Zealand for shooting sport were totally different from the docile domestic rabbits, already present and useful only for food. From their first arrival in the 1860s wild rabbits over-ran the newly established pastoral runs in Otago, Marlborough and Hawke's Bay. From the 1870s, rabbits spread across all open country in plague numbers, ruining the pastures and thereby devastating the wool clip and lambing rate of malnourished ewes, and the profits of previously wealthy runholders. Nineteenth-century science was based on the traditional idea of 'the balance of nature'; therefore, the obvious reason why rabbits in New Zealand behaved so differently from their less abundant British ancestors was that they had been imported without natural enemies. The colonial government passed anti-rabbit legislation, and in 1881 appointed Benjamin Bayly to administer it and to organise the importation of suitable natural enemies. For eight years Bayly fulfilled his mandate with vigour, despite minimal effect on rabbits and repeated criticism from naturalists at home and overseas. New income from rabbits could not replace lost profits from wool, so many runholders were bankrupted, but it did keep some struggling farmers on their own land until conditions improved.

Notes

1. Among the various citations of this famous witticism, this seems to be closest to the first version published in a documented source. See https://quoteinvestigator.com/2016/07/17/solution/
2. For a more complete open-access list of the very many primary sources underlying this chapter, see King, C. M. (2017). The chronology of a sad historical misjudgement: The introductions of rabbits and ferrets in nineteenth-century New Zealand. International Review of Environmental History, 3, 141–175. doi: http://press-files.anu.edu.au/downloads/press/n2519/pdf/08_king.pdf

3. Prickett, N. (2002). The archaeology of New Zealand shore whaling. Department of Conservation, National Historic Heritage Workshop, Wellington.
4. Guthrie-Smith, H. (1969). Tutira. The Story of a New Zealand Sheep Station, 4th edition. Wellington: A. H. & A. W. Reed.
5. Peden, R. (2011). Making Sheep Country: Mt. Peel station and the transformation of the tussock lands. Auckland: Auckland University Press.
6. McDowall, R. M. (1994). Gamekeepers for the nation: The story of New Zealand's Acclimatisation Societies 1861–1990. Christchurch: Canterbury University Press.
7. Sinclair, F. R. J. (1996). High Street Quaking: A history of Dunedin's "Inner Circle". (PhD thesis), University of Otago, Dunedin.
8. Thompson, H. V., & King, C. M. (Eds.). (1994). The European Rabbit: The History and Biology of a Successful Colonizer. Oxford: Oxford University Press.
9. Dunlap, T. R. (1999). Nature and the English diaspora: Environment and history in the United States, Canada, Australia and New Zealand. Cambridge: Cambridge University Press.
10. Hamilton, G. A. (1952). History of Northern Southland. Invercargill: Southland Times. P. 34.
11. Beattie, J. H. (1979). The Southern Runs. Gore: The Gore Historical Society. Pp. 252–253.
12. Sharrard, J. M. (1966). Kaikoura: A history of the District. Kaikoura: Kaikoura County Council. Pp. 184–185, 188, 126.
13. Clark, A. H. (1949). The invasion of New Zealand by people, plants and animals: The South Island. New Brunswick: Rutgers University Press, citing the 1940 history of Marlborough by historian Sir Alister McIntosh.
14. Burdon, R. M. (1938). High Country: The evolution of a New Zealand sheep farm. Auckland: Whitcombe and Tombs.
15. Gibb, J. A., & Williams, J. M. (1994). The rabbit in New Zealand. In H. V. Thompson & C. M. King (Eds.), The European Rabbit (pp. 158–204).
16. Weazels for New Zealand. (1876). Otago Witness, 9 December p. 21.
17. Kricher, J. (2009). The Balance of Nature: Ecology's enduring myth. Princeton, NJ: Princeton University Press.
18. Dunlap, T. R. (1999). Nature and the English diaspora. P. 73.
19. Jefferies, R. (1878). The Gamekeeper at home (Reprinted by Oxford: Oxford University Press, 1978).
20. Holland, P., & Figgins, G. Environmental disturbance triggering infestations of gorse, rabbits and thistles in southern New Zealand: 1850 to 1980. International Review of Environmental History, 1(2015), 41–79.
21. Midgley, M. (1992). Science as salvation. London: Routledge. P. 78.

22. Bishop, W. (1878). Letter 13 July to Colonial Secretary asking permission to land 2 dozen ferrets. Wellington: Archives New Zealand R24273700.
23. King, C. M. et al. (2015). The Drama of Conservation: The history of Pureora Forest. Hamilton NZ: Department of Conservation and Springer Verlag.
24. Buller, W. L. (1877). On the proposed introduction of the polecat into New Zealand. Transactions & Proceedings of the New Zealand Institute, 9, 634–635.
25. Pinney, R. (1981). Early Northern Otago Runs. Auckland: Collins, Collectors Edition.
26. Pinney, R. (Unpublished). Notebook M1: Extracts from Benmore Letter Books, 1877–1887. Dunedin: Hocken Library MS-3178/013.
27. Our Kaikoura Letter [Stoats & weasels needed to save NZ wool]. Marlborough Express, 25 July 1877, p. 7.
28. Anon. (1881). Report of the Rabbit Nuisance Committee [1880–81]. Appendix to the Journal of the House of Representatives, 1881 I I-06.
29. Multiple correspondents. (1886). Correspondence protesting introduction of stoats and weasels. Wellington: Archives New Zealand R24456501.
30. Ibid.
31. Teschemaker, C. d. V. (1883). The Rabbit Nuisance. In. Auckland: Auckland City Library, Sir George Grey Special Collection 636.9 T33.
32. Hill, S., & Hill, J. (1987). Richard Henry of Resolution Island. Dunedin: John McIndoe Ltd.
33. McDowall, R. M. (1994). Gamekeepers for the nation.
34. Coman, B. (1999). Tooth and Nail: The story of the rabbit in Australia. Melbourne: The Text Publishing Co. P. 79.
35. Clark, A. H. (1949). The invasion of New Zealand by people, plants and animals: The South Island. New Brunswick: Rutgers University Press.
36. Rolls, E. C. (1969). They all ran wild. Sydney: Angus and Robertson. P. 78.
37. [Editorial]. Oamaru Mail, 30 June 1887.
38. Wall, A. (1999). A song of Brer Rabbit of Otago. In J. Gordon (Ed.), Out of Town. Writing from the New Zealand Countryside (pp. 189). Christchurch: Shoal Bay Press.

CHAPTER 8

Ferrets (1852–1892)

> We may infer from these facts [on the tameness of the birds on Galapagos] what havoc the introduction of any new beast of prey must cause in a country, before the instincts of the inhabitants have become adapted to the stranger's craft or power. Charles Darwin (1845)[1]

Ferrets are semi-domesticated animals, subject to a long history of artificial breeding, either as pets or as trained working companions for a rabbit-catcher, so are unused to living independently of human care. In Britain, the use and management of ferrets to control and harvest rabbits was a well-established rural skill,[2] widely practised there and easily imported into New Zealand with experienced immigrants. The ferrets themselves often suffered high mortality in transit, but the hardiest survivors could be used to establish local ferret breeding stud farms to supply the accelerating demand from New Zealand landholders. For at least 50 years after 1870, tens of thousands of ferrets were bred and released on the rabbit-damaged pastoral lands of New Zealand.[3]

Polecats, the ancestors of ferrets, evolved in Eurasia as specialist predators of small rodents and rabbits. Their long, thin bodies and short legs give them a smooth tubular shape that moves easily through rabbit burrows. They have sharp, triangular faces, small rounded ears folded back against the head, and a short excitable tail. There are no true polecats in New Zealand, but feral ferrets, as domesticated animals that have escaped

C. M. King, *Invasive Predators in New Zealand*,
Palgrave Studies in World Environmental History,
https://doi.org/10.1007/978-3-030-32138-3_8

human control, are still generally less aggressive than the other wild carnivores with which they have to compete. Live-caught feral cats and wild stoats have to be anaesthetised to prevent them from tearing a trapper's hands to shreds, but a feral ferret can be handled conscious, and are large enough to carry a radio transmitter (Fig. 8.1).

The idea of employing ferrets to kill rabbits was not a new one. The Roman author Strabo describes how ferrets were specially bred for the purpose and muzzled before being sent into the rabbit burrows. When the skill of using ferrets to flush rabbits out into the open became known during the Middle Ages, the practice became widespread in Europe. Rabbits and ferrets both arrived in England, possibly together, some time between the eleventh and fourteenth centuries.

The introduction of invertebrate natural enemies is a standard technique for management of insect pests, a task for which invertebrate predators can sometimes be spectacularly successful. But attempts to employ

Fig. 8.1 A wild ferret fitted with a radio transmitter. (Photograph by Author)

vertebrate predators to manage vertebrate pests seldom work so well, or at all, except under special conditions, and most turn out to have disastrous side effects (Chap. 12). The story of mustelids in New Zealand is not only one of the best known illustrations of this generalisation, but also a genuine piece of social and biological history. Because the importations of the mustelids had to be organised deliberately, and were well documented at the time, the historical contexts of their introductions are much better known than for the accidental arrivals of rats and cats.

Early Private Importations of Ferrets, from 1852

In the early nineteenth century, there was a strong demand in Australia for ferrets to take part in the Antipodean equivalent of the medieval 'sport' of bear-baiting. Live-captured Norway rats—the bigger and meaner, the better—were placed in a pit surrounded by tiered seats for eager spectators. Ferrets, terriers or mongooses were released by turns into the pit for an agreed time, and the audience placed bets on the number of kills each made. There were several such pits in Melbourne and at least one in Sydney.[4]

A few ferrets, perhaps already accustomed to this gruesome game, found their way to New Zealand even before rabbits became a nuisance—as is demonstrated by an advertisement of 1852 offering three pairs for sale, and another in 1860 offering four ferrets at 35 shillings a pair. In 1867, the *Bella Mary* brought ten ferrets from Hobart, 'which will be parted with on reasonable terms'.[5]

Even if rat-baiting was never practised in New Zealand, private imports had been going on since the late 1850s. Ferrets had long been useful for purposes other than rabbit control, for example, clearing ships and buildings of rats, so the Canterbury Acclimatisation Society brought in five ferrets in 1867, plus another one in 1868, and kept them at the Society's Gardens in Christchurch.[6]

'Working ferrets' were trained to hunt rabbits alongside a team of dogs (Fig. 8.2). The long slim body of the ferret was inserted into the burrow to flush the rabbits out into the open, where they were caught by the dogs, or by men stationed at the holes with clubs or nets. Good working ferrets were valuable property and well cared for because they were part of their owners' toolkits. Farmers advertising for labourers favoured rabbiters

Fig. 8.2 A rabbiter working with a co-operating team of dogs and ferrets. (From *Auckland Weekly News*, 11 December 1924, courtesy Keith Giles, Auckland Central City Library)

with their own ferrets, and immigrants skilled in such work were much in demand. These ferrets were not intended for release, and any that got lost were not expected to be able to survive for long independently of human partners.

The area of ground that a ferreting team could cover was limited, and the need to control rabbits over vast areas of seldom-visited mountain country was overwhelming. So it was the runholders in those districts who were the first to try a new idea, importing private shipments of ferrets for deliberate release on their own land, and to persuade the government to do the same for Crown lands on an industrial scale. Inevitably the emphasis shifted to importing ferrets in huge numbers, specifically to be liberated into the wild.

The chances of the released ferrets surviving long enough to establish a wild population varied with the habitat and climate of the distribution area, and with their own ability to adapt to an independent life. One objection to releasing captive-bred ferrets was that, after generations of selection for the docile qualities preferred in tame domestic animals (Fig. 8.1), they might be unable to make a living independently. That was certainly an issue at first, when young ferrets were turned out straight from a comfortable hutch into the unprotected wild.

One vociferous runholder, Charles de Vere Teschemaker, was among the ferrets' keenest supporters, but he conceded that domestic ferrets were rather delicate, and did not survive the English winter if left unprotected. New Zealand was different, he added, because there was so much food available for them, and they could survive well here especially if they were trained to hunt and kill before being turned out. He was right only up to a point.

Otago and Southland

Soon after rabbits arrived in Otago in the mid-1860s, the economic viability of the back-country sheep stations, including two of the largest, Earnscleugh and Galloway (Fig. 8.3), began rapidly to decline. By June 1875, the runholders were pinning their hopes on the idea of ordering supplies of ferrets for their stations.

This first 'supply' of ferrets for Earnscleugh amounted to only five animals in 1875, but then the station staff started breeding their own ferrets for release in the 1880s.[7] The ferrets did not save Earnscleugh; it was resumed by the Crown (i.e. abandoned by the leaseholder) in 1895, by which time it could carry only 12,000 sheep, half the number it had supported in 1879.

Cuthbert Cowan, a leading Southland runholder, organised a private shipment of ferrets from Britain, which arrived in late 1876. The *Southland Times* reported:

> One of the passengers by the *Arawata* on her last trip from Melbourne, was a young gentleman [named in another source as Allen Bowler] who brought 36 ferrets in the *St. Osyth* from England to Melbourne and thence to this place, for rabbit killing purposes. He has proceeded to Mr C. Cowan's station.[8]

The report adds: 'In conveying his ferrets from Glasgow to London by rail, a distance of about 400 miles, the charge made for the journey on account of the ferrets was 14s; but on arriving at the Bluff and having them conveyed by rail thence to Invercargill, a distance of 20 miles, the

Fig. 8.3 Map of ferret release sites. (Cartography by Max Oulton, from Author's data). Solid lines enclose open country habitats favoured by rabbits

charge he had to pay for their carriage was £1 9s 9d! This struck the stranger "all of a heap"'.

In evidence to a later Parliamentary enquiry, Cowan was asked if he had ever tried natural enemies as a remedy against rabbits. He replied that he had landed 16 ferrets [maybe Bowler lost 20 on the way] and put them out on his own property. Now the country was 'overrun with ferrets', which were breeding in the wild and doing no harm to lambs.

Life at Benmore Station, in the Waitaki Valley in northern Otago, is well recorded in Robert Pinney's classic book[9] based on a series of diaries and letters written by the station manager Thomas Middleton and summarised in research notes kept in the Hocken Library. Ferrets were first taken there in 1875, and by 1881 they were 'all over the run'. Middleton attributed his moderate early success in reducing the numbers of rabbits in part to the work of ferrets, but regretted that they were so vulnerable to the effects of the highly infectious and lethal disease canine distemper and of eating the carcases of poisoned rabbits.

Many business enterprises were seriously challenged by rabbits, but at first, faced with a completely unknown dilemma, they had no way to predict the size of the threat to their profits. For example, the National Mortgage and Agency Company of New Zealand Ltd. (NMA), in Dunedin, was a leading rural supply company, supplying livestock, seeds and equipment to pastoral runholders, and handling the sales of the wool and tallow they produced. They were among the first commercial entities at risk, but their directors saw no reason to be alarmed. In 1875, NMA's founder J.M. Ritchie wrote: 'I hear that Johnson has been spending upwards of £15 a week to try to keep down the rabbits. He is using men and dogs. Far better to get some ferrets'.[10]

NMA was quite prepared to put its advice into practice. The unpublished ledgers of their Dunedin Office include the following entries for consignments of mustelids to J.C. Buckland at Tumai, Waikouaiti: 27 November 1882: 'Ferrets from Invercargill £6-17-6'; 23 December 1882: 'Ferrets from Invercargill £24-10-0'; 9 April 1884: 'Exp[ense]s re stoats, ferrets etc, 13s 0d'. These animals were presumably imported into Bluff (probably from Melbourne) and forwarded to the Dunedin office for distribution to client landholders.

W.S. Davison was one of the architects of the frozen meat trade and the early dairy export industry, and a member of the NMA Board. In 1888, 13 years after Ritchie's complacent statement, he travelled to London to speak at a meeting of NMA shareholders. He told them: 'I have not the slightest doubt that we shall very easily keep the rabbits down in New Zealand. It only requires an increase in the numerous stoats and weasels which have been sent out to keep the pests in check. In a few years we shall be little troubled by rabbits'. Gordon Parry, the author of the centenary history of the NMA company, commented: 'Here were two intelligent, prudent and successful men, quite sure that the rabbit problem was only a passing phase … in the interval between the two statements [1875 to 1888] ... the Government had lost more than £315,000 [a year] in rentals from leasehold [pastoral] lands, and the annual national loss [including lowered yields of wool lambs] was being estimated at £2,000,000'. Parry's figures differ slightly from those given previously by Teschemaker (Chap. 7), but the message is the same.

Marlborough

In 1878, Joseph Ward of Wairau and some of his neighbours ordered a large consignment of 600 ferrets to be dispatched from England to New Zealand on the sailing ship *Rialto*. On the voyage, distemper broke out among them, and only three ferrets survived, so 'it was a rather expensive affair'. Another lot of 600 ferrets was imported by George Bullen of Kaikoura, but every one of that consignment died. He started again with 700, and got two ferrets out of the 1300. The two shipments cost some £800, for virtually no result. A later and presumably better-managed shipment was more successful in keeping ferrets alive, at a cost of 10s 8½d per ferret, in part because by then it was known that ferrets had to be kept well away from any dog with distemper.

Besides disease, another likely reason for these failures was inappropriate feeding and housing on board ship. Gamekeepers with long experience of keeping ferrets described them as 'shivery creatures' that need warm dry hutches,[11] and are always thirsty, so they must be given plenty of water. Shipments accompanied by couriers unaware of their animals' special needs might well have been at risk of high mortality in transit.

Bullen also encouraged British ferreters to emigrate to New Zealand, bringing their trained working ferret and dog teams with them. These shipments recorded many fewer losses, not least because these animals were accompanied by experienced ferret-keepers. For example, the ship *Warwick* left Plymouth on 30 November 1878, arriving at Lyttelton on 3 March 1879. The *Nelson Evening Mail* announced that the *Warwick* carried two rabbit-catchers accompanied by their ferrets and dogs, and engaged to work for 'a run holder at the Kaikouras'. They were to be paid between 4s 6d and 2s 6d per dozen rabbit skins, depending on the time of year. The *Warwick*'s passenger list names all the men on board with their trades: two of them, David Cook and David Lang, both 23 and both from Fife, were listed as rabbit-catchers. Bullen is named in another report as the runholder engaging them.

Other Marlborough runholders tried a different tactic in 1881. Messrs. Fell Bros, agents for the Starborough Station near Blenheim, offered to pay stewards and others working on the wool ships 12 shillings a head for all ferrets landed alive and in healthy condition. The adverts were repeated at least five times over the next two months. If any ferrets came in this way, they could only have been in small numbers.

North Island

Wild rabbits liberated in the southern Wairarapa near Carterton had greatly increased in numbers by 1870,[12] and by 1875 they had reached William Beetham's station Brancepeth, southwest of Masterton, about 15 km away.[13] Further south, the Huangarua station near Martinborough was owned from 1869 to 1878 by the Hon. George Waterhouse, a Wellington politician and pastoralist. He had had rabbits on his property since about 1871. Waterhouse was one of the speakers at the Legislative Council debate on whether to impose a ban on introduced predators in 1876 (Chap. 7). He must have been relieved at the decision that private imports were not to be prohibited, because during the debate he revealed that 'he had, within the last three or four months, turned loose a considerable number of ferrets'.

The result was apparently good. Runholders in Southland read, perhaps with a touch of envy, reports that the ferrets imported by the Hon. Mr. Waterhouse, released on North Island pastures thick with rabbits, had cleared them out completely by January 1877, right down to the last one in the neighbourhood. Such unequivocal statements are now impossible to confirm, but seem to modern eyes incautious: absence of observations (of rabbits surviving the arrival of ferrets in a locality) does not mean that none did.

Official Ferret Importations, 1882–1886

As the advancing horde of rabbits surged northwards from Southland into Otago, the government planned to establish resident populations of natural enemies to stand in their way. The need to do this was considered especially urgent where labour-intensive methods of rabbit control were impossible, such as on the vast unoccupied Crown lands and thick tussocks of the remote back country. They expected to be able to do this by importing and releasing thousands of ferrets on the mountain lands dividing the headwaters of the main east-flowing rivers, and on the western shores of the lakes. This policy was not understood at the time, or later—especially by those concerned about the protection of native birds. Edward Melland held two pastoral leases in far western Otago, the Eglinton and Cheviot (Fig. 8.3). He was also a scientist, and was twice president of the Otago Branch of the New Zealand Institute. Melland criticised the government's 'incredible folly' in turning out ferrets on the west shore of Lake Manapouri, where kākāpō, then still common, are now extinct.

Benjamin Bayly had been appointed as the government's superintending rabbit inspector in November 1881 (Chap. 7). By the time of his first report to the House of Representatives in 1882, he was able to say that several shipments of ferrets had been procured from Melbourne, and small lots were expected regularly by sailing vessels from England. The Inwards Letter Books kept at the Colonial Secretary's office in Wellington recorded a long series of correspondence with London in that year concerning shipments of ferrets. Most of these letters are now lost, so it is now difficult to tell where the animals came from, but one letter of 5 October 1882 from the Agent-General mentions a shipment sent in September from a dealer 'from the Clyde'. The public accounts for 1881–1882 list expenditures of £122 10s for purchasing ferrets and £28 15s for their hutches.

The annual volumes of *Statistics of the Colony of New Zealand* include comprehensive listings of imports and exports by item and origin. Under 'Animals, living', a total of 1972 animals listed as 'Livestock: other kinds' are recorded as imported from Australia and the UK between 1870 and 1880 inclusive (Table 8.1), that is, before the arrival of the first official imports of stoats and weasels (Chap. 10). Another 590 animals were recorded as imported in the four years 1882–1885, listed as 'Ferrets and weasels', mostly ferrets from Australia where there were no stoats or weasels.

Bayly's early enthusiasm was quickly tempered by unpleasant surprises. A serious problem with transporting ferrets around the world was that, as George Bullen and other private importers had already discovered, these shipments were often disastrous failures. A letter from the Colonial Secretary in Wellington to the Agent-General in London, dated 9 September 1882, included a memorandum listing the very high mortality rates of shipments of ferrets from Britain. The Colonial Secretary had apparently not worked out how to mitigate losses in transit, due partly to inexpert handling and partly to the great susceptibility of ferrets to canine distemper.

Table 8.1 Official records of the numbers of live animals other than domestic stock imported into New Zealand, 1870–1886; many more came in unrecorded. Most of the 1972 unspecified arrivals of 1870–1880 would have been private shipments of ferrets, especially those from Australia. In 1882, ferrets were given their own import category until ferret imports ended in 1886, and 590 arrivals were recorded before locally bred stock became more economic. Imports from the UK in 1883–1886 (in parentheses) were probably mostly stoats and weasels

Year	*Import category*	*Numbers of animals*				*Total*
		NSW	*Victoria*	*Tasmania*	*UK*	
1870	Livestock: Other kinds		1		342	343
1871	Livestock: Other kinds	3	10		670	683
1872	No records					0
1873	Livestock: Other kinds					0
1874	Livestock: Other kinds	51				51
1875	Livestock: Other kinds	310				310
1876	Livestock: Other kinds	8	421		9	438
1877	Livestock: Other kinds	4	17		6	27
1878	Livestock: Other kinds	13	23		29	65
1879	Livestock: Other kinds	4	20	1	12	37
1880	Animals: Other kinds	4	12	2		18
1881	No records					0
1882	Animals: Ferrets		204	28	10	242
1883	Animals: Ferrets and weasels		66	20	(15)	101
1884	Animals: Ferrets and weasels			25	(6)	31
1885	Animals: Ferrets and weasels		23	41	(152)	216
1886	Animals: Ferrets stoats and weasels				(284)	284
	Total	397	797	117	1535	2846

King, C. M. (2017). Pandora's Box down-under: Origins and numbers of mustelids transported to New Zealand for biological control of rabbits. Biological Invasions, 19, 1811–1823. Data from the annual issues of *Statistics of the Colony of New Zealand*

Among the suppliers of live animals for export from England was a well-known dealer, Mr. Cross of Liverpool. He sent 120 ferrets, accompanied by 750 pigeons as food for them, 10 sacks of biscuits and 375 tins of Nestle's milk, to 'a large wool farmer' in New Zealand. Cross quoted his client as believing he was facing ruin unless he could prevent the rabbits from eating up the pasture on which his sheep ought to be feeding, so he told Cross to send the order regardless of the expense.

Curiously enough, at about the same time Mr. Cross was also commissioned to send 2000 live rabbits to British Columbia. News editors in New Zealand were quick to point out the irony of Mr. Cross supplying rabbits to be introduced alive into one country and carnivores to help the extermination of introduced rabbits in another country. Why, they asked, have not the western Canadians learned from what rabbits have done to New Zealand and Australia? Such questions, now prominent among the many *post hoc* sorrows of the story, apparently did not arise at the time.

Bayly's 1883 report to the Rabbit Nuisance Committee listed the losses recorded from 25 shipments of ferrets landed between March 1882 and 30 June 1883:

> Of a total of 1217 ferrets shipped from England, 178 were landed alive, at a cost of £953, plus 241 shipped from Melbourne between March and April 1882, of which 198 landed alive, at a cost of £224 – to which must be added add 122 for natural increase, less 157 that died from distemper. In total, 376 ferrets were landed alive by Government.[14]

The final straw was the voyage of the sailing barque *Lastingham*, commissioned in London by the Agent-General in April 1883 to carry a shipment of ferrets to New Zealand under the care of Francis Court. All the ferrets died on the way. The Colonial Secretary directed the Agent-General to dispatch all future shipments on steamships and to ensure that all consignments be accompanied, reasonably enough, by someone 'with knowledge of the animal'.[15]

On top of losses in transit, competition with private interests was making it increasingly difficult for the government to meet its strategy of obtaining large cohorts of ferrets for liberation on Crown lands. Bayly found that the private demand for ferrets meant that the government had been able to buy only two or three lots during the year 1883–1884. Charles Teschemaker complained: 'I lately sent to Dunedin for 20 pairs of ferrets … [the dealer] could only supply me with six pairs and I was told he could have sold them ten times over'.[16]

By the end of 1883, it had become clear that the policy of importing ferrets was never going to be economic. Fortunately for the landholders (but not for the native fauna), local breeding was fast becoming a feasible and more attractive alternative, so Bayly shifted to a different strategy, the purchasing of ferrets bred in the colony for sale to the government at fixed prices per head. His 1883–1884 annual report on the rabbit nuisance announced the end of official importations of ferrets, although private shipments continued for several years more.

Ferret Breeding Programmes, from 1870s

The earliest private ferret breeding programme was established in the mid-1870s in Christchurch by the Canterbury Acclimatisation Society. The Curator of the Association's gardens (A.M. Johnson) kept ferrets there for Mr. Morton, a member of the Society, on whose behalf he sold some of them; two were still there in 1874.[17] A larger breeding programme began there in 1882, when 20 ferrets were brought in for the government, which was to pay for their food and upkeep, at a daily cost of 3¼ pence per ferret. A large hutch was built for them costing £46 12s. By December 1882, they had increased to 60, and their hutch was enlarged.

In 1882–1883, the government established three official depots for breeding ferrets, at Waimata and Masterton in the North Island, and near Christchurch in the South Island, and many other establishments followed, as listed by local rabbit inspectors contributing to Bayly's annual reports. Most of these animals were bred for distribution and release on Crown lands, but private owners, notably George Bullen in Marlborough, ran their own programmes, turning out up to an additional 500 ferrets in one season on his own land.

The total numbers of ferrets released (both imported and home-bred) by all parties soon reached extraordinary levels. In the three years of 1884–1886, the official tally was 4400 in 1884, 6024 in 1885, and 7539 in 1886, and no doubt there were many other releases unrecorded. Within two years of establishing organised ferret studs, all imports ceased, and releases of locally bred ferrets far exceeded the numbers of previously imported stock.

The Joint Committee of the Canterbury Chamber of Commerce and the Canterbury Agricultural and Pastoral Association feared the prospect of rabbits invading Canterbury in myriads. They were confident that natural enemies judiciously placed throughout the back country would kill off the rabbits as fast as they arrived, and in December 1887 the Christchurch *Press* summarised a public meeting held to consider their report. The committee urged the government to accelerate the production processes by offering a bonus for ferret breeding.

Among the large stations establishing their own ferret breeding facilities was Benmore, which had a breeding stock of 112 doe and 23 buck ferrets at the homestead, and also bought ferrets bred at Quail Burn and Mount Royal (Fig. 8.3). In April 1888, 689 ferrets were turned out on the station, most locally bred. Still, rabbit numbers remained astronomical: 40,000 were killed on Benmore in February–March 1889 alone.[18]

While the demand lasted and official encouragement continued, the ferret breeding business paid some enterprising operators very well. In September 1888, the *Otago Witness* visited a private ferret stud run at Wairuna Hills (inland from Balclutha) by Messrs. Allen and Riggs. They had a contract to supply the colonial government with 10,000 ferrets a year for three years, at 7s 6d a head. The breeding stock numbered 200 full-grown ferrets, of which 40 were males. They ate 30 rabbits a day, plus the milk of three cows, and every detail of their diet, housing, mating and nursing of offspring was carefully managed to keep them healthy.

Ferret breeding was a long-term and chancy operation, so the government had to offer private breeders a guaranteed purchase price for their ferrets in order to get them to invest in it. Such a system created an unforeseen problem: an attractive opportunity for dishonest profiteering. In districts where large numbers of ferrets had already been turned out, the high price offered for live ferrets gave rabbiters an incentive to poach wild ferrets and offer them for sale. In order to ensure that sheep inspectors were not buying back their own (unmarked) ferrets previously released, the government had to change the rules to authorise purchases only from known breeders.

In turn, local sheep inspectors had to check if more ferrets were needed before ordering any. For example, in February 1886, W.A. Smith wrote to E. Clifton, the sheep inspector at Kaikoura, offering ferrets for sale at 10s each, 'do you want them?' Bayly confirmed that ferrets were needed for turning out on Crown lands, so the Colonial Secretary authorised Clifton to buy as many young, healthy and properly seasoned ferrets as were offered, at not more than 10s each, or less if possible.

In the southern South Island, the local rabbit inspectors believed that ferrets had proved their worth, because in some areas the rabbits were disappearing, and in their place ferrets could be found in considerable numbers. Bayly's 1887 annual report on the rabbit nuisance lists a total of 1922 ferret releases during the year (800 in Southland alone), plus others unrecorded. The sheep inspector for Gore, Angus Macdonald, reported that ferret breeding had become an industry in his district and agreed that breeders should be licensed to reduce the temptation to illegal traffic in government stock.

The following year's annual report claimed with pride that ferret breeding was now being done on a large scale; 3600 ferrets had been bred and 2755 released in Otago, 367 in Canterbury, 480 Marlborough, plus >4000 by private owners. A separate report lists the number of current contracts for a total 21,760 ferrets to be supplied to the Stock Department from 1 January 1887 to 8 June 1888. All the consignees were South Island rabbit inspectors, although a footnote adds that fewer than half of the contracts were expected to be met.

The End of the Official Strategy, 1889–1891

Eventually there had to be an end to this programme, and it came during 1888–1889. In 1889, the government bowed to ongoing criticism of the idea of using official funds to supply natural enemies to control rabbits,[19] and withdrew all support for it. Bayly was transferred to the Auckland regional inspectors' office, and responsibility for writing the annual report to the Rabbit Nuisance Committee in May 1889 was taken over by some unnamed official.

The independent Joint Committee on Livestock and Rabbits produced its 1889 annual report in September. It strongly criticised the change in policy connected with the removal of Bayly's influence, in the most uncompromising terms:

> Ferrets stoats and weasels [are] agreed by all to be the cheapest and most self-acting mode of dealing with rabbits … Committee are at a loss to conceive on what grounds the Government have decided to discontinue the purchase of ferrets bred in the colony, and also to cease the importation of stoats and weasels … we believe this course will prove a great national calamity, and they strongly urge the Government to reconsider the matter.[20]

The government did not change its policy, but the ferret breeders did. The rabbit inspector for Otago, Alfred Douglas, had reported that 2230 ferrets had been produced for purchase by the government for turning out on the high country in 1890. When some runholders found that the government had stopped all contracts, they simply released all or most of their breeding stock.

From then on, private enterprise was left to meet the demand for natural enemies by direct negotiations with purchasers. A private shipment of 185 ferrets intended for the South Island and dispatched on the *Aorangi* from London on 9 January 1891 did nothing to help the shortage, since only two arrived alive. But the newly established regional Rabbit Boards started placing orders to supply ferrets from private breeders, which gradually helped to recover confidence in the ferret breeding business, and the local ferret production industry continued long after all official shipments of mustelids ended in 1892 (Chap. 10).

For example, the North Wairarapa Rabbit Board turned loose about 1000 locally bred ferrets in 1892–1893, and in Marlborough hundreds of ferrets were liberated in that spring and summer, with many more following. The extent of the potential supply is illustrated by a file of official correspondence in the archives, including a letter written in February 1899 by E. Field, the sheep inspector at Gore. Field reassured Ritchie that he could get local breeders to commit to producing any number of ferrets up to 1000 at 5s each, provided the government guaranteed to purchase them in September. Ritchie recommended that Field order 1000–1500 for liberation on Crown lands, where Ritchie considered that the natural enemy were doing excellent work in the high country.

In all, the 50-year history of attempts to use ferrets to control rabbits in the South Island involved tens of thousands of animals. Clark[21] cites an unpublished estimate by the historian R.M. Burdon that overall a total of 'upward of 75,000' ferrets were released in the South Island alone. The written records are full of astonishing figures that are too fragmentary and scattered, and too often overlapping, to add up to an actual count, but they are consistent with Burdon's figure taken over the long term.

Spread and Impact of Ferrets

Ferrets soon spread from local release sites throughout the open country of both main islands of New Zealand, and, contrary to expectations, they survived the Otago winter easily in warm, dry burrows, provided there was enough food—the only conditions they could not tolerate were being wet, cold and hungry for too long. They killed rabbits by the thousands, and followed the first surge of rabbits into areas of unsuitable habitat where both are now uncommon, for example, the wet forests of Westland. Charles Douglas, well aware of the damage already being done by Norway rats and cats (Chaps. 4 and 6), reported in despair:

> Ferrets and stoats are now up to the Waiho Forks, and will soon be over the northern districts. They will finish the birds, and no doubt will get all the blame; but the process of extinction had commenced long before they arrived in the country.[22]

George Bullen, the first to try controlled releases of ferrets on a large and systematic scale on his run at Kaikoura, was also the first to find the formula required to make ferrets serve his interests as well as their own.

He was successful enough to be able to suspend all other forms of rabbit control on his land. He understood, as few others did, that once the increase in rabbit populations passed a certain point, ferrets could make a difference only given certain conditions. These were:

> Rabbit numbers have to be first reduced so far as possible by other means. The quickest way of doing this was by poisoning the rabbits in winter, but the timing had to be correct, to avoid the risk of ferrets eating the poisoned rabbits; otherwise they would be found 'dead all over the run'.
>
> Ferrets must be turned out in early spring and in sufficient numbers; those liberated only a few at a time on thickly infested ground could not be expected to cope with the number and increase in rabbits.
>
> No rabbit-trappers can be employed anywhere that ferrets are working, since rabbit traps kill ferrets as well as rabbits. As run-holders reported: 'So long as we employed rabbiters, the rabbits increased.' 'If you want to breed rabbits, keep rabbiters'. '[We] tried [rabbit] trapping and caught 70 ferrets'.
>
> Domestic ferrets should be trained to kill live rabbits before release into the wild.

After eight years, Bullen could report in 1881 that he had many hundreds of ferrets on his property of >100,000 acres. He routinely supplemented their numbers by breeding and releasing about 200 a year, and proudly told the Rabbit Nuisance Committee that he needed no other method of rabbit control.

Contemporary research on the effects of predation by foxes on the numbers of rabbits in Australia[23] supports Bullen's view that ferrets (like foxes) could potentially keep the rabbits down only *if* the breeding stock of rabbits was already reduced by some other factors and *if* there were enough ferrets present at or just before the start of the rabbit breeding season. In addition, ferrets have an extra advantage not available to foxes: they can search the burrows and kill whole litters of young rabbits before they emerge. Timing and numbers are critical. If these are got right, this policy can be the most effective way of preventing a later increase in numbers of adult rabbits, but it is usually not possible unless ferret numbers are intensively supplemented in early spring.

Several other runholders in Marlborough were inspired by Bullen to use the same strategy. Walter Gibson summarised the results for the Joint Committee's 1884 report:

> I purchased all the ferrets I could get and turned them out. Since then I have been working with ferrets and have been breeding a large number. … The whole of that country has been pretty well cleared of rabbits … when fresh detachments of rabbits [arrive from higher ground] we send ferrets out. They kill all the young ones, consequently the whole country is now practically free of them. … Kaikoura district was once almost ruined by rabbits … it was quite a desert. It is now covered with grass. I shall breed 1,000 ferrets this year.[24]

The problems were that, first, in most other places, ferrets tended to be released in too small numbers, or too late in the breeding season, when the advantage of their ability to clean out whole nests of small young was already lost. Second, there was no way to make owners of neighbouring properties agree to act together, or to protect treated properties with rabbit fences, so as to prevent land that had been cleared from being immediately reinvaded from adjacent uncontrolled lands.

The result was that Bayly's general policy of releasing ferrets onto huge unfenced runs failed, because hard-won local benefits were quickly reversed by unstoppable reinvasion. Over the country as a whole, although wild ferrets left to their own devices certainly did survive and breed, and they did kill an awful lot of rabbits, ferret numbers were and are determined by rabbit numbers rather than vice versa.[25] Rough and variable though the records undoubtedly are, they hardly support the government's early confidence that ferrets would be the salvation of the pastoral industry. No one understood why until the development, decades later, of deeper understandings of predator–prey relationships (Chap. 12).

Ferrets and Weka

One of the clearest and saddest stories concerning the effects of introduced mustelids on native birds concerns the weka, a large flightless rail endemic to New Zealand. This bird had long been an important item of diet for the Māori, and early Europeans working in the back country also often depended for food on the bird they knew as the Māori hen or woodhen (Fig. 8.4). Weka are bold, curious, predatory birds with a dagger-like beak and omnivorous feeding habits. They are capable of attacking potential prey ranging from insects to small vertebrates, supplemented with fruit and carrion. G.M. Thomson[26] cites an observation of a weka killing a weasel, in which it 'circled round the weasel watching a chance to spring

Fig. 8.4 The weka is a native ground predator formerly widespread on both main islands. Thanks to ferrets, the buff weka, a distinct subspecies once common on open country in the South Island, is now extinct except for an introduced population on the Chatham Islands. (Photograph by Grant Morriss)

in and strike it, which it did, always on the head, finally stretching its opponent out'. During an irruption of forest rodents in Fiordland, weka could be seen eagerly snapping up mice and swallowing them head first.

In Southland (Fig. 8.3), J.H. Beattie's detailed history of the southern runs describes the local abundance of weka in the 1860s. The Otakarama run, east of the present town of Gore, was

> quite treeless in the 1850s, but littered with charred logs from ancient Māori fires. The ground was dissected with creeks and gullies full of weka, which were abundant until 1888 or 1889, and then they seemed to disappear completely.[27]

A few kilometres west, on the Hokonui Hills, weka were so numerous in 1868 that everything moveable had to be put up out of their reach or they would be dragged off. Māori tradition recognised that district as a good place for catching fat weka, and they made annual foraging trips to collect them for winter storage. Further north, on the Fairlight run east of the Eyre Mountains, there were no rabbits in 1865, [only] luxuriant cattle feed, with weka abounding, and other native birds simply swarming. Rabbits arrived in 1873.

The European settlers were well aware of the potential value of the weka as a resident predator of the rabbit, perhaps even capable of stemming the invasion of pastoral lands (Chap. 7). For example, in 1876, the president of the Otago Institute suggested that the settlers should preserve the woodhen as a useful ally in the destruction of rabbits. Another report added his opinion that the rabbit would never have attained its present pest proportions if the southern settlers had been less destructive of the woodhen. As late as 1880, Thomas Middleton was still assuring Hon. Robert Campbell: 'I am certain that the weka is the best friend we have in keeping down the rabbits. I have no fear of the sort of rabbit we have at present ever becoming dangerous'.

Despite this early optimism, over the next five years, first ferrets and then stoats and weasels arrived at Benmore, and Middleton's opinion changed. His comments ranged from anxious prediction in October 1881 ('Ferrets are to be found all over the run … rabbiters say they will kill the wekas'), through confirmation in May 1882 ('Ferrets … will help keep down the rabbits but are destroying wekas') to anger in September 1882 ('It is to be deplored that the ferret and wekas are enemies. The former will kill the weka in a hole and the weka again is supposed to [help] the ferret [by killing rabbits] in the open') to sadness in February 1885 ('The wekas are very plentiful at the low end of Benmore but I fear they won't last long') and 1886 ('No wekas due to work of ferrets').

Richard Henry suggested one reason why the normally pugnacious weka was losing the battle against ferrets on the mainland:

> The poor weka was once very plentiful in New Zealand, but before the ferret, it disappears like a shadow. It used to kill mice, rats, and young rabbits, and when it saw a ferret, it chased it like a rat, and of course gave up the ghost.[28]

Middleton and Henry were not alone in their opinions. At the Orari Gorge run, inland from Geraldine, C.J. Tripp joined a long list of people writing to the Colonial Secretary protesting against the deliberate introduction of mustelids and their effects on weka. In February 1886, he wrote:

> I am quite convinced in my mind that the Government have made a very great mistake. … In every rabbit hole on my run in which I try a ferret to turn out the rabbits out comes a weka; now these wekas destroy not only rabbits but rats – I have seen a weka chase a rat in and out of a thicket and across a stream and seen the rat run for his life from these birds … we can

> with poison and dogs keep the rabbits well under with the assistance of the weka which goes into holes in the ground and kills every rabbit small as well as big.[29]

Tripp was convinced that 'the vermin have killed off every weka in the District'. The weka may not have disappeared quite as fast as he claimed, but Tripp's concern was valid and widely shared.

Others were more worried about protecting introduced game birds than native birds, even to the extent of hunting weka with greyhounds, in order to make room for the pheasants. Yet others agreed that weka were a better weapon against rabbits than introduced mustelids or poisoned bait, both of which also killed game birds. The correspondence columns of the Otago newspapers preserve a flavour of the debate. For example:

> I fully endorse the popular opinion … We have a better remedy, and more suitable to the Colony, in wild cats, wekas, and hawks, all of which are deadly enemies to 'bunny.' The wekas … pay the penalty through the poison laid for rabbits … [The Otago Acclimatisation Society viewed] with deep regret the steps which the Government has seen fit to take in connection with the introduction of … vermin … these animals will prove of little use in destroying rabbits, and will on the other hand, in all likelihood, destroy Māori hens and other native birds, as well as domestic poultry.[30]

By far the most passionate and sustained rear-guard campaign to protect the weka was staged by William Rees, the South Canterbury sheep inspector based at Timaru. A long correspondence between Rees and the Colonial Secretary in Wellington preserves the detailed arguments in one archives file.[31]

In October 1883, Rees complained to the Colonial Secretary that weka were being wantonly destroyed, and requested that they be legally protected as a natural enemy of the rabbit, in the same way as mustelids and cats already were. The official reply pointed out that weka were an important article of food for Māori, so it would not be advisable to prohibit the killing of weka by them. Rees retorted that it would be better to take off every man and dog now employed in rabbiting, because they systematically destroy weka, and let the weka deal with the rabbit alone. The Colonial Secretary took refuge in semantics, stating that 'the law will not allow protection of the weka since the word used in Section 28 of the Rabbit Nuisance Act 1882 is "animal", and Government is advised that in law that word includes only mammals'.

Rees was not put off. Bayly's annual report to the Rabbit Nuisance Committee published in June 1884 includes Rees' statement, expressed as strongly as permitted, that 'I earnestly wish the weka to be protected, as we have no better natural enemy to the rabbit'.

The archives file includes a very detailed report of the distribution of rabbits in the South Canterbury district, submitted by Rees in June 1886 complete with maps and a five-page memo. He pointed out that, after a concentrated three-year campaign, many properties had few rabbits and no rabbiters or mustelids. Over most of the district, weka were swarming, except on the Te Akatarawa run on the north bank of the Waitaki River, where a professional rabbiter ran traps that killed many weka. Rees pleaded to be heard, because

> I feel quite certain that to the weka we principally owe at this moment the marked decrease of the rabbits in S Canterbury where three years since they were in thousands. … Every one who has had experience in country life where there are rabbits and wekas will confirm what I say. I have seen them kill full grown rabbits – and as to the young ones they never leave one alive when once they find the nest. … Nothing but my extreme conviction of the necessity of giving immediate protection to the weka causes me again to urge this matter.

Rees added letters from some of his neighbours confirming their agreement, including one from J.W. Miller of Hakataramea Downs, that he had

> three ferrets on the property for rats but should decidedly not think of turning them out as they would destroy the wekas which are better rabbit exterminators than all the other [introduced] vermin … The weka is protected here and therefore numerous and which I believe is the cause of the rabbits not now increasing on this property.

Rees ended by

> drawing your attention most earnestly to my statements as regards the weka – and the great necessity that exists for its immediate legal protection (at all times) from destruction as being one of, if not the greatest, natural enemy we possess to the rabbit.

The Colonial Secretary forwarded Rees' report to the Joint Committee on Rabbit and Sheep Acts then sitting. The Joint Committee's report published on 20 July 1886 does not mention Rees or the weka, and the official position that the weka could not legally be protected remained unchanged.

The lack of any official protection meant that for the following decades weka had to contend with human hunters as well as predation by mustelids. The file of correspondence ends in 1889, with a note from J.M. Scott, the sheep inspector at Queenstown: 'Sorry to report weka being wantonly destroyed in the back country near Mt. Aspiring … At all the rabbiters huts they are killed by the dozen … they are not poisoned but cured and smoked for cooking'. Scott asked his superiors for advice on 'what steps I am supposed to take to punish these offenders'. Without legal authority, nothing was or could be done.

In the 1850s, the buff weka, the subspecies endemic to the South Island, had been widespread and abundant throughout the southeastern districts. Inevitably, it was extinct on the mainland by or before 1920. Contemporary observations help to explain why the buff weka, once one of the most common birds in the south, was gone within the period of a few decades. It now survives only on Chatham Island and Pitt Island, where by wonderful good fortune a group of 12 buff weka were introduced in 1905.

At this distance in time, it is not possible for us to distinguish the effects of ferret predation on weka in isolation from the many other changes of the time (e.g. in snow cover or drought, land-use, pastoral management, fire regime, rabbit control and weed invasion). These and other environmental disturbances engineered by the pastoral runholders to favour their sheep regardless of consequences for wildlife (Chap. 7) could all have helped to disadvantage weka in the southeastern South Island at the same time. It may well be that some combination of these factors depressed the productivity of weka below replacement level; and after that, predation by rabbiters plus the recently arrived ferrets could have been the last straw. The observations reported here illustrate the last stages of that doleful process and help to explain why the mere knowledge of the danger was not enough to save the mainland population of buff weka. Sadly, the same or a similar story must have been repeated but unrecorded in many other places as ferrets spread around the country.

Summary

English-domesticated ferrets were traditionally trained to hunt rabbits in a team along with a man and his dogs. They were easy to purchase and too small to damage lambs, so they were the first natural enemies to be imported. Early official importations from dealers in England and Australia beginning in 1882 were often unsuccessful. Shiploads of ferrets were purchased, up to 600 at a time, but canine distemper and poor management during a long voyage often destroyed many or even all of them. Instead, ferret breeding studs were established in New Zealand after 1884, and the progeny released onto rabbit-sick pastures. The government's Stock Department bought tens of thousands of ferrets from private stud owners for up to 10 shillings each, and supplied them to local sheep inspectors for release on Crown lands. Government participation in the ferret business ended in 1889, but private organisations and Rabbit Boards continued for decades; at least 75,000 ferrets were released in the South Island alone. Mortality among them was very high, especially in high country. The effect on rabbits could be useful under certain restricted conditions, not widely practised. Open-country native fauna such as the buff weka were severely damaged or extinguished.

Notes

1. Darwin, C. (1845). Journal of Researches into the Natural History and Geology of the Countries Visited during the Voyage Round the World of HMS Beagle under the Command of Captain Fitzroy RN. London: John Murray (Reprinted from 1860 edition by Folio Society, London, 2003), p. 400).
2. Marchington, J. (1978). Pugs and Drummers: Ferrets and Rabbits in Britain. London: Faber and Faber.
3. For more detailed lists of page citations to the primary sources underlying this chapter, see King, C. M. (2017). The chronology of a sad historical misjudgement: The introductions of rabbits and ferrets in nineteenth-century New Zealand. International Review of Environmental History, 3, 141–175. doi: http://press-files.anu.edu.au/downloads/press/n2519/pdf/08_king.pdf; and King, C. M. (2017). Contemporary observations of predation on the buff weka (*Gallirallus australis hectori*) by ferrets in the South Island during the nineteenth century. Notornis, 64, 52–55.
4. Rolls, E. C. (1969). They All Ran Wild: The Story of Pests on the Land in Australia. Sydney: Angus and Robertson. Pp. 113–114.
5. Importation of birds etc. New Zealand Herald, 1 March 1867.
6. Thomson, G. M. (1922). The naturalisation of animals and plants in New Zealand. Cambridge: Cambridge University Press. P. 70.
7. Peden, R. (2011). Making Sheep Country: Mt. Peel station and the transformation of the tussock lands. Auckland: Auckland University Press.
8. Southland Times, 15 November 1876. [Ferrets for Cowan on the Arawata].
9. Pinney, R. (1981). Early Northern Otago Runs. Auckland: Collins.

10. Parry, G. (1964). N.M.A. The Story of the First 100 Years. The National Mortgage and Agency Company of New Zealand Ltd. 1864–1964. Dunedin: The National Mortgage and Agency Company of New Zealand Ltd. P. 122.
11. Jefferies, R. (1878). The Gamekeeper at home. Oxford: Reprinted 1978 by Oxford University Press.
12. Gibb, J. A., & Williams, J. M. (1994). The rabbit in New Zealand. In H. V. Thompson & C. M. King (Eds.), The European Rabbit: The History and Biology of a Successful Colonizer (pp. 158–204). Oxford: Oxford University Press. P. 162.
13. McLean, W. H. (1966). Rabbits Galore. Wellington: A.H. and A.W. Reed.
14. Bayly, B. P. (1883). Annual Report on the Rabbit Nuisance [1882–83]. Appendix to the Journal of the House of Representatives, 1883 H-18.
15. Colonial Secretary. (1883). Letter 1 August to Agent General [Ferrets to be carried on steam ships]. In Outwards Letterbook – Agent-General (Vol. Folio 356–57). Wellington: Archives New Zealand R20557923.
16. Teschemaker, C. d. V. (1883). Letter to Major Atkinson [NZ Premier] on obtaining information on stoats and weasels. Wellington: Archives New Zealand R24413234.
17. Lamb, R. C. (1964). Birds, beasts and fishes: The first hundred years of the North Canterbury Acclimatisation Society. Christchurch: North Canterbury Acclimatisation Society. P. 19.
18. Pinney, R. (1981). Early Northern Otago Runs; Pinney, R. (Unpublished). Notebook L: Extracts from Benmore Diaries, 1866–88. Dunedin: Hocken Library MS-3178/012.
19. Martin, H. B. (1885). Objections to the introduction of beasts of prey to destroy the rabbit. Transactions & Proceedings of the New Zealand Institute, 17, 179–182. doi: http://rsnz.natlib.govt.nz/volume/rsnz_17/rsnz_17_00_003550.html
20. Acland, J. B. A. (1889). Report of Joint Committee on Live stock and Rabbits. Appendix to the Journal of the House of Representatives, 1889 I-11.
21. Clark, A. H. (1949). The invasion of New Zealand by people, plants and animals: The South Island. New Brunswick: Rutgers University Press. P. 266
22. Douglas, C. E. (1894). Report on the Westland Alps. Appendices to the Journal of the House of Representatives, 1894 C1, 71–74.
23. Pech, R. P., Sinclair, A. R. E., Newsome, A. E., & Catling, P. C. (1992). Limits to predator regulation of rabbits in Australia: evidence from predator-removal experiments. Oecologia (Berl.), 89, 102–112; Parkes, J. (1995). Rabbits as pests in New Zealand: A summary of the issues and critical information (LC9495/141). Landcare Research, Lincoln.

24. Randall Johnson, G. (1884). Report of the Joint Committee on Sheep and Rabbit Acts. Appendix to the Journal of the House of Representatives, 1884 I–5.
25. Norbury, G., Heyward, R., & Parkes, J. (2002). Short-term ecological effects of rabbit haemorrhagic disease in the short-tussock grasslands of the South Island, New Zealand. Wildlife Research, 29(6), 599–604; Norbury, G., & Jones, C. (2015). Pests controlling pests: Does predator control lead to greater European rabbit abundance in Australasia? Mammal Review, 45, 79–87.
26. Thomson, G. M. (1922). The naturalisation of animals and plants in New Zealand. Cambridge: Cambridge University Press. P. 74.
27. Beattie, J. H. (1979). The Southern Runs. Gore: The Gore Historical Society. Pp. 161, 218, 351.
28. Henry, R. (1887). The New Zealand Rabbit and its prey. In. Lyttleton Times, Christchurch: Auckland Institute Museum, QL737.L32 HEN.
29. Multiple correspondents. (1886). Correspondence protesting introduction of stoats and weasels. Wellington: Archives New Zealand R24456501.
30. Otago Witness 10 July 1882. Maniototo – from our own correspondent.
31. Rees, W., & Colonial Secretary. (1883–86). Re protection of the weka as a natural enemy of the rabbit. Wellington: Archives New Zealand R17620862.

CHAPTER 9

Stoats and Weasels (1876–1883)

> It was an outrage that any individual or local body should have been allowed to attempt to correct a blunder by a crime. H. Guthrie-Smith (1969)[1]

From the mid-1870s, the grey blanket of rabbits advanced over the South Island mountain lands, sweeping both the tattered tussock sward and the starving sheep away together.[2] The runholders had to stand by and watch the investments and returns of 20 years whirled downstream in the maelstrom of rising debts and collapsing profits. Wool clips and lambing percentages fell year by year as the losses undermining the entire pastoral industry escalated into a complete disaster.

The runholders' gloomy view of their future is illustrated by an 1877 item in the *Marlborough Express*, predicting that only stoats and weasels could save the New Zealand wool industry. Stoats and weasels are often confused for one another, because both are small, brown and white, fast-moving tube-shaped animals, and the distinguishing black-tipped tail of the stoat is not always noticed, then or now. Most of the primary sources cited in this chapter refer to both together, often under the general family name 'weasels' (variant spellings 'weasle', 'weazel' are given here as in the originals).

C. M. King, *Invasive Predators in New Zealand*,
Palgrave Studies in World Environmental History,
https://doi.org/10.1007/978-3-030-32138-3_9

It was widely believed that, since no amount of human effort could undo the damage already being done by rabbits, only the importation of natural enemies in large numbers could restore the balance of nature and prevent the complete collapse of the wool industry (until then earning 30–40% of the colony's export income; Chap. 7) within the next ten years.

This deep-seated philosophical faith in the ability of predators to control rabbits seemed to be immune to the evidence, not only that the most obvious natural enemies already present (cats and ferrets, plus native harrier hawks and weka; Chaps. 6 and 8) had not prevented the rabbit invasion, but also that it was already too late to stop rabbits from ravaging the high country. Hard experience had already confirmed that the 'Importation of ferrets has not been attended with the success that was anticipated' so runholders clung to the belief that 'To be thoroughly serviceable, a hardier animal is required, especially in the colder portions of Otago'.[3]

From the runholders' point of view, there was nothing unreasonable in continuing to draw vast and gratifying conclusions about the natural control of rabbit populations directly from such very slender arguments. The arguments themselves were seldom scrutinised because, in philosopher Mary Midgley's memorable phrase, they peeped out only briefly, like very early mammals, from a protective thicket of assumptions.[4] Midgley was writing a critique of unrealistic proposals about the human colonisation of space, which some might see as the modern equivalent of the nineteenth-century programme of introducing the natural enemies of rabbits. There is no shortage of comparably unsupported ideas in wide circulation today (Chap. 12).

So runholders lobbied the colonial government to approve an official programme of mass importations of wild stoats and weasels. Vehement objections to the plan began at the first mention of the idea in 1874 (p. 151), both in prominent newspapers such as the *Otago Witness* and in scientific journals,[5] and continued long after it was too late to stop it. The strong language used in these sources illustrates the intensity of the debate better than any subsequent summary.

So why were the environmental consequences of these and other European activities for endemic flora and fauna so often dismissed? In part, because of simple self-interest, and in part because it was widely believed that the New Zealand endemic biota was already doomed to extinction. Charles Darwin himself saw, as early as 1835, that 'The endemic productions of New Zealand, for instance, are perfect one compared with another; but they are now rapidly yielding before the advancing legions of plants and animals introduced from Europe'.[6]

The historic details of these events illustrate how the powerful economic interests and dominant cultural assumptions of the runholders authorised a programme of biological introductions that ran against the clearly expressed interests and views of many local residents and well-qualified independent advisers. This finding suggests that acclimatisation should be viewed in much wider than biological terms, a conclusion with implications for thinking about contemporary ecological problems (Chap. 12)—especially predator–prey relationships.

1874–1880: The First Proposals

In contrast to the docile ferrets, stoats and weasels are small, agile and intractably wild carnivores, which evolved in the colder parts of the northern hemisphere as specialist predators of small rodents—voles, mice and lemmings.[7] Stoats native to England had the advantage that they are large enough to specialise on rabbits, but are too small to pose a serious threat to lambs, unlike the foxes that were already established in Australia (Chap. 11). Stoats are untameable once their eyes are open, and their rigid once-a-year birthing season makes them difficult to manage and impractical to breed in captivity. But they are common in Britain, and easily trapped by skilled gamekeepers using tools well developed by the late nineteenth century (Fig. 9.1).

Fig. 9.1 (**a**) English gamekeepers developed the Fenn trap, the most widely used 'humane' trap available for 40 years, although now unable to pass more stringent animal welfare standards. The stoat's distinguishing mark is a bushy black tail-tip. (**b**) Weasels are similar but proportionately smaller and with short brown tails. (Photographs by **a**: Jim Mills; **b**: Andrew Thompson)

The immediate problem to deal with was how the proposed imports of such hostile wild carnivores native to the other side of the world could be organised. The first suggestion came from Macrorie and Cuthbertson, a firm of stock and station agents and auctioneers established in Invercargill in 1875 (Chap. 7), who saw a possible commercial opportunity. They anticipated a demand for weasels for liberation on the rabbit-sick pastures of Southland clients of their new business, but wisely first checked that importing them was not already illegal.

Reassured, they wrote on 8 May 1876 to Frank Buckland, an English naturalist and leading proponent of uninhibited acclimatisation. They proposed a plan to offer £5 per pair for weasels, delivered alive to Bluff, and asked Buckland whether he thought such imports would be useful and feasible. The incentive offered was about five times the average weekly wage for an English gamekeeper at the time, but Buckland replied that shipping weasels round the world to New Zealand would be impossible. He recommended ferrets instead.

In fact, some private individuals were already trying to import weasels on their own accounts. One was the Hon. Dr. Daniel Pollen of Auckland, who, during a bad-tempered debate in the Legislative Council in 1876, confessed to having 'some weasels in his possession'. The main problem was that, even if the difficulties of collection and transport of these aggressive little carnivores could be solved, bringing them in would be wasted effort if they could not be landed in numbers sufficient to establish a breeding population. Cuthbert Cowan told the Rabbit Nuisance Committee in 1876 that 'they are bad travellers ... [but] I have some coming out. They were to have been dispatched in July'. Almost a decade later, Cowan stated that 'I landed 16 ferrets and 3 weasels and put them out on my own property. Now the country is overrun with ferrets. Nothing has been seen of the weasels'.[8]

The Hon. Robert Campbell's extensive estates in New Zealand (Chap. 7) were indirectly managed from his family base at Buscot Park, in England. This link may explain why several English newspapers announced in October 1876 that 'a large number of live stoats collected from various English counties' were about to be shipped to New Zealand by Robert's older brother William Campbell. The notices do not say that the stoats were destined for the Hon. Robert's rabbit-infested runs, but it would be surprising if they were not.

The idea of exporting weasels to New Zealand was widely reported in the English press. It provoked a passionate response from leading ornithologists concerned with the threat to New Zealand's endemic flightless native birds, represented in England by Professor Alfred Newton of Cambridge University, and endorsed with fury in New Zealand by Walter Buller. For different reasons, opposition was also strongly expressed by landowners who had been investing heavily in acclimatising game birds for sport. In December 1876, both sides of the debate were summarised in a long article in a British naturalists' journal (*Land and Water*, edited by Frank Buckland) and reprinted in the *Otago Witness*. It ended by Buckland reluctantly conceding Newton's point and withdrawing his support for the idea.

Meanwhile, the colonial office in London was sufficiently alarmed by William Campbell's proposal to address a separate official enquiry to Professor Newton 'as to the necessity of proceeding to such extreme measures'. Newton repeated his opposition in forceful terms, and instead forwarded plans for a method of rabbit trapping that he predicted should be

> sufficient to remove the injury without … letting loose ferrets, stoats or other predatory beasts, which cannot fail eventually to produce far greater mischief than that which they are intended to remedy … there is no gamekeeper in this country who does not wage incessant war against them.[9]

This response was duly conveyed to the Colonial Secretary in Wellington. Such firm official disapproval of an action that could be linked to the Hon. Robert, a serving member of the Legislative Council, could explain the absence of any mention in Papers Past, the Bluff Maritime Museum records or any other Southland archive of William Campbell's supposed cargo arriving at Bluff in the late 1870s.

Even if it did, the chances of establishing a new population of stoats and weasels were not good unless enough of the animals survived the journey and were released together. Successful introduction of any exotic species depends largely on the size of the colonising group,[10] which suggests that the few weasels obtained by Cowan and Pollen may not have survived for long. Neither, apparently, did two weasels landed by Joseph Ward of Wairau (Marlborough) among a consignment of ferrets. It seems that no one at that time had worked out how to bring such highly strung and aggressive wild animals alive and healthy around the world, in numbers sufficient to establish a viable population. Meanwhile, rabbits continued to ravage the South Island runs.

Disappointed runholders in Marlborough continued trying to make their own arrangements. J. Stace had reported killing 84,700 rabbits on the Starborough run (Fig. 9.2) in the previous nine months. An 1881 motion to a meeting of landowners, proposing a subscription levied on all sheep owners between the Opawa and Clarence rivers to finance a large collective importation of the natural enemies of rabbits, was fiercely debated, and lost. A month later, on 14 December 1881, a Blenheim stock agent placed an advertisement announcing:

> To Stewards and others of Wool Ships etc – Wanted to be known, that the undersigned is prepared to give £1 apiece for all Stoats and Weasels … landed alive, and in healthy condition, the same to be shipped over as soon as possible after arrival to care of Messrs Fell Bros, Blenheim, who will act as agents and settle payment. H. Stace, Starboro', Blenheim.[11]

The chances of any unfortunate stoats or weasels surviving the journey in such inexperienced hands were probably close to zero.

Another enterprising party, signing himself only 'Colonist', wrote a letter in 1882 to *The Field*, the premier magazine of English country life, asking if any readers could supply information as to the means of procuring stoats, polecats or weasels, with suggestions for their care during transport to New Zealand. He explained that, once the approaching plague of rabbits arrives and becomes firmly established, there will be no remedy but large-scale poisoning, which he described as 'an objectionable method'. Rather, he planned to meet them with their natural enemies and, thus, 'prevent their attaining to unmanageable numbers'.[12]

The editor appended a warning that such animals would not confine their attentions to rabbits; much mischief had already been done, and more would follow unless 'well-intentioned gentlemen' considered more carefully the likely results of their experiments. The letter provoked several replies, including one from Newton, agreeing on the danger and advising 'Colonist' to think again. 'Better', one said, 'to send out professional rabbit catchers with plenty of long nets … than adding more enemies to the poor kiwis. Will not the authorities prohibit the introduction of these vermin before it is too late?' On the contrary, before long it would be the authorities who were promoting them.

So, although there may well have been some individual weasels or stoats in New Zealand before 1883, it seems unlikely that they had actually established a breeding population. Public opposition had already persuaded Buckland to back down from his advocacy of the proposed imports, and both Macrorie and John Cuthbertson died within a few years of establishing their partnership. John's own brother Robert Cuthbertson stated in evidence to the Rabbit Nuisance Committee: 'As Chairman of the West Oreti Trust, I sent Home some years ago, carte blanche, to get out £500 worth of vermin, but failed to get any at all'.[13]

Meanwhile, professional rabbiters and gamekeepers in England soon recognised the new employment opportunities offered by the demand for their skills in New Zealand. For several years (at least 1876–1878), William Campbell recruited or encouraged potential emigrants with advertisements in many British newspapers, for example, 'Wanted, a few more Rabbit Trappers for New Zealand. Apply, giving full particulars and references, to W. H. Campbell Esq, Buscot Park, Lechlade, Gloucestershire'. James Sellwood, a gamekeeper from Gloucester, emigrated with his family on the *Oamaru* in 1876, and there must have been many more.

1880–1882: The Colonial Government Creaks into Action

The colonial government's first response was to toughen up earlier and less effective anti-rabbit legislation, and pass the Rabbit Nuisance Act 1881, quickly followed by several amendments (Chap. 7). The new rules provided legal protection for all enemies of the rabbit already present, and there were no exemptions other than for the weka, because it was traditional food for Māori (Chap. 8). Defaulters could be fined up to £10 for killing a ferret or weasel, and the rule applied over the whole country, causing particular outrage when stoats and weasels spread into areas where there were no rabbits. Complaints from the settlers, that this illogical provision prevented them from destroying predators to protect their cherished poultry and carefully nurtured game birds, produced only a terse official notice from John McKenzie, minister for agriculture:

> Attention is drawn to the Provisions of 'The Rabbit Nuisance Act 1882' for the protection of the natural enemy of the rabbit, and it is hereby notified that the animals mentioned hereunder have been Protected, as provided for in Section 29 of the abovementioned Act: ferrets, cats, stoats, weasles, mongoose.[14]

In 1897, the Auckland and Tauranga Acclimatisation Societies tried several times to persuade the minister to grant exemptions under the act, to permit local removal of protection for stoats and weasels in districts not infested with rabbits. He replied that 'we cannot exempt a single district'. It took until 1903 before amended legislation finally allowed targeted removal of natural enemies of game birds and poultry, strengthened in 1910 but reversed in 1923. Only in 1936 was all protection for mustelids finally removed.[15]

Benjamin Bayly

The first report from the government's new and energetic superintending inspector, Mr. Benjamin P. Bayly (Chap. 7), was published in June 1882. It described the establishment of his administration and the ongoing importations of ferrets, and adds cautiously: 'Other enemies, such as the stoat and weasel may be introduced, if found practicable'. The careful wording shows that the committee was not yet convinced that the difficulties likely to be caused by the public reaction, not to mention the sheer practical techniques required, could ever be resolved. Indeed, almost at once, the minister faced a question in the House, asking whether, 'In view of the conflicting opinions expressed as to the introduction of ferrets, weasles, &c, to overcome the rabbit nuisance, [would] Government cause the matter to be more fully considered'. The formal reply stated that 'Government had not sufficient reason to doubt [their] expediency'.[16]

For years afterwards, Bayly and his supporters ignored the ongoing protests, arguing that 'The introduction of the stoat and weasels is also a legitimate function of the Legislature, seeing that it would be the means of giving a good money value to millions of acres of Crown lands now abandoned as worthless'. Bayly was much encouraged by proponents of the idea who cited cases in Britain where rabbits (a valued animal there) could not be kept on ground not cleared of stoats. The problem was, how could that favourable situation be recreated in New Zealand?

The prospects of successfully introducing wild stoats and weasels depended on one important condition: the recruitment of a skilled trapper with sufficient knowledge of these aggressive little animals to collect them from Britain, then feed and house them for long enough to convey them alive to New Zealand. The trade was a new one, because there had never before been any demand for such new and unusual articles of commerce. Fortunately, there was one man who was already familiar with New Zealand, and was also able to facilitate the right contacts.

The Brigg Connection

In 1880, two English gentlemen farmers, Samuel Grant of Healing and John S. Foster of Ludborough (Fig. 9.5, inset), made a study tour of the farming districts of both main islands, on behalf of Lincolnshire tenant farmers considering emigration as an escape from increasingly difficult conditions at Home. Grant and Foster made a detailed assessment of the main factors that an emigrant farmer has to consider, which they listed as the climate, soil, market, and labour resources of the Colony.[17]

It is a remarkable example of selective reporting, that the English visitors did not mention rabbits at all until they got to Southland. There they admitted (a massive understatement; see Chap. 7) that rabbits were becoming a serious nuisance. Nevertheless, they concluded that the prospects of New Zealand as a profitable site for emigration were so favourable that it 'will not fail to become a great producing and exporting country of all the chief food staples'; and it is 'suitable for sheep of every kind … The runholders have, in times past, made very large sums of money, and no doubt will continue to do so'.

Grant and Foster's report was sponsored by the colonial government, with whom they negotiated a 'special settlement' for purchase of land near Te Aroha, provided they brought in a certain number of Lincolnshire farmers as permanent settlers. Rabbits were indeed scarce in the North Island at the time, but one cannot help wondering if one or other of the parties had some vested interest in underplaying the threat of future rabbit invasions. Contrary to Grant and Foster's cheerful prediction, the rabbits spreading north from their early liberation sites in Southland were already forcing many formerly wealthy runholders of Southland and Otago into bankruptcy. Eventually, the deal fell through, and Grant was himself bankrupted in 1884.

Meanwhile, Grant played a critical role in the story of stoats and weasel introductions. Among the pastoralists in distress were several with whom the two Englishmen had stayed during their 1880 tour, and with whom they afterwards kept in touch. Charles Teschemaker, a prominent South Island runholder (Chap. 7), and his brother Thomas in England, had been conducting a long correspondence with Grant about their common interests in rabbit control. Grant had certainly visited the Teschemakers' New Zealand run at Kauru Hill and stayed at their homestead at Taipo (Fig. 9.2). He probably also visited Bushey Park, the property of Francis Dyer Rich, 48 km from Taipo. Rich and Charles Teschemaker were both members of the Dunedin Club, the sort of place that runholders would

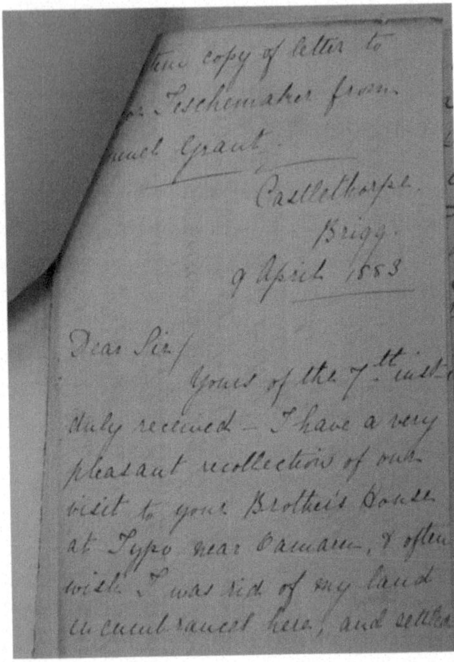
ue copy of letter to
r Teschemaker from
muel Grant.

Castlethorpe.
Brigg.
9 April 1883

Dear Sir/
Yours of the 7th inst
duly received — I have a very
pleasant recollection of our
visit to your Brother's House
at Typo near Oamaru, & often
wish I was rid of my land
encumbrances here, and settle

Fig. 9.2 This letter from Samuel Grant confirms his 1880 visit to Taipo, and that he was subsequently responsible for arranging the first shipment of stoats and weasels to New Zealand. (Grant, S. (1883). Letter to Major Teschemaker. Wellington: Archives New Zealand R24413234)

naturally entertain like-minded gentlemen visiting from England, such as Grant and Foster. Two years later, Rich asked Grant to supply a shipment of mustelids for Bushey Park, with an experienced man to accompany them on the journey from England. Grant was one of few English gentlemen farmers who would know what to do with Rich's request.

In nineteenth-century England, rabbits were an important economic harvest,[18] so they were provided with artificial warrens to encourage breeding and protected by networks of traps set to remove the stoats and weasels that competed with the farmers for rabbits. The small town of Brigg, in Lincolnshire (Fig. 9.3), was a local centre of the rabbit-warrening and fur dressing industry.[19] Warrens were especially common around Brigg and on the nearby Lincolnshire Wolds.

Samuel Grant owned a farm at Castlethorpe, 2 km from Brigg, where local lads helped to work the traps. The late nineteenth century was also the heyday of the sporting estate, where gamekeepers were employed to remove stoats and weasels (and other predators) that destroyed many game birds.[20] One such estate was owned by the Earl of Yarborough at Brocklesby Park, 16 km from Brigg.

Fig. 9.3 Display in Brigg Heritage Centre, Lincolnshire, celebrating the important local rabbit fur dressing industry. Rabbit fur traded through the Brigg market was used to trim gloves and muffs, and to make the felt needed for trilby hats. (Photograph by Author)

The Allbones family were professional vermin destroyers based in Brigg. Mathew Henry Allbones (1836–1909) had worked at Brocklesby Park, and his eldest son, Walter Allbones (1863–1948), had learned the skills of trapping on Grant's farm at Castlethorpe. Both were especially knowledgeable about stoats and weasels, so Grant commended Walter ('a first-class rabbiter') as well qualified to travel with Rich's animals (Fig. 9.4). Henry organised the business end of the commission, which proved to be the first of many he received over the next seven years. (One of the acute ironies of this story is that the Allbones' expertise in vermin trapping, recruited to help New Zealand farmers *remove* introduced rabbits, had been gained in the course of their work in *protecting* English rabbits!)

1882: The Allbones System

Henry Allbones was described as 'in appearance and manner the very "spit" of the professional vermin-catchers to be found in most of the English agricultural counties' of the time. His proposed system for managing large numbers of stoats and weasels on such a long journey assumed that the diet of these active small predators required fresh blood as much or more than meat. This was a common perception at the time, derived from widely reported observations that a stoat or weasel entering a hen-house or chick run would kill every bird in sight, leaving nearly all of them as uneaten carcases with needle-point punctures in the neck.

Fig. 9.4 Walter Allbones. (Photographer unknown. From the Allbones family archives, courtesy Janice Salmon)

Such behaviour had been an important adaptive strategy for the ancestors of small mustelids that evolved to live on fluctuating populations of voles and lemmings in the frozen north. It allowed them to collect and cache surplus food during years when hunting was easy, which in that cold climate stayed fresh enough to eat during the starvation years that followed.[21]

In other contexts, the same bloodthirsty behaviour seemed either inexplicably wasteful, or just what was needed, depending on the point of view. Ornithologists and keepers of game birds condemned them as the most 'destructive of our four-footed vermin'. Runholders desperate for an ally against the rabbits praised their 'love of slaughter having apparently no connection with the mere vulgar craving for food. The … weasel is like a conqueror, who slays his thousands for glory, and goes on slaying till he is too old to slay any more'.[22]

Among the early objections to attempts to import stoats and weasels from the other side of the world was the prediction that such small, active animals could not endure close confinement for months at a time, because 'they knock themselves to pieces about the sides of the cage'. But the Allbones' system proved that the little animals were remarkably tough, and if supplied with good bedding kept scrupulously clean and dry, well fed on fresh meat still containing blood (carried on the ships as live pigeons) plus plenty of water, the mortality rate averaged (with rare exceptions) under 10%. Between 1882 and 1890, Walter undertook at least five round-the-world deliveries, and Henry at least seven (Chap. 10).

At first, Henry collected stock in England while Walter accompanied the shipments. On at least two occasions Walter took along an assistant, his friend Fowler Metcalf, a gamekeeper employed at Broughton Lodge near Brigg (Fig. 9.5). After nearly four years of back-and-forth voyages, Walter married in October 1886, and settled down in Brigg to produce a large family, swapping roles with his father. Henry Allbones last appeared in the passenger records in May 1890, when he arrived on the *Tongariro* with his third son Henry junior.

A journalist was sent to interview Henry in Lyttelton in 1887, and his account, widely syndicated, included some very revealing details about his business methods, including a first-hand description of its fully developed form. Henry told him:

> 'You will get nothing better for rabbits than stoats and weasels Every time they want a feed they kill a fresh rabbit, and when they are hunting, if they light upon a nest of ten young ones, they kill them all. They eat nothing dead, not even a pigeon that dies itself, though they are very fond of pigeons. ... There won't be any overcrowding with them. If there were, they are very soon trapped off, if you want to trap them off. But they don't increase very fast. For one thing their life is but three years, for they wear out their pin teeth cracking the rabbit's skull. They strike just behind the ear, so as to cut the jugular vein, and their teeth strike the bone and wear away quickly, for they are a regular needle point. And then the young ones are six weeks before they can see. No they'll never get too plentiful ... You ... ought to get more stoats and weasels for rabbit infested runs. They would keep the rabbits down. Mr Swailes, of Rawby [Wrawby], in Lincolnshire, had a stoat among his rabbits last year, and he had hardly a rabbit to shoot.'
>
> 'That was among burrows, Mr Allbones; will they be as successful with the rabbits here, that nest in a tussock on the hillside?'
>
> 'That they will; they are always travelling, are stoats, and they will catch a hare. The hare can never get away from them, though she runs through hedges, and they bite down through the top of the head till they kill her, and they never take more than just the blood.'
>
> 'For a shipment of 300 you want 4,000 pigeons to be used on the voyage, and that would give you an idea of the way they would destroy the rabbits where they were thick. I had to go to Antwerp for the pigeons – couldn't have got them in England for anything like a reasonable price – and I bought them there 8d a piece easily enough. ... They would come to about 10d a piece with the shipping charges. I paid 4s 6d apiece for weasels,

and 8d carriage, reckoning an average, as I paid the carriage whether they came from far or near ... I paid 6s a piece for stoats, so they had to be in good condition for me to accept them.'

'I used to look at every one—catch him behind the head and hold his hind feet, and sometimes he'd have lost an eye or a leg, and sometimes his teeth would be nearly done ... I built a great shed to put them in, fitted all round with shelves. Boxes had to be lined with zinc, for they'd soon gnaw a hole big enough for them to get out of, and the Company would not ship them without. The first lot of boxes cost 16s; afterwards I bought the material and got it made up, and they came down to 1s 6d. They have to be made with fixings so that they can be kept clean. ... There are not many men in the business, and it isn't everybody could manage it. You see, it takes a bit of scholarship to buy all that material and make all the arrangements.' And Mr Allbones shows our reporter with pride a letter signed by Mr Walter Kennaway, Secretary of the Agent-General's office, testifying to the faithfulness and ability with which his work has been done. 'I have been all my life at this sort of work,' he continues. 'I am a vermin destroyer for a number of large estates, including Lord Yarborough's estate, Brocklesby Park'.[23]

This detailed account of the amount of trouble Henry had to take to select and care for his fierce little passengers helps to explain why previous attempts to manage the shipments probably did not succeed.

Allbones' expert services did not, of course, come cheap, but the animals he brought were valuable to runholders, and competition for them was stiff. When 20 weasels and 6 stoats from the *Ionic* were sold in January 1885 at a Christchurch auction, the weasels fetched £3 each, and the stoats £5 5s each. Such high prices created quite a sensation: some English newspapers reported with astonishment that 'these animals, which have hitherto been looked upon as vermin, now ... possess a marketable value'; 'farmers are making 6s and 4s a head, so the trade is likely to become quite an opening for the distressed agriculturalist'. The *Naturalist*, a periodical of natural history for the North of England, merely predicted that 'Doubtless we shall hear in a few years that the New Zealanders are at their wit's end to devise means for getting rid of weasels'.[24]

Allbones also sometimes had to defend his operation against public sympathy for the pigeons—'Poor pigeons', sighed the *Leamington Spa Courier* in 1888—and especially against sensational and inaccurate reports that the pigeons were fed to the weasels alive, not freshly dead. As Allbones pointed out, that would be completely impractical, as they would flap violently about the cages, upset the water tins and everything else. There was no more cruelty involved than in the eating of a pigeon pie, which at that time was a common dish in rural England. Perhaps the local taste for pigeon pie explains why a local lad stole two of Allbones' pigeons from the Woolwich docks, where they were awaiting their ship.

Origins of the Animals Transported

The great majority of the animals shipped to New Zealand were collected from farms around the Allbones' home base in northern England. To start with, handbills were circulated throughout north Lincolnshire, and the potential trade was publicised by local newspapers such as the *Lincolnshire Chronicle*: 'About 100 stoats and weasels, which have been caught with considerable difficulty in different parts of Lincolnshire by a man named Allbones, a vermin-catcher, of Brigg, were yesterday conveyed by him to London, en route for New Zealand'. Allbones received offers of animals from all over the country but, at least in mid-1885, was still finding it cheaper to collect them himself in Lincolnshire. Walter got many of them from a farm at Croxby Top (Fig. 9.5), near Rothwell in the Lincolnshire Wolds, which is still a working farm complete with the thick hedges and scattered copses that have long been the gamekeepers' favourite trapping sites.

An account of Walter's work was written decades later for the *Lincolnshire Times*, 2 April 1976, under the heading 'Rambler's Notebook: Exported weasels to New Zealand'. It is valuable because it is the only source that mentions Croxby Top, but it also includes several red herrings. The worst was the mention of an unnamed four-masted ship, which (after hours of searching for likely sailing vessels) turned out to be the steamship *Doric* (early steamers retained masts, but did not rely on them) (Fig. 10.1).

As the demand grew, Henry began to source animals from elsewhere. He advertised in other county papers

> such as the *Sheffield Daily Telegraph*, *Fife Free Press* and the *Yorkshire Post*, so as to get it known everywhere, and someone put it in *The Field* … At first people thought it was a hoax, but by-and by they began to send them to me … they'd shove them into an old box, or watering pot, or kerosene tin, or anything handy … many of [the animals] were maimed or damaged [but] they had to be in good condition for me to accept them. I built a great shed to put them in, fitted all round with shelves.[25]

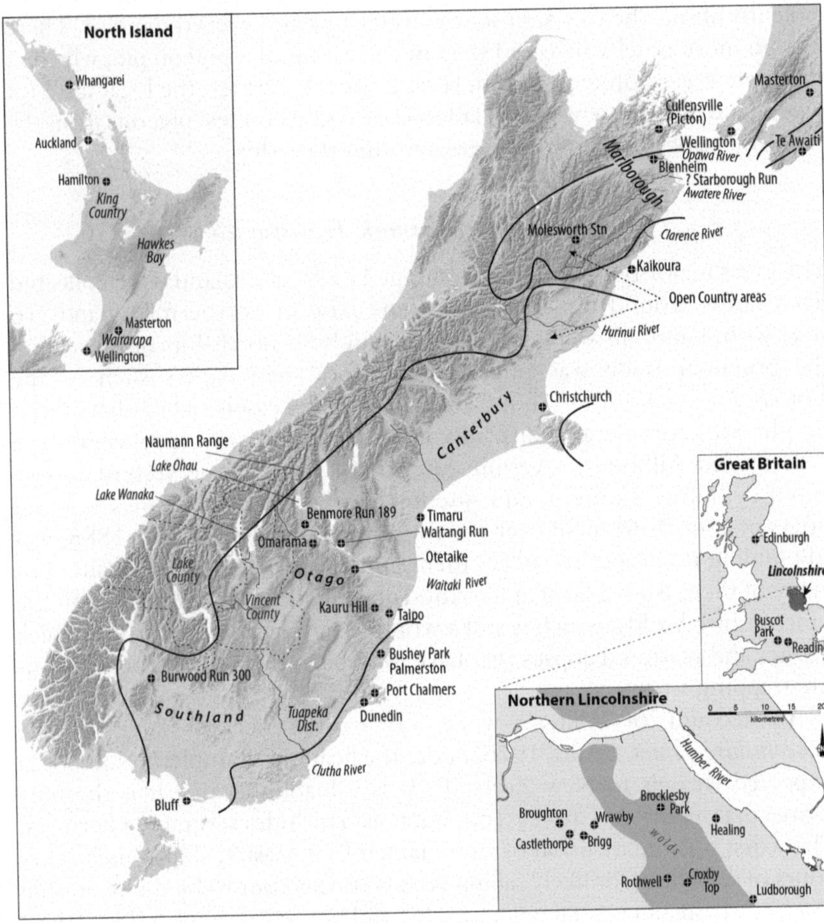

Fig. 9.5 Map of the pastoral runs first receiving imported stoats and weasels in the late 1880s, and (inset) the location of Brigg in Lincolnshire, the source of the early collections. Dashed lines enclose Lake and Vincent Counties, which were among the first and worst affected by rabbit damage. (Cartography by Max Oulton, from Author's data)

Throughout his operation he still continued advertising in Lincolnshire—offering to supply traps at 3s 6d and instructions if needed.

At one point, Allbones was offered 'a full supply from Ireland', but no New Zealand stoats sampled so far have showed any genetic connections with Ireland, or from continental Europe.[26] So the conclusion remains that the ancestors of most stoats now in New Zealand came from northern England, plus perhaps some from lowland Scotland.

1882–1883: The First Trial Shipment

By late 1882, when the details of Allbones' system were decided, Samuel Grant and Henry Allbones supplied a trial consignment of 25 stoats and weasels plus 8–10 ferrets for transport to F.D. Rich, to be accompanied by Walter Allbones and Fowler Metcalf. The animals and their two caretakers were shipped aboard the sailing vessel *Waitangi* (Fig. 9.6) on 22 December, along with their live pigeons, sacks of pigeon food and a variety of other livestock and passengers.

Captain Friston of the *Waitangi* reported later that in early January the ship met a hard WSW gale and fearful sea in the English Channel, which increased in force for several days. Then on 8 January a heavy sea 'broke on board, smashing the sheep pens, cowhouses, and pigstyes, killing and

Fig. 9.6 The very first stoats and weasels to be landed successfully in New Zealand on 26 March 1883 were the survivors of a trial private shipment carried on the deck of the sailing ship *Waitangi*. (ATL G16642–1/2, Alexander Turnbull Library)

drowning sheep, poultry, ferrets, and weasels, together with a number of pigeons, &c.'. The ship recovered, and eventually arrived at Port Chalmers on 26 March 1883, with the first few stoats and weasels to have a real chance of establishing themselves on New Zealand soil.

An eyewitness account of the voyage is given in a shipboard newspaper, preserved in the Alexander Turnbull Library. It describes the cabin occupied by Allbones and Metcalf, and their ordering of waggons at Port Chalmers to carry their surviving animals to their destination. Curiously, Walter's first letter from New Zealand to his fiancée Annie (a copy is still held by his family in Brigg) mentioned the storm and the loss of the sheep and pigeons, but nothing about his lost mustelids. Later reports agree that only 10 of the consignment of 33–35 mustelids were landed, but these did 'extraordinary execution' when released. Rich promptly engaged Walter to work on his station for a year.

Rich informed Bayly of his recent import, and the Rabbit Nuisance Committee reported to Parliament on 31 July 1883. Bayly agreed that 'the stoat and weasel, if they can possibly be imported, would be best agents of destruction ... A good bonus has been offered to anyone who will introduce a certain number of these animals in a healthy condition ... one man has proceeded to England for the purpose of making an attempt'. Bayly does not name the 'one man', but at the date of Bayly's report, it could not have been Walter Allbones. Walter's letter to Annie in September 1883 states that 'I dont intend to stop in New Zealand very long however I will stop my year out'. Neither could it have been Fowler Metcalf, who, according to Walter, had by that time been for the last three months 'out on a wild run among the wild pigs rabbiting'. The 'one man' already on his way back to England in July 1883 must have been someone else, possibly engaged to bring out another private shipment for Rich.

Walter's success on behalf of Rich greatly encouraged Bayly in his hope that stoats and weasels would be better able to stand the voyage and less vulnerable to distemper than were ferrets. Although he agreed that the runholders who were most impaired by the rabbit pest should combine to meet the cost, 'the State should be first to move in the matter ... once the fact of this importation was established there is little doubt that private enterprise would follow'. Bayly urged the colonial government to import stoats and weasels both officially and by offering 'liberal support' to private importers. Bayly's views were summarised in the *York Herald* of 23 November 1883, so the potential profit to be made from this new export trade was already known in England.

On 4 December 1883, the Colonial Secretary wrote to New Zealand's Agent-General in London the key letter that committed the government to take part in what had hitherto been an entirely private game:

> I have the honour to inform you that a man named Walter Allbones sails for England as a steerage passenger per 'Tongariro' under engagement to the Government to procure and import to New Zealand a number of weasels and stoats. Allbones has been selected for this service as possessing special knowledge of the animals, their habits and management, having proved his aptitude for it by a successful importation which he has made for Mr. F. D. Rich of Bushy Park. … I enclose an estimate of the cost of importing 100 animals although it is to be understood that the experiment is not to be confined to that number if more can be procured without involving undue delay. Allbones will call upon you and will explain the details of his intended movements and of the way in which the arrangements are to be carried out; and I shall be glad if you will furnish the necessary funds out of your general imprest account.
>
> Mr. Samuel Grant of Castlethorpe, Brigg, Lincolnshire, is acquainted with Allbones, who was brought up on his farm or in its immediate neighbourhood, and I have to request that you will correspond with this gentleman and bespeak his good offices, which it is believed that he will willingly render to the Government in the enterprise they are undertaking. It will be advisable to ask Mr. Grant whilst doing this, to see that the young man uses all diligence in his work that no time may be lost in his return to the Colony.
>
> It is of course unnecessary that I should explain to you the reason why the Government are about to make this experiment. The rabbit pest has attained such dimensions, the effects of the ravages of the animal are so disastrous, and every means hitherto adopted has proved so unequal to anything approaching eradication, that it has been determined to import considerable numbers of the natural enemies of the rabbit in the hope that the nuisance may thus at last be abated. If this experiment proves a success as, with the rapid communication now existing [by international telegraph, since 1876] there is every reason to hope it will, it is the intention of the Government to push on the enterprise until the extermination of rabbits in New Zealand has been accomplished.[27]

The equivalent pest extermination ambition of the twenty-first century, the Predator Free New Zealand 2050 (PFNZ 2050) programme, clearly has long roots in history.

Summary

After ferrets proved to be more fragile and less successful in controlling rabbits than expected, the colonial government searched for hardier predators better able to survive high country winters. Stoats and weasels evolved in the cold boreal regions and, although smaller than ferrets, are still effective killers of young rabbits. But they are also intractably wild and hard to handle, and initial private attempts at transporting them in 1876–1882 failed. F.D. Rich from Bushey Park (South Island) had met Samuel Grant from Brigg, Lincolnshire, during Grant's study tour in 1880. Rich commissioned Grant to supply a trial shipment of animals and an experienced gamekeeper to accompany them. Grant chose Walter Allbones, assisted by his father Henry. They developed a successful transportation system by which the animals were collected from the wild in live traps, kept in dark, scrupulously clean boxes and fed daily on freshly dead pigeons. Rich's trial consignment, accompanied by live pigeons with their own food supplies and escorted by Walter, reached Bushey Park in March 1883 on the sailing vessel *Waitangi*. The colonial government immediately contracted Walter to return to England and bring back an official shipment of 100 animals in 1884, the first of many over the next decade.

Notes

1. Guthrie-Smith, H. (1969). Tutira. The Story of a New Zealand Sheep Station, 4th edition. Wellington: A. H. & A. W. Reed, p. 354.
2. For an open access, more detailed list of page citations to the many primary sources underlying this chapter, see King, C. M. (2017). The history of transportations of stoats (*Mustela erminea*) and weasels (*M. nivalis*) to New Zealand, 1883–1892. International Review of Environmental History, 3, 51–87. https://doi.org/10.22459/IREH.03.02.2017.04
3. Bayly, B. P. (1883). Annual Report on the Rabbit Nuisance [1882–83]. Appendix to the Journal of the House of Representatives, 1883 H-18.
4. Paraphrased from Midgley, M. (1992). Science as salvation. London: Routledge. P. 16.
5. Buller, W. L. (1877). On the proposed introduction of the polecat into New Zealand. Transactions & Proceedings of the New Zealand Institute, 9, 634–635; Martin, H. B. (1885). Objections to the introduction of beasts of prey to destroy the rabbit. Transactions & Proceedings of the New Zealand Institute, 17, 179–182.
6. Darwin, C. (1859). The Origin of Species [London: Folio Society 2006 reprint of 1st edition of 1859]. P. 160.

7. King, C. M., & Powell, R. A. (2007). The Natural History of Weasels and Stoats: Ecology, Behavior and Management, 2nd edition. New York: Oxford University Press.
8. Randall Johnson, G. (1884). Report of the Joint Committee on Sheep and Rabbit Acts. Appendix to the Journal of the House of Representatives, 1884 I-5.
9. Agent-General. (1877). Methods for diminishing the rabbit nuisance. Appendix to the Journal of the House of Representatives, 1877 I, H-02.
10. Blackburn, T. M., Lockwood, J. L., & Cassey, P. (2015). The influence of numbers on invasion success. Molecular Ecology, 24, 1942–1953.
11. Evening Post 14 December 1881. Advertisements: To Stewards and others of wool ships &c.., p. 3.
12. 'Colonist'. (1882). Stoats and polecats for New Zealand. The Field, 60, 536.
13. Anon. (1881). Report of the Rabbit Nuisance Committee [1880–81]. Appendix to the Journal of the House of Representatives, 1881 I I-06.
14. Otago Witness 25 October 1894. Natural enemy of the rabbit protected. P. 27.
15. Wells, P. (2009). The Fall and Fall in the Legal Status of Mustelids in New Zealand. Environment and History, 15, 343–368.
16. Grey River Argus 8 July 1882. [Question to the Minister]...
17. Grant, S., & Foster, J. S. (1880). New Zealand: A Report on its Agricultural Conditions and Prospects. London: G. Street.
18. Thompson, H. V., & King, C. M. (Eds.). (1994). The European Rabbit: The History and Biology of a Successful Colonizer. Oxford: Oxford University Press. Pp. 64–67.
19. Beastall, T. W. (1978). The Agricultural Revolution in Lincolnshire. Lincoln: History of Lincolnshire Committee. Pp. 143–144.
20. McDonald, R. A., & Murphy, E. C. (2000). A comparison of the management of stoats and weasels in Great Britain and New Zealand. In H. I. Griffiths (Ed.), Mustelids in a modern world (pp. 21–40). Leiden: Backhuys Publishers.
21. King, C. M., & Powell, R. A. (2007).
22. Southland Times 1 July 1882 p. 2. Natural enemies of the rabbit.
23. Mataura Ensign 24 May 1887 p. 6. Stoats and Weasels: an interview with an expert.
24. Marlborough Express 21 April 1885 p. 3. An Anti-ferret Testimony.
25. Weekly Press 20 May 1887. Stoats and weasels [Allbones interview].
26. Veale, A. J., Clout, M. N., Holland, O. J., & Gleeson, D. M. (2015). An invasive non-native mammal population conserves genetic diversity lost from its native range. Molecular Ecology, 24, 2156–2163.
27. Colonial Secretary. (1883). Letter 4 December to Agent General [re Allbones commission]. In Outwards Letterbook – Agent General. Wellington: Archives New Zealand R20557923.

CHAPTER 10

Stoats and Weasels (1883–1892)

> Nothing in connection with the naturalisation of wild animals into New Zealand has caused so much heart-burning and controversy as the introduction of these bloodthirsty creatures. G.M. Thomson (1922)[1]

The 1874 settlers' meeting in Southland, which first strongly recommended the introduction of weasels to prevent the spread of rabbits, was followed by almost ten years of raging arguments and private, mostly failed, solutions[2] (Chaps. 7, 8 and 9). The afflicted pastoralists, having tried everything else they could think of, eventually managed to persuade the Colonial Secretary that *government-sponsored* introductions of stoats and weasels were the only remaining option to curb the rabbit menace.

The fatal die was cast in 1883, when the colonial government accepted the advice of F.D. Rich and Benjamin Bayly to employ Henry Allbones and his son Walter to manage an official programme of importations of stoats and weasels (Chap. 9). From then for the next ten years, a long series of organised shipments pursued the largest ever attempt to achieve biological control of a vertebrate pest—a deliberate translocation experiment of small predators intended to prove the capability of natural enemies to suppress plague numbers of rabbits.

C. M. King, *Invasive Predators in New Zealand*,
Palgrave Studies in World Environmental History,
https://doi.org/10.1007/978-3-030-32138-3_10

1884–1886: The First Official Government Imports

Bayly must have arranged for Rich to release Walter Allbones early from his job at Bushey Park, so as to undertake his new commission. Walter returned to England in December 1883, and left again for New Zealand on the *Doric* on 24 April 1884, with his first official consignment of stoats and weasels for the government (Fig. 10.1; Table 10.1). The whole business of colonial shipping was a matter of great interest to newspaper readers, so this and subsequent shipments were keenly observed and commented on. His departure on the *Doric* was widely reported in England, in *The Field*, and in half a dozen English and Scottish newspapers.

June 1884

When Bayly's annual report to the colonial government's Rabbit Nuisance Committee for 1884 was published on 6 June, Walter Allbones was still on his way back to New Zealand with his cargo. Unable to report his arrival yet, Bayly could only confirm the link to Rich's pioneering effort:

> an endeavour is being made to introduce the stoat and weasel … and an agent sent Home to procure them. It has been proved by Mr Rich of Palmerston South, that they can be imported, and the same man who

Fig. 10.1 The first official shipment of stoats and weasels to New Zealand arrived on the steamship *Doric* on 28 June 1884. (ATL 1/2-150418-F, Alexander Turnbull Library)

Table 10.1 Minimum numbers of known consignments of British stoats and weasels to New Zealand, 1883–1891.[4] The two species are easily confused and are often listed together under the general term 'weasels', 'weasles' or 'weazels'. See Table 10.2 for a subset list of the 15 shipments where the two species are listed separately, although there is no guarantee of correct identification. Names of ships and dates of arrival at the first port of call in New Zealand are taken from the online Comber Index until it ends in 1889. '# Animals landed/y' is compiled from the annual total number of animals arriving on reported shipments until 1886, and then from import data recorded in the annual *Statistics of the Colony of New Zealand* until weasel imports ended in 1892. The average mortality rate during a typical six- to seven-week voyage was about 10%, except during three disastrous voyages in early 1891. The total costs listed here exclude the *Waitangi* trial shipment; the ultimate cost of the programme is unknown

Year	*Ship*	*Date arrived NZ*	*Courier*	*Client*	*#Animals landed/y*	*Cost, £*
1883	*Waitangi*	March 26	WA, FM	Francis Dyer Rich	10	245
1884	*Doric*	June 28	WA	NZ Govt	90	505
1885	*Ionic*	January 20	WA	NZ Govt	919	475
	Rimutaka	July 12	HA	E J Riddiford, C Goulter		803
	Ionic	October 28	UN	NZ Govt		837
	Rimutaka	November 6	WA, FM	E J Riddiford		659
1886	*Doric*	January 19	HA	NZ Govt	303	
1887	*Rimutaka*	October. 9	HA	NZ Govt/subsidy	433	
1888	*Tongariro*	January 5	HA	Awatere Rabbit Bd	1453	
	Kaikoura	April 23	HA	NZ Govt/subsidy		
	Doric	May 9	UN	Awatere Rabbit Bd		
	Ionic	July 3	GS	Awatere Rabbit Bd		799
	Ruapehu	August 12	HA	Awatere Rabbit Bd		755
	Ionic	November 21	UN	UN		
1889	*Kaikoura*	June 18	UN	Awatere Rabbit Bd	1238	
	Aorangi	August 13	HA	Awatere Rabbit Bd		607
1890	*Tongariro*	May 9	HA & (3rd) son	Awatere Rabbit Bd	2292	
	Ruapehu	August 6	UN	N Wairarapa Rabbit Bd		
	Rangatira	October 3	UN	UN		
1891	*Tongariro*	January 30	UN	N Wairarapa Rabbit Bd	1100	
	Aorangi	February 26	UN	NZ Govt		
	Rimutaka	February	UN	N Wairarapa Rabbit Bd		
	Ruapehu	April 21	UN	N Wairarapa Rabbit Bd		
	Kaikoura	May	UN	N Wairarapa Rabbit Bd		
	Aorangi	July 16	UN	N Wairarapa Rabbit Bd		
1892	*Unknown*	Unknown	UN	UN	838	
		Totals			**7838**	

Abbreviations: *WA* Walter Allbones; *HA* Henry Allbones (Walter's father); *GS* George Storey; *FM* Fowler Metcalf; *UN* Unknown

> brought his out has been engaged by the Government; and as a certain amount of success attended Mr Rich's efforts where a sailing vessel was the means of transit, better success may be anticipated by utilising one of the large steamers.[3]

By this time, steamships could make the voyage in half the time taken by the *Waitangi* and were equipped with larger accommodation, interior heating and refrigeration of stores. The success of the stoat and weasel import programme is undoubtedly linked to its coincidence with the beginnings of regular steamer schedules.

The Colonial Secretary in Wellington informed the Agent-General of Walter's arrival on 28 June with 23 stoats and 67 weasels. Henry was happy to receive word of his safe arrival, commenting that Walter 'has been more fortunate this voyage than he was in 1882' (on the *Waitangi*; Chap. 9). The local Rabbit Inspector Blair Fullerton reported that when the boxes were opened on the rabbit-infested grasslands of Otago, the little animals ran out of them and appeared quite as lively as if they had never experienced a long sea voyage and close confinement.

When Bayly later visited the Wanaka release area, he was pleased to find rabbits to be scarce. His assumption that this was due to the success of the weasels released from the *Doric* fired up the neighbouring runholders as little else could have done. Conditions on the Wanaka peninsula were not necessarily typical, and elsewhere the outlook was much less favourable, but no one noticed that. 'Scarcely a rabbit is to be seen, so 8,000 acres of badly infested country was cleared for about £500', he reported smugly. Runholders took literally this optimistic assessment of the success of the weasels and stoked up an ever-increasing demand for more. Bayly reported in June 1885 that a syndicate of runholders was immediately formed to employ Allbones to bring out a private shipment, all expenses paid. By November, Walter and Henry back in England had collected a batch of 80 and expected to complete a shipment of 150 within a month.

January to July 1885

During the 13 months from January 1885 to January 1886, five separate shipments of stoats and weasels arrived, totalling 1222 animals. Three lots were commissioned by the government for turning out on Crown lands, mainly in the South Island, and two were private orders for properties in the southern North Island.

First to arrive, on 20 January, was Walter Allbones on the *Ionic*, with his second official consignment (his third in total, counting the trial shipment for Rich on the *Waitangi*), of 158 stoats and weasels shipped in November 1884, 3 in a box, along with 2400 pigeons, served out at the rate of a pigeon per box per day. Presumably, each box in which three small carnivores were transported together must have been divided into separate internal compartments. Of the 126 weasels and 33 stoats loaded, 116 weasels and 32 stoats were landed, totalling 148 animals at cost of £475 0s 2d.

Part of this consignment was dispatched to the Hon. Robert Campbell's Benmore Station. Campbell's manager reported on 13 February 1885 that he had

> Got up safely with the weasels. Turned them out at 'the junction'. In 2 lots, 6 each. We put them where rabbits bolted under rocks, and they started work at once. We could hear the rabbits singing out. While we were watching them, a young weka rushed out from under a rock with a weasel hanging on, but it fell off with a mouthful of feathers. The wekas are very plentiful at the low end of Benmore but I fear they won't last long.[5]

No such considerations for native fauna had any bearing on the urgent desire of other landholders to get in on the act. Within days Walter was on his way home with an order for another consignment. This time he was commissioned to bring a private shipment to be divided between Edward J. Riddiford, a prominent Wellington politician and runholder of the Wairarapa, and Cyrus Goulter of Marlborough. Riddiford paid Allbones' £21 passage home on the *Ionic*. As soon as Walter got back to Brigg with the contract, Henry set about collecting the animals from local suppliers. Henry proudly claimed that he had been permanently appointed by the Government of New Zealand to collect weasels and stoats for export.

Henry must have filled Riddiford's order early, because he set off on 21 May on the *Rimutaka* and reached Wellington on 12 July 1885. The *Daily Telegraph*'s report of his departure added hopefully that 'Bunny has a rough time in store'. On this occasion, 240 animals (167 weasels and 55 stoats) were loaded into their special maritime lodgings of 80 zinc lined boxes, accompanied by 4000 live pigeons and 42 quarters of avian shipboard rations. Mr. Stenhouse, a reporter sent to interview Henry on board the *Rimutaka* in Wellington harbour, was impressed. When invited to have a look at the consignment, Stenhouse reported that 'the animals came out in fine condition ... and only 14 died on the passage'.

Riddiford's first shipment had been organised through the Colonial Secretary's office and the Agent-General in London. The Agent-General charged £674 13s 4d for collecting the animals and organising their transport and feeding on the voyage, on top of which the Colonial Secretary added wages for Allbones and his assistant, plus a government bonus of 5s for every weasel and 7s for every stoat landed alive. The total came to £803 16s 8d, an eye-watering NZ$148,000 in today's money. Hardly surprising, then, that Riddiford was slow to pay it. Yet, even as these animals were being distributed across Riddiford's Te Awaiti station in July 1885, Riddiford and Henry sat down at the homestead and signed another contract, commissioning Riddiford's second private consignment, on the same terms as the first. The first page of this contract is reproduced in a previous, detailed report.[6]

Despite the cost, word was getting around the runholders, and they besieged Bayly with requests for allocations. The heartfelt pleas of William Stronach of Lake Ohau Station illustrate his deepening fears that the rabbits would soon overrun the whole of Otago. He begged for a supply of natural enemies, his only hope of stopping the rabbit invasion in such wild country:

> if you find you can spare me some of your next shipment of stoats and weasels I should be very much obliged … the rabbits [are] much more numerous there [on the Naumann Range] than I even dreamed of … in a short stretch of time I am afraid they will from thence over-run the whole of our country … nothing but the natural enemy can check them in such wild country … let us have them in sufficient numbers.[7]

Stronach's idea of 'sufficient numbers' was certainly in the thousands rather than the hundreds distributed so far.

October and November 1885

The *Ionic* set off back to New Zealand again on 10 September and arrived on 28 October 1885, with 89 stoats and 196 weasels shipped, of which 82 stoats and 167 weasels were landed, making 268 animals at cost of £836 16s 7d. This shipment was the fourth of five to arrive within a few months (Table 10.1) and the first that was probably not escorted by either of the Allbones. At the time the *Ionic* left England, both were committed to collecting animals for other contracts. Beaten by the competition, the government had to employ 'another agent for similar purpose'. Meanwhile, Walter left England on 26 September 1885 with his next consignment of 350 on the *Rimutaka*, arriving on 6 November, again accompanied by Fowler Metcalf.

This was the last trip for both Walter Allbones and Fowler Metcalf. Metcalf accompanied Walter and the animals to Riddiford's station in late 1885, but then married and settled in Dunedin and produced two sons. He eventually moved to Napier, where he died in 1938. Walter returned to England, married his sweetheart Annie in October 1886 and settled down in Brigg to produce a large family. He took over organising the Brigg end of the business, leaving Henry to continue the job of shepherding the shipments around the world.

January 1886

The fifth of the five shipments sent during this busy period is much less well documented than the others, but it was probably Henry's 'further order' expected to follow Walter's shipment on the *Rimutaka*. If so, it would be the one brought by Henry on the *Doric*, which left on 4 December 1885 with 101 boxes and arrived on 19 January 1886. It is significant because it was the last known consignment to arrive for almost two years.

A correspondent to the *Auckland Star* described it as Allbones' '7th or 8th shipment', which from later information is either an exaggeration, or evidence of missing data. Only five Allbones shipments can be confirmed from known records, totalling 1322 stoats and weasels between March 1883 and January 1886 (Table 10.1), arriving on the *Doric* (June 1884), *Ionic* (January 1885), *Rimutaka* (July and November 1885) and *Doric* (January 1886), plus the ill-fated consignment for Rich on the *Waitangi* in March 1883, not counting the *Ionic* of October 1885 which was accompanied by someone else. There may have been others, but if so, they missed the observant press reporters on the docks.

1886: An Interruption

In January 1886 came an unexpected halt to the importation programme. The Agent-General forwarded a letter to the Colonial Secretary from the Shaw, Savill and Albion Line, who chartered the *Ionic* and the *Doric* for the New Zealand trade, stating they would not be able in future to convey stoats and weasels to New Zealand because of passenger complaints received after the recent voyage of the *Ionic*. At least one man who had travelled on such a ship denied that his fellow passengers were even aware that there were any animals on board until the captain told them; nevertheless, the government was obliged to search around for alternatives.

The need to find one was painfully clear to the Joint Committee on Rabbit and Sheep Acts. Their 1886 Report pointed out that

> the colony is not only suffering a heavy annual loss from the very great numbers [of rabbits] still existing in infested districts, but that the infested area is constantly increasing ... rabbits are making their way into ... Canterbury from Amuri in the north and from the Mackenzie country in the south and west; while in the North Island they are spreading from the Waikato into the King Country and from the Wairarapa into Hawke's Bay ... [we] should without delay take steps to provide for a large and constant supply of stoats and weasels.[8]

So the Committee was no doubt relieved to receive a letter on 15 June from the Christchurch stock agents Cuff & Graham. This firm offered to provide stoats and weasels to the government in the required 'large and continuous supply'—in any numbers up to 2000 over two years, using the cargo ship *Elderslie*. Their terms were seen to be very reasonable: 40s per head, in the ratio of 2:1 in favour of stoats (more efficient rabbit killers), with the first shipment of say 200 to arrive by January 1887.

The Committee's decision to accept the offer was supported both by Bayly and by Cuthbert Cowan and was probably influenced by a letter from R. Maunsell, chair of the Wairarapa Rabbit Association. He offered to pay £500 for 250 stoats and weasels by the end of January if 75% of the shipment were stoats and all were landed in good condition, plus 2s 6d per stoat. They were even willing to add another £500 for a later shipment.

Bayly also sent one of his staff, Blair Fullerton, to make his own enquiries about the *Elderslie*. He was told it was designed to ship frozen mutton from any port, spending only about 48 hours in each, and its schedule was too irregular and unpredictable to make the organising and delivery to port of shipments of live animals reliable. Fullerton and Bayly agreed it would not do. Unfortunately, these enquiries caused a delay which, by the end of August, had exhausted Cuff & Graham's patience. They notified the Colonial Secretary that, since they had received no definite commitment, the offer was withdrawn, although it could be renegotiated later. The end of 1886 passed with the stalemate over shipping still unresolved.

Frustration at the ongoing delay was mounting rapidly among the runholders of the upper Waitaki valley. The rabbit fences erected along the north banks of the Waitaki and Hurunui Rivers were slowing the rabbits' advance, but could not be a permanent solution.[9] In February 1887, a formal petition was composed and signed by all the owners and managers of the famous stations described in Pinney's classic history of the early northern Otago runs.[10] The Hon. Robert Campbell submitted the petition on 11 March 1887 to Patrick Buckley, the Colonial Secretary:

> We the undersigned ... would most earnestly beg to draw the attention of the Government to the urgent necessity of taking immediate action of introducing large numbers of the natural enemy of the rabbit ... Judging from the sad experience of those who at one time held runs in the upper Clutha district, unless the plague which is moving steadily from there upon us is checked by efficient means this District will suffer the fate of the back runs in the Lake and Vincent Counties. [Fig. 9.5] The runs there a few years ago yielded good rents; now they are in the hands of the Rabbit Department which is spending enormous sums (about £100 per 10,000 acres) annually in keeping the pest from overwhelming the inner runs ... [We] are certain that in the natural enemy alone remains the only chance of saving the outer runs in this District ... the only means to avert the calamitous result of a wholesale abandonment of runs ... at no very distant period.[11]

To make sure nobody missed the point, Thomas Middleton added a letter to John Wilson (Campbell's London agent) that 'We are urging the Government to get out stoats *in thousands*' [my italics].

The government's own estimates backed up the runholders' concerns. As of 1 March 1887, £32,803 of government revenue had been lost from the 1,346,554 acres (544,931 ha) of Otago sheep country which had been surrendered since it became infested with rabbits. In Southland, the losses must also have been enormous, but the rentals there were so modest, 2d/acre, that none had been abandoned. Without rabbits the same runs would have been re-let when the leases expired (mostly in 1883) at 4d/acre, so the losses to the government in Southland were not less than £15,000 a year.

Under such pressure, the Colonial Secretary agreed in April 1887 to resume the importations of stoats and weasels, and instructed Bayly to take up Cuff & Graham's offer to resume the negotiations. Bayly's cautious enquiry to them ('It having been again determined to resume the importation of stoats and weasels at an early date ... [we now] request whether your firm is prepared to re-open negotiations … as mentioned in yours of 30 Aug [1886]') met with a brief and frosty reception: 'Circumstances have changed, [we are] not now prepared to reopen negotiations'. The file does not record Bayly's reaction.

By May 1887, Bayly's next report on the rabbit nuisance still could not list any new imports of stoats and weasels. Yet, an April 1887 official report by J.H. Baker, commissioner for Crown Lands, Canterbury, to the undersecretary for Crown Lands, Wellington, found rabbits massed on the Otago frontier ready to make a raid on Canterbury. Omarama was 'literally eaten out', by rabbits 'in denser numbers even than in Southland'. Baker feared that rabbits will inevitably overrun the whole hill country, costing the government £10,000 a year in lost revenue, plus the loss to wool producers of £35–40,000 a year. Such numbers must have given Bayly nightmares.

On top of the disputes with shipping companies, the problem of financing the imports was another serious brake on Bayly's programme, which the Hon. Robert Campbell discussed with Patrick Buckley, the Colonial Secretary. According to Campbell, Buckley made a verbal promise that the government would subsidise private imports of stoats and weasels at the rate of £2:1. Campbell then wrote to the minister, William Larnach, from his homestead at Otekaike on 11 July, that

> I have by this mail sent home an order to send out about 300 stoats and weasels … in accordance with a verbal promise made to me by Mr Buckley that the Govt would give £2 to £1 towards the expense of the consignment … the introduction of these, followed by more as I can manage, will keep the rabbits out of Canterbury.[12]

Larnach, greatly surprised, asked Buckley: 'Did you promise Mr Campbell the subsidy referred to herein?' Larnach replied on 15 July that he remembered a conversation at which Hon. Mr. Campbell was present, and he would support the idea, but such a large expenditure would have to be approved by the House, which had not been done, and, besides, the owners of the direct steamers had refused to introduce any more of the rabbits' enemies because of the complaints of passengers. Larnach added that Mr. Campbell then said, 'I can get them out by the Southern Steamers' (presumably using the Campbell family shipping connections in Sydney; Chap. 3).

On 21 July, the minister's office confirmed to Campbell that, 'with the view of encouraging private enterprise in importing stoats and weasels, I will authorise the payment of half the expense of consignment'. Rabbit inspectors around the country were quick to apply for permission to inform runholders of this arrangement. The rabbit inspector in Dunedin (E. Clifton) reported that private individuals were so anxious to obtain these animals that, in the Tuapeka District alone, half the cost of importing 1000 stoats and weasels would be at once guaranteed. On 28 November, Clifton forwarded orders to purchase 1475 animals at the subsidised rate.

Henry Allbones must have been much vexed by the interruption to his business caused by the Shaw, Savill and Albion Line's ban on carrying stoats and weasels. He travelled to New Zealand on the *Ionic* in March 1887, without any animal companions. He must have failed to persuade them to reconsider, because he turned to the rival shipping line, New Zealand Shipping Company. Their ships *Rimutaka*, *Tongariro* and *Kaikoura* carried the Allbones' next three consignments (Table 10.1).

During this trip, Henry was interviewed by a reporter from the *Press*, whose account of his operation (Chap. 9) was widely syndicated in different versions around the South Island newspapers. When asked about the number being imported, he replied, 'My son has brought over five shipments, making 1160 stoats and weasels, and there is another shipment for Mr. Rich, of Bushy Park, I think, of which most were drowned'. Henry added: 'At present there are no shipments coming out, but I had an enquiry whether I could get more, and I said I could'. The fruit of that promise reached New Zealand on the *Rimutaka* on 9 October.

1887–1889: Back to Business

October 1887

Henry's foresight and family connections ensured the fastest possible resumption of the trade. Back in Brigg, Walter must have been notified of Henry's new arrangements by telegraph, because advertisements immediately appeared in several British newspapers: 'Wanted, by May 29, 150 Weazels and 200 Stoats for New Zealand. 4s each will be given for Weazels and 6s each for Stoats, and carriage will be paid. Must be sound in condition. Apply Henry Allbones, Brigg'. Walter and his younger brothers collected and maintained the stocks of animals awaiting shipment in the purpose-built shed on Henry's property. The next lot of 200 animals was already on its way to London by train from Brigg in early June, well before Henry could have got back himself.

By the time Henry reached London, Walter's June consignment from Brigg was awaiting him there. They therefore did not have to sit long on the docks before Henry was loading them onto the *Rimutaka* and setting off back to New Zealand on 27 August, arriving on 9 October.

The colonial government was obviously impressed, because on 12 October, while on board the *Rimutaka* back in Wellington harbour, they offered Henry a new contract. He accepted on the usual terms:

> Return steamer fare to New Zealand and back to be provided by the New Zealand Government;
> Wages twenty five shillings a week to be paid me from the date I sail from England to the date I return to England;
> If the animals can be obtained, the shipment to consist of not less than three hundred;
> A bonus of seven shillings to be paid for every stoat and five shillings for every weasel landed alive in the Colony;
> The assistance of one passenger to be allowed during the passage out at twenty shillings a week.
> I will do my utmost to get the animals here in as short a time as possible.
> I expect the Agent-General to pay all expenses for bringing the animals, pigeons for their feed, corn for the pigeons, boxes for the animals and the birds and other necessities of that kind.[13]

January 1888

Henry's system must have been running like clockwork to clear the backlog, because he returned to England, collected another lot and got back to New Zealand on the *Tongariro* with 285 of them on 5 January 1888. The *Tongariro* shipment was probably one of several delayed orders for private clients. The Awatere Rabbit Board, in the Marlborough province, was certainly among the clients most actively importing stoats and weasels at this time.

The *Tongariro*'s load was the first instalment of an ambitious order, placed by the Awatere Rabbit Board in October 1887, for 1000 animals, to be divided between the station owners of the Awatere district in the proportion of one stoat or weasel to 2000 sheep, at a cost of about £2000. The *Marlborough Express* reported that 56 stoats and 229 weasels were transferred to the coastal steamer *Penguin* for the final passage to Picton.

Under instructions from Messrs Goulter and Vavasour, both Awatere Rabbit Board members, other consignments were delivered to the Dumgree, Starborough, Richmond Brook and Blairich Stations on 15 July, at an average cost of £3 7s a head.

Unlike government shipments, private shipments did not have to be reported to Parliament. This one might have included the group of 250 which started their journey from Brigg in September 1887, described by the *Grantham Journal* as 'the ninth shipment of these animals Mr Allbones has forwarded, and he has orders for one thousand more'.

A couple of months later, the *Leeds Mercury* marvelled that the New Zealand government has

> spared no expense … nearly 2,000 exported, still the demand continues for all we can supply … 4s for weasels, 6s for stoats … when sold the shillings become guineas … another cargo of over 600 is on its way to NZ, under the experienced guardianship of an agent in Lincolnshire … [it is] no easy task … the latest consignment needed 7,000 pigeons.[14]

By May 1888, the cumulative tally of 12 cargoes had reached more than 4000 stoats and weasels, and about 50,000 pigeons. On average it was calculated that each stoat or weasel had eaten about 100 pigeons during the voyage. The stoats and weasels thrived on this diet, although it was 'Rough on pigeons', commented the *Sheffield Daily Telegraph*.

April to November 1888

The year 1888 was turning out to be the busiest yet, although a report in March that 'stoats and weasles are arriving almost monthly from England' was overstating the case (six shipments are known for that year; Table 10.1).

Kaikoura's load of 351 animals, of which 319 survived, left England on 10 March and arrived on 23 April. It was probably the government shipment for which Henry had signed up in October 1887. Bayly told the Rabbit Nuisance Committee that a shipment of 94 stoats and 225 weasels had been recently imported, adding that 'besides this shipment several have arrived for private owners', which presumably includes the October and January consignments.

The *Doric* delivered 137 stoats and weasels in May; two more shipments, on the *Ionic* and *Ruapehu*, arriving in July and August, plus two otherwise unknown additional shipments on the *Rimutaka* and the *Amuri*, made the five shipments needed to complete the Awatere Rabbit Board's first order.

Such frequent consignments needed the help of additional agents. One was a local man, George D. Storey from Cullensville, a small town west of Picton. After the report of a Board meeting in February 1889 published its decision to import a second lot of stoats and weasels, Storey wrote in to remind the Board of his previous service:

> I am taking steps to place my services before the Board, and in doing so may state that I have been in Government employ importing these vermin, and have done so as lately as July last [July 1888]. I landed in Dunedin per *Ionic* some 302 alive of 319 shipped in London. I left Wellington in February to fetch that consignment. … Mr Allbones and myself, who collected same and also a further consignment per *Ruapehu* which followed me, will be glad to again contract with your Board.[15]

Storey had indeed been employed to accompany the 312 animals shipped from London to Dunedin on the *Ionic*, arriving on 3 July 1888. The New Zealand and Australian Land Company's Clydevale Station received 81 stoats and weasels; 84 had been shipped, and only 3 died during the voyage. The Awatere Rabbit Board received two consignments in 1888, 197 on 11 July, followed by 74 on 17 July.

The *Ionic* and *Ruapehu* shipments were confirmed in mid-1888 when the chief inspector of Stock in Sydney addressed an official enquiry to the Wellington government as to the cost of importing the natural enemies of the rabbit into New Zealand. According to these more official records, 257 of 273 weasels survived, plus 38 of 39 stoats, of a total of 295 landed. The average cost was £2 14s 2d per head, totalling £798 19s 2d. The same source quoted 192 of 208 weasels, and 120 of 128 stoats landed from the *Ruapehu*, arriving on 12 August 1888, at an average cost of £2 8s 5d a head (£755 5s 9d).

The last shipment of 1888 was loaded onto the *Ionic*, which arrived on 21 November bringing 'a hundred stoats and weasels for Wellington'. They were then forwarded to Lyttelton on the coastal steamer *Rotorua*.

1889–1891: Rabbit Boards Take the Lead

For the hitherto indefatigable Chief Inspector Benjamin Bayly, 1889 was a bad year. He had pursued his strategy of importing natural enemies against frequent protests from those who did not agree with his view that the risk to native birds was a price worth paying for saving the wool industry. But as stoats and weasels spread into native forests and began to kill more birds than rabbits, the accumulating pressure of objections against his policies precipitated an important change in official government policy. Bayly was demoted to the office of an ordinary district sheep inspector in Auckland, and all official imports of stoats and weasels ended abruptly. The southern pastoralists who continued looking to the government for supplies of stoats and weasels were disappointed. The April monthly report was from E. Clifton, the Wellington area stock inspector, who objected to the change in policy as strongly as he dared:

> the natural enemy has had the most encouraging effect and it is only unfortunate that this introduction has not been made more general … Government should resume the introduction of stoats and weasels and their liberation on Crown lands bordering rabbit nuisance country.[16]

In May 1889, the annual report on the rabbit nuisance, which Bayly had authored for years, no longer carried his name.

The Joint Committee on Livestock and Rabbits produced its own independent annual report in September 1889, compiled by leading runholder J.B. Acland. They protested against the sudden change in official policy connected with the removal of Bayly's influence. They pointed out that a recent shipment of 300 had arrived, and another was already on the way. They cited the success of Mr. Riddiford, who had turned out a large number of stoats and weasels where two to three years ago rabbits had been thick: now hardly one is seen, because all the young ones had been eaten.

Bayly had been the public servant with most influence over official anti-rabbit policy through the 1880s, which was much influenced by the pastoralists' lobbying. He had been long accustomed to frequent interaction with the powerful inhabitants of the Colonial Secretary's office and the House of Representatives, and took his demotion hard. His experience as the architect of the national mustelid introduction strategy was no use in the Auckland area where rabbits were few and stoats and weasels had not yet reached. He gloomily reported the difficulty caused to him and his sub-inspectors in Whangarei and Hamilton by a prohibition against using poison for rabbits, because of the risk of secondary effects on wild pigs. 'The only natural enemies in this district are [harrier] hawks and [feral] cats. The latter are pretty numerous and are of great service'. He added rather wistfully: 'I should like to see ferrets liberated in the King Country'. It was his misfortune that history has proved his strategy wrong.

Worse was to come. In a parliamentary report of June 1890, 'Mr Bryce presented a petition from Benjamin Bayly of Hamilton, complaining of his dismissal from the position of sheep inspector for the Auckland district, and asking for relief'. It seems the request was denied, because from then on, Bayly's name disappears from the archives. The Hamilton Electoral Roll has Benjamin Peddie Bayly registered as an auctioneer in 1890.

Without Bayly's support, local communities had to organise further importations themselves. The government had not banned the shipments; it had only refused to spend further government funds on them, pointing instead to the subsidy on private importations of stoats and weasels proposed in 1887 by the Hon. Robert Campbell (p. 224). So the Otara and East Coast Rabbit Association, still vexed at having failed to persuade the government itself to turn out a sufficient supply of natural enemies to keep down the rabbits in the South Wairarapa, requested official confirmation of the subsidy scheme, and details of whether a voluntary body would qualify and on what conditions.

The official reply of 4 October 1889 was that, where the settlers establish a Rabbit Board formed under the act, it will be entitled to a pound for pound subsidy. If they do not constitute themselves into a Rabbit Board, they will be entitled to the bonus of £1 per stoat and 10s per weasel imported by them into the colony. In this and many other rabbit-infested districts, the affected landholders resolved to form a new Rabbit Board and go in strongly for the private importation of stoats and weasels. Rabbit Boards ordered some very large consignments over the next few years.

In Hawke's Bay, the debate about how to organise efficient rabbit extermination was conducted for much of the winter of 1890 in the pages of the local papers. Most correspondents were in favour of continuing to

> introduce the stoat and weasel. It is the only rational way of dealing with the rabbit pest. If we had taken the steps we are now taking two years ago, this district would have been richer by hundreds of thousands of pounds.[17]

Not everyone agreed, and a month later a public meeting voted against any imports until 'a majority of the stock-owners had expressed approval of the introduction of stoats and weasels, [otherwise] the Board would not be justified in spending public money on importing them'. Other Rabbit Boards continued undeterred.

June and August 1889

The Awatere Rabbit Board agreed in February 1889 to place a second order for £1200 worth of stoats and weasels (enough for a 'supply of 600 vermin'), to be landed in Blenheim in September/October 1889, including an extra 5s for stoats. The premium on stoats was a response to advice from Acton Adams of Molesworth Station, pointing out that £900 on a mixed cargo of stoats and weasels would be better spent only on stoats, since the cost to buy them in England was only an extra 5s, and the freight was the same, whereas one stoat was worth two weasels in killing rabbits. Allbones responded by increasing his purchase price to 7s 6d for stoats, and some advertisements specified stoats only: 'No more weasels wanted'. Storey wrote in offering his services again.

The *Kaikoura* loaded 300 animals, of which 270 landed on 18 June 1889, announced as part of the second order confirmed by the Board on 12 April, the remaining 300 to be procured through Messrs. Holmes and Bell. Another 227 (of 254) lot arrived on the *Aorangi* on 13 August, a month ahead of the deadline. Holmes & Bell, the Board's agents in Blenheim, informed them that the *Aorangi* carried Henry Allbones' 17th shipment, and his aggregate total now exceeded 5000. If so, not all of his previous cargoes were reported, even though the Allbones' uncommon name has made searching for his movements relatively easy.

The official import data confirm that there must certainly have been many more, because they list a total of 1238 stoats and weasels landed in the year 1889 alone (Table 10.1). But most shipping records do not mention the animals, and often not the names of steerage passengers. Tracking down the many unspecified consignments would be nearly impossible to find without the name of the accompanying agents.

At the April 1890 meeting of the Awatere Rabbit Board, Holmes & Bell submitted a memo from Henry Allbones requesting payment of £202 5s 3d, the balance on his stoats and weasels account. The Board resolved to pay it on production of their agreement with Allbones.

May to October 1890

Import records list a total of 2292 stoats and weasels arriving in 1890, but newspapers mention only three shipments. It seems that by this stage the press in both countries had almost lost interest in the matter.

The *Tongariro* arrived on 9 May with the third of three shipments for the Awatere Rabbit Board. The May 1890 meeting received another memo from Allbones, dated 20 August 1889, for payment of an invoice for 182 weasels and 45 stoats, total gross £607 10s for 225 animals at £2 14s each, plus a £10 gratuity and his passage money home. He requested this payment to be made to the London office or to his account at the Lincoln & Lindsey Bank (England). By mid-1890, another 1081 stoats and weasels had been liberated by the Board. *Ruapehu* arrived on 6 August, with a consignment for the North Wairarapa Rabbit Board, and *Rangatira* followed in October (Table 10.1).

The Awatere Rabbit Board's policy of extensive mustelid imports was part-financed by the government's commitment to a pound for pound subsidy paid against evidence of rates collected by the Board for destruction of rabbits, of c. £2000/year. The file is full of applications for subsidy payments, but they were apparently not enough—the Awatere Rabbit Board was abolished in 1893.

The ongoing disagreements on rabbit control policy within government circles are well illustrated by the report of the Joint Committee on Livestock and Rabbits, dated 11 September 1890. Long after Bayly's departure, the Committee repeated its 1889 criticism of the change in official policy. As it pointed out, the rabbit nuisance continued to spread, at an annual loss to the colony 'probably equal to the interest on our national debt'. The Committee seriously complained of 'the short-sightedness of a policy which allows session after session to pass' without providing the means for bringing in stoats and weasels. The government 'should encourage the breeding of ferrets in every possible way, and continue the introduction of stoats and weasels in large and continuous numbers'. Nothing changed: the renamed Live-stock and Rabbits Committee's report for 1891 does not mention natural enemies at all.

January to July 1891

After years of successful shipments smoothly organised by Henry Allbones and Benjamin Bayly, things began to go wrong. Six shipments were delivered to the North Wairarapa Rabbit Board during the first seven months of 1891. Three of them recorded unusually high mortality of stoats and weasels. Only 42 (11%) of the 374 stoats and weasels loaded on to the *Tongariro* were still alive by the time the ship arrived on 30 January. Of 306 animals shipped on the *Aorangi*, only 13 of 93 stoats (13%), plus 2 of

185 ferrets and none of 28 weasels survived until the ship docked on 26 February. The *Kaikoura's* cargo arriving in May 1891 suffered 'extraordinary mortality' en route, attributed by the receiving inspectors attending the May meeting of the Board to 'want of cleanliness'. High mortality from distemper among ferrets was common (Chap. 8), but stoats and weasels were normally much less vulnerable to it, certainly while being maintained by Henry or Walter Allbones in clean travelling boxes and fed on freshly dead pigeons.

Henry Allbones last appeared in the passenger records in 1890, when he arrived on the *Tongariro*. His methods were expensive, but he had shown that he could handle large consignments with minimal losses, so it is hard to believe that Henry, with all his experience of successful deliveries over almost a decade, would have been responsible for any of the disastrous voyages of 1891. It seems more likely that the Wairarapa Rabbit Board was cutting costs by employing other, less expert agents and/or supplying inferior rations. Perhaps someone tried Acton Adams' recommendation of 1889. He had suggested that the animals in transit should be fed mutton not pigeons, which is much cheaper, making the landed cost less than 30s a head.

Three other 1891 shipments had mixed or unrecorded results, so far as the records show. *Rimutaka* left England with 356 stoats and weasels and arrived in February. *Ruapehu* loaded a shipment of 228 weasles and 117 stoats and arrived on 21 April with 250 of the 345 animals alive (72% survival rate); 165 weasles and 85 stoats were delivered to the North Wairarapa Rabbit Board in Masterton. *Aorangi* made up for past mistakes by landing in July a creditable total of 328 of the 349 animals loaded (a 93% survival rate). In the 1892 annual report on the rabbit nuisance, the rabbit inspector for the Wellington district (J. Drummond) says: 'I am satisfied these are increasing and doing good work'.

Even so, the long procession of shipments of stoats and weasels to New Zealand continued to fascinate newspaper readers in England. The influential London paper *Pall Mall Gazette*, perhaps unaware of recent disasters, commented admiringly:

> An increasing business is being done in the exportation of stoats and weasels to New Zealand ... the catching of these natural enemies, not only of the rodents, but of everything furred or feathered which comes within their power, has taken the dignity of a very profitable occupation for adepts in catching them alive. So far as New Zealand is concerned, the supervision of

> this trade is undertaken in London by the Loan and Mercantile Agency of that colony, who are still able to make a good profit after as much as five or six shillings a head has been paid to the trappers, in addition to the not inconsiderable cost of transport. The London correspondent of the *Sheffield Telegraph* has seen at Waterloo station a consignment of 320 of these little animals intended for shipment to Wellington in the *Kaikoura*. This is the fourth lot received at the South Western terminus alone within a month, each party is accompanied by an experienced man, whose chief duty is to provide its members with a liberal supply of fresh-killed fowl on the voyage, for the stoat … disdains food from which the aroma of the warm blood has departed. He has, in fact, altogether a good time of it, for not only is he transported in a well ventilated and comfortably bedded box … but goes to a land where the problem of over-population and food supply can never have a painful interest for him.[18]

But by this time, mid-1891, the business was no longer increasing. The Wairarapa Rabbit Board was developing doubts about the wisdom of continuing the importation programme, and it was not alone. Ratepayers attending Board meetings complained about the great expense of the importations from Britain, especially when half of any loss came out of their pockets. The May meeting decided that all further importations of stoats and weasels should be stopped until further notice, except those that were already en route and those that had already been collected in England, and the local manager of the Loan and Mercantile Agency was asked to cable their London agents to that effect. The *Wairarapa Daily Times* reported that, despite the success of the last consignment (on the *Aorangi*) the Board agreed that it 'should seriously consider the question as to whether it is advisable to import more', since ferrets 'can be bred in the district at about one third less cost than stoats and weasels can be imported'.[19]

Local newspapers continued to report the settlers' despair at the spread of stoats and weasels into country without rabbits, but did not mention any new arrivals. It seems likely that the *Aorangi*'s July shipment was the last. After December 1891, the import category for stoats and weasels disappeared from the annual statistical report (Table 10.1).

Less than two decades later, but at opposite ends of the world, Henry Allbones and Benjamin Bayly both died in 1909. Neither of their gravestones mentions the huge contributions they each made to changing the environmental history of New Zealand. Walter died in 1948; his descendants still live in Brigg, and the Brigg Heritage Centre (Fig. 9.3) actively remembers its New Zealand connections.

Total Numbers and Costs of Imports

Clearly, the records reviewed above do not tell the complete story—there are too many gaps and blanks through which quite a few shipments could have slipped through. Consignments arriving up to January 1886 were widely observed and commented on, especially those ordered through the Colonial Secretary and reported in Parliamentary documents. After 1887, government switched to subsidising private importations, on which the news reports became increasingly vague, if existing at all, leaving many cargoes unaccounted for.

Fortunately, the annual official volumes of *Statistics of the Colony of New Zealand* include comprehensive listings of imports and exports by item and origin. From 1887 to 1891, the annual lists of imported living animals distinguished stoats and weasels as a separate category, totalling 6516 animals, until the category was abolished in 1892. So the *bare minimum* number of stoats and weasels known to have been landed must have been at least 7838 individuals, carried on many more than the 25 known shipments listed in Table 10.1. The total known cost of the organised programme (not counting Rich's trial shipment) was *at least* £5428, or (allowing for the number of gaps in the data, more likely twice that by the end of 1892), up to NZ$ 2 million in today's money. Now, the Department of Conservation spends far more than that every two to three years, to protect beech forest birds from post-seedfall irruptions of stoats (and rats) (Chap. 12).

In July 1900, the New Zealand government received an official enquiry from Sydney as to the advantage or otherwise of introducing weasels for the purpose of keeping down the rabbit pest, and, if advisable, how to obtain them. The formal reply provides an unprecedented *post hoc* assessment of official thinking, formulated only a few years after the New Zealand imports had ceased:

> Both the Stoat and Weasel have certainly been of great assistance in checking the rabbit pest on our high pastoral country. Many runs which a few years ago were swarming with rabbits are now practically clean. This is however not altogether due to Stoats and Weasels. The country was first thoroughly poisoned thereby giving the enemy an opportunity of coping with what was left. Many of the Runs now receive but little attention in the way of poisoning. I should not however like to say that the introduction of Stoats and Weasels is advisable.

> I doubt if a supply could be got from here. They are now to be found in nearly every part of the colony and in many places where there are no rabbits. The Department has offered to pay for any secured in non-infested districts for removal to infested districts, but none have been brought in.
>
> The cost of importing is about £2 a head [estimated from the 1888 figures included in the file]. The risk of loss is not great. The best way is to advertise in Great Britain for delivery at some centre, as in the Colony. Our Agent-General has all the particulars of our shipments, and no doubt would be pleased to render your London office any assistance.[20]

The Australians wisely took the hint.

Release Areas

As soon as each eagerly awaited shipment arrived, the animals were transported to the worst-affected runs. The new arrivals were liberated only on rabbit-infested pastures, but they did not stay there. Detailed analyses of these liberation records, and of subsequent public comment if available, can be used to reconstruct the pattern of spread of stoats and weasels, and the approximate dates of their arrival into districts far from where they were first released against vehement public opposition (Table 10.2; Fig. 10.2).

Within 20–30 years, stoats and weasels were becoming common throughout both main islands. Both are small enough to be vulnerable to attack by larger predators (ferrets, cats and hawks) resident in the open country favoured by rabbits, but they are both good climbers, and readily find food and cover in forests.[21] Hence, the later distributions of stoats and weasels bear little relationship to their original release areas.

The demand from landholders required Bayly to distribute the animals in much smaller lots than needed to maximise their chances of establishment. For example, only 30 weasels and 8 stoats were taken to Queenstown for C.C. Boyes on 6 February 1885. Middleton's detailed records show that the Hon. Robert Campbell managed to buy 7 stoats and 25 weasels for his Benmore run via his Dunedin agent, A.C. Begg. The archives of the Loan and Mercantile Agency in Dunedin include an entry: 3 February 1885 'To stoats and weasels £90-05-00: Robert Campbell & Sons, Ltd. A.C. Begg, Dunedin'. Not everyone got as many animals as they wanted, or even any at all: Bayly told the disgruntled landowners that 'there were so many applications that if I had had ten times the number I could have disposed of them'.

Table 10.2 Subset of 15 named shipments in which stoats and weasels imported into New Zealand were recorded separately (plus one record possibly referring to more than one shipment), and the locations where they were released, 1884–1892. These records mentioning the species composition of successful consignments amount to 3585 animals, less than half the minimum estimate of 7838 stoats and weasels landed. In another nine shipments, and in all import records, the species were not distinguished (Table 10.1). Numbers of animals are only approximate, and sources for each entry do not always agree

Arrived	*Ship*	*Stoats sent*	*Stoats landed*	*Weasels sent*	*Weasels landed*	*Distributed to*
1884.6.28	*Doric*	27	23	74	67	Otago
1885.1.20	*Ionic*	33	32	128	116	Marlborough/ Wairarapa, Otago
1885.7.12	*Rimutaka*	55	55	181	167	Wairarapa
1885.10.28	*Ionic*	89	82	196	167	Otago
1888.1.5	*Tongariro*	NR	56	NR	229	Marlborough
1888.4.23	*Kaikoura*	NR	94	NR	225	Clarence, Canterbury/ Otago boundary
1888.7.3	*Ionic*	39	38	273	257	Marlborough/Otago
1888.8.12	*Ruapehu*,	128	120	208	192	Marlborough
1889.8.13	*Aorang*	NR	45	NR	182	Marlborough
1891.1.?30	*Tongariro*	150	NR	224	NR	Wairarapa
1891.2.26	*Aorangi*	98	13	28	0	South Island
1891.2.12	*Rimutaka*	120	NR	236	NR	Wairarapa
1891.4.21	*Ruapehu*	117	85	228	165	Wairarapa
1891.5.?	*Kaikoura*	104	NR	428	NR	Wairarapa
1891.7.16	*Aorangi*	91	83	258	245	Wairarapa
1891–1892	?	NR	237	NR	610	Wellington
Totals		**1051**	**963**	**2462**	**2622**	

NR Not recorded

Stoats and weasels rapidly became numerous in the Waitaki Valley, introducing a conflict of interest between station owners and managers on the one hand and the men employed to control rabbits by traditional trapping methods on the other. James Sutherland, by then the manager of the Benmore run, reported with regret in April 1907 that stoats and weasels were often caught in traps set for rabbits. Rabbit trappers regarded them as enemies threatening their means of living, so they killed every stoat and weasel they could—ignoring their legal protection. The trappers' antipathy is understandable from their point of view: one setting of 20 rabbit traps near Lowburn collected 11 weasels and 3 rabbits.

Fig. 10.2 Locations of places where stoats and weasels were first liberated and later spread unaided. (Cartography by Max Oulton from Author's data)

An official 1893 survey found that large numbers of stoats and weasels had drifted into the Marlborough Sounds country, which was unfortunate as settlers there had no rabbits and little sympathy for the 'natural enemy' that was brought in to save the inland runholders. Rather, the Sounds settlers destroyed stoats and weasels whenever possible, to protect their poultry.

On the other hand, E.J. Riddiford must have been pleased with his investment of hundreds of pounds into his two private shipments (Table 10.1) because less than two years later Bayly's 1887 annual report included a statement from the local rabbit inspector: 'Stoats and weasels are spreading over the district; 300 were recently turned out on Te Awaiti by Riddiford, with excellent results'. J.B. Acland's 1889 annual report of the Joint Committee on Livestock and Rabbits states that stoats and weasels had been seen 30–40 miles away from Te Awaiti.

In 1890, five years after the releases, C.R. Caverhill, Riddiford's then manager, stated:

> I am of the opinion that these animals have done good work since Mr Riddiford turned them out ... [proved by] the small number of young rabbits seen on the place ... [and] very few does in full milk ... having lost their young.[22]

Juvenile survival is one of the strongest influences on the population dynamics and local numbers of fast-breeding, short-lived species like rabbits, rats and mice, whose numbers depend on their very high potential net productivity.

No weasels, stoats or ferrets were deliberately released in Westland, where the forest was wet and thick, rabbits were few and runholders absent. But by the late 1880s, Gerhard Mueller, an explorer and surveyor working in south Westland, began to observe

> the invasion of stoats, weazels, etc. from Otago, which had first driven the rabbit plague over the mountain to our side and then followed, and now finding the birds more attractive than the rabbits, are leaving them to multiply and meanwhile destroying ... the birds the kiwi, etc.[23]

The surveyor Alfred Douglas reported on 8 May 1890 that

> several of those (or their progeny) which were turned out at Lake Wanaka some five years ago have lately been seen in various [Westland] places, one ... at the mouth of the Okuru River.[24]

Northern Hawke's Bay was invaded by weasels moving up from the Wairarapa and Hastings in the south. Weasels thrived in great numbers. Their northern migration is vividly described by Guthrie-Smith:

> By 1901 [they] had reached Tangoio [on the coast north of Napier], by 1902 a specimen had been seen on Tutira … For a short period weasels overran like fire the east coast between Tutira and Poverty Bay, and then like fire died out … Nowadays on Tutira I do not hear from shepherds or fencers of the weasel once in six years. There is something ridiculous in the fact that the weasel should have arrived at the station before the rabbit, and that later, when rabbits had become numerous, weasels should have practically passed out of the district – that the cure, in fact, should have preceded the disease.
>
> [Weasels] passed through the district and disappeared … none of any age, young or old, remained. The three years' irruption of weasels through Tutira … resembled the progress of a comet through the heavens … the invasion rolled itself up like a scroll; it came and went like a thunderstorm; … then it rained weasels; they poured along the roads … [now] the weasel is a thing of the past, and [has] remained so for many years.[25]

Weasels are now by far the least widespread of the three mustelids in New Zealand, and are seldom common anywhere.[26] Assuming that Guthrie-Smith, a careful and observant naturalist, knew the difference between weasels and stoats, the puzzle now is to explain why weasels once invaded Tutira in such unprecedented numbers, and then simply disappeared.

Ratio of Stoats to Weasels

Since the average of all shipments was strongly biased towards weasels (Table 10.2), and after release they were at first reported to be more abundant than stoats, how is it that, depending on habitat, region and other conditions, many broad-scale predator control programmes of the past have caught vastly fewer weasels than stoats? If the reported numbers of each species imported were usually more-or-less correct (a large if), there seems to have been a substantial historic change in the proportional abundance of the two species since the nineteenth century. Two possible explanations operating at different geographic scales can be proposed.

(1) In their native environments in the northern hemisphere, weasels evolved as specialist predators of small rodents such as mice, voles and lemmings averaging <50 g body weight. The only rodent that small which is present in New Zealand is the wild house mouse (Chap. 5), and now rarely in any numbers except during a temporary pulse in food resources after a beech seedfall. But in the nineteenth century, the New Zealand landscape offered weasels many

acceptable dietary substitutes for small rodents, including ground-nesting birds, bats, lizards, frogs, weta and other large invertebrates. One man observed a weasel rearing a family of nestlings in a rat hole and watched her carrying food to them every day: 'it was simply lizards every time'. So long as such unsuspecting small prey were abundant and widespread on the mainland, weasels could increase in numbers very rapidly. That is because, in good years, well-fed adult female weasels can produce a second, summer litter, and female young can breed in the year of their birth, and between them they can generate a very high population rate of increase.

In modern New Zealand, there are many fewer small prey available on the ground compared with the nineteenth century. The resources offered by most New Zealand habitats are now marginal for the thin and energy-hungry weasel body, preventing adults breeding again in summer, and making first-year litters impossible. Weasels still eat lizards, eggs, birds and very young rabbits if available, and can respond quickly to a local irruption of mice. A clear example of that was observed in Pureora Forest Park in 1984, when the number of weasels collected by a fixed array of traps in thick grass under a young pine plantation briefly increased *more than eight-fold* in six months, following a sudden irruption of wild house mice.[27] But in most years, the absence of voles and lemmings means that potential prey in the range of body sizes preferred by weasels are now at best scarce and patchy in New Zealand, and their population rate of increase is marginal. By contrast, stoats can take ship rats, possums and rabbits as well as smaller prey when available. Hence, one possible explanation is that the relative numbers of the two species may now be correlated primarily with the present distribution of their preferred prey, not with the numbers originally released.

(2) An alternative but not mutually exclusive idea is that the smallest members of a carnivore guild are always vulnerable to interference competition from the larger members. Wherever stoats are common, weasels suffer that disadvantage; the same applies to stoats in competition with cats and ferrets.[28] Effective removals of stoats after the arrival of myxomatosis in England, and of American mink in Belarus, were both followed by significantly increased numbers of weasels caught in the same traps. In New Zealand today, reports of a similar effect are emerging in places where stoats have become less common, at least locally and temporarily, in, for example, kiwi sanctuaries and similar protected areas after years of stoat trapping.[29]

Consequences for Birds

It seemed self-evident to early observers that the loss of native birds from untouched standing forest must have been linked to the first arrival of mustelids. According to B.H. Wilmot, weka, kiwi, and kākāpō used to be plentiful in the Makarora Valley before 1890, but 'since the advent of the stoats and weasels they are very rare, and yet rabbiting tallies have not depreciated'. J. Donovan's description of the abundance of native birds at Okarito until about 1894 was firmly linked by him (and many others) to the time when 'we first saw the stoats and weasels ... [then] the birds disappeared'.[30]

Sir Walter Buller repeatedly criticised the introductions and continued campaigning against them long after the shipments had ceased in 1892. 'The vermin that every farmer on the Old Country was trying to extirpate as an unmitigated evil our wise Government bought up by the hundred and imported into this country in the vain hope that [they] would change their habits and take to a rabbit diet, to the exclusion of everything else!'

The correlations pointed out by Donovan, Buller and others certainly contained some truth, but did not allow for the less obvious damage being done by previous and contemporary new arrivals in the forest, the three species of rats (Chaps. 2 and 4). It would be unfair to criticise nineteenth-century observers for not understanding what even scientists of the time (and often, even now) did not appreciate: the already-well-advanced consequences for nesting birds of hundreds of years of rat damage. Stoats and weasels live at the top of the food chain, so are relatively few, but they are active during the day. That is when their predatory activities can be observed, and easily confused with the far greater unseen damage done at night by hordes of rats. The only way to disentangle different causes of the same effects is by double-blind reversible experiments, which were not done at the time.

The grief expressed by local naturalists was genuine, then and now, but their explanation falls into the logical trap of arguing *post hoc, ergo propter hoc* (after this, therefore because of this): mustelids arrived, birds have gone, therefore mustelids are the cause of their loss. With hindsight we can point out, without denying the severe effects of predation, that bird faunas are also sensitive indicators of all sorts of overlapping environmental changes that are very hard to disentangle, and much of the original habitat to which they were closely adapted has simply been swept off the map. For the mourning colonists of the time, and for tens of thousands of members of the Forest and Bird Society now, that is no consolation.

By contrast, the politicians and landholders most closely involved in the decision to import mustelids were less concerned, because they calmly expected the native fauna to disappear naturally in the face of competition from the superior products of the northern hemisphere. European zoologists from Darwin onwards assumed that most of the native fauna was already doomed. Even the Māori anticipated their own demise, displaced by European settlers in the same way as the kiore had been displaced by the Norway rat (Chaps. 2 and 4).

The colonists did indeed express some regret about the losses of introduced game birds, but native birds were not seen as so important at the time, so the warnings were ignored. 'Like everything else which is native to the place, it [native forest] seems unable to bear the advance of civilisation', concluded Samuel Grant and John Foster after their 1880 survey (Chap. 9). Unfortunately, long before Bayly lost his job in 1889, he had succeeded in importing enough stoats and weasels to ensure both species established and spread far outside their original release areas. By 1909, Sir Joseph Ward, the New Zealand premier, considered that they had spread pretty well all over the Dominion: by the early 1920s, stoats and weasels were common in nearly every part of New Zealand, and in some were 'enormously abundant'.[31]

Reactions in England

By the end of 1886, even some English landowners were beginning to wonder what effect the continued removal of native predators were having on 'the balance of nature' on their own estates. The numbers of animals removed for export were probably only substituting for the numbers normally killed by gamekeepers, but people began to ask whether local increases in rats and field voles could be due to the losses of their natural enemies. The landowners of six British counties agreed to prohibit the export of stoats and weasels to New Zealand. Fluctuations in numbers of field voles had been known for centuries, but now the zoologist J. E. Harting[32] cited a shepherd attributing an irruption of voles in 1892 to the trapping of weasels and stoats for sending out of the country. Ironically, the same argument was used to support the policy of importing mustelids into New Zealand: 'The remonstrances of the Lincolnshire farmers against the continual export of these animals should be a convincing proof of their usefulness'.[33]

A different reaction to the shipments illustrates the conflict of human interests between the British landowners and the trappers working where the animals were being collected. Gamekeepers in England were no longer so keen to destroy stoats and weasels as before, since they discovered that money could be made from them when captured alive, but landowners wanted them destroyed to protect game birds, not kept as valuable for exports. Men with the right contacts and able to supply animals to purchasing agents would 'pass word around the county for 50 miles [80 km] that no stoats should be killed but all saved to be trapped after they have had young'.

On the other hand, the massive sample of British stoats exported to New Zealand also had a surprisingly positive consequence. Geneticists tell us that the founder effect—the minimal sampling of an ancestral population by a colonising group—normally ensures that the genetic diversity of a population of introduced animals is less than that of their ancestors. But New Zealand stoats, contrary to expectation, carry five haplotypes, only one of which survives in Britain. This surprising result seems to be because Henry Allbones' collecting of English stoats for transportation to New Zealand in the 1880s sampled the genetic diversity of the species that was then present in his part of Britain. That native diversity was lost when British stoats barely survived a catastrophic population decline that followed the removal of their main food supply by the rabbit disease myxomatosis in 1953–55.[34]

British stoats are not in any danger of extinction, but if ever authorities there wanted to retrieve their former genetic diversity, pest managers in New Zealand would no doubt be happy to supply some shipments of stoats to travel back to their ancestral homeland.

Rethinking the Mustelid Introductions

Nineteenth-century Acclimatisation Societies are often criticised for their ill-advised importations, but in fact, most of the provincial Societies, except Southland, had always been against the introductions of any alien predators, if only because they would also compete with their Society's members for game birds. The real responsibility lay with the colonial government, for reasons that seemed valid to them at the time (Chap. 7).

In 1909, the Council of the Otago Acclimatisation Society went so far as to declare the importations of 'ferrets, stoats, and weasles … a disastrous mistake, owing to game [birds] being destroyed'. To contemporary eyes it is remarkable, even shocking, that their concern was still mainly for introduced game, rather than for native fauna. But it is important to understand the decisions and opinions of the past in the contexts in which they were made. At that time, arguments based on conservation of native wildlife always ran a poor second to the colonial ambition to create a 'Britain of the South',[35] complete with abundant game to provide shooting sport for everyman.

Attitudes in colonial New Zealand on how to control rabbits were based on the simple assumption that every rabbit killed by a 'natural enemy' would be one less rabbit in the population. The colonists reasoned that, because stoats were certainly good at killing individual rabbits, more stoats doing more of the same could remove a whole population of rabbits. Hence, stoats and weasels were widely assumed to be actually capable of preventing the spread of rabbits from their earliest release areas and reversing the damage in areas where rabbits were established.

This assumption was reinforced by the confident prediction of European authorities, from Darwin onwards, that the endemic biota of New Zealand will inevitably be replaced by plants and animals from Europe (Chap. 4). So, in contrast to the declining and inferior native predators, stoats and weasels had the advantage: they were British.[36] Eventually, as David Young, Paul Star and others show, attitudes began to change.

The bitter criticism of Bayly and his colleagues by later writers is understandable but unfair. On the contrary, like the American mink freely admitted to Britain to establish fur farms, the introduction of mustelids to New Zealand was actively encouraged by officialdom all the while these animals were seen as having an economic value. Mink farms in the UK were a reliable source of luxury fur, and the risk of escapees breeding in the wild was ignored. Stoats and weasels in New Zealand were seen as self-perpetuating allies in the battle against the rabbit pest and a source of rural employment. Repeated warnings of the risk of these new arrivals becoming an irreversible serious threat to their new environments were simply overruled. Similar warnings are now appearing during heated debates about the use of gene drives to control pest populations.

The surviving documentation describing events in colonial New Zealand is a sobering record of the urgency and optimism under which the responsible authorities chose to deploy a potent new tool, against fervent opposition and without any suspicion that things could turn out so differently. Self-perpetuating weapons are still a dream for new generations of pest control authorities (Chap. 12). History provides many cautionary tales and much to think about.

Summary

Over the decade 1883–1892, at least 7838 British stoats and weasels were landed in New Zealand. At least 25 shipments are known, on named ships, with their consignees and destinations. The number of animals landed according to official customs import data adds up to many more than are listed in the shipping records, so these are minimum estimates. The programme was driven against all objections by Benjamin Bayly, the government's supervising rabbit inspector, until he was dismissed in 1889 and the government withdrew support for the programme. After 1889, all consignments were commissioned by regional Rabbit Boards. The Allbones' system of carrying live pigeons 1882–90 was expensive but generally successful (survival rates c. 90%). Three of six shipments in 1891 recording catastrophic mortality caused serious financial losses to Rabbit Boards, which in 1892 switched to cheaper, locally bred ferrets. More weasels than stoats were landed (2622:963), and both spread widely throughout both main islands, but weasels are now rare. The two species were nearly the last alien predators to be introduced, and their effects on the native fauna are often less than those of rats, but kiwi chicks, rock wrens and beech forest endemics are especially vulnerable to stoats.

Notes

1. Thomson, G. M. (1922). The naturalisation of animals and plants in New Zealand. Cambridge: Cambridge University Press, p. 70.
2. For a more complete list of the primary resources underlying the first part of this chapter, see King, C. M. (2017). The history of transportations of stoats (*Mustela erminea*) and weasels (*M. nivalis*) to New Zealand, 1883–1892. International Review of Environmental History, 3, 51–87. https://doi.org/10.22459/IREH.03.02.2017.04; and for the second part, King, C. M. (2017). Liberation and spread of stoats (*Mustela erminea*) and weasels (*M. nivalis*) in New Zealand, 1883–1920. New Zealand Journal of Ecology, 41, 163–177. https://newzealandecology.org/nzje/3311
3. King, C. M. (2017). History of transportations
4. Bayly, B. P. (1884). Annual Report on the Rabbit Nuisance [1883–84]. Appendix to the Journal of the House of Representatives, 1884 II, H-02.
5. Pinney, R. (Unpublished). Notebook M1: Extracts from Benmore Letter Books, 1877–1887. In. Dunedin: Hocken Library MS-3178/013.
6. King (2017), History of transportations…p. 71.
7. Multiple authors. (1885). Correspondence concerning natural enemies for Lake Ohau Station. In. Wellington: Archives New Zealand R24418111.

8. Anon. (1886). Report of the Joint Committee on Rabbit and Sheep Acts. Appendix to the Journal of the House of Representatives, 1886 I, I-05.
9. For a vivid description of how rabbits managed to get around the fences in Australia, see Rolls, E. C. (1969). They all ran wild. Sydney: Angus and Robertson.
10. Pinney, R. (1981). Early Northern Otago Runs. Auckland: Collins, Collectors Edition.
11. [Runholders of Upper Waitaki]. (1887). Correspondence on the state of the rabbit scourge in upper Waitaki. Wellington: Archives New Zealand R17620769.
12. Campbell, R., & Colonial Secretary. (1887). Correspondence concerning the promised subsidy for importing the natural enemy. Wellington: Archives New Zealand R17620769.
13. Allbones, H. (1887). Letter 12 Oct to Minister in Charge of Livestock, Wellington: Acceptance of contract terms. Wellington: Archives New Zealand R17620769.
14. Leeds Mercury 9 November 1887. Politics and Society, pp. 4, 5.
15. Marlborough Express 5 February 1889. p. 3 Awatere Rabbit Board.
16. Clifton, E. (1889). Monthly report for Wellington area to Stock Inspector. Wellington: Archives New Zealand R17620841.
17. Hawke's Bay Herald 5 June 1890. Rabbit extermination.
18. Pall Mall Gazette 2 April 1891. Stoats and weasels for the Antipodes.
19. Wairarapa Daily Times 13 August 1891. Rabbit Board.
20. Multiple authors. (1896–1898). [Complaints about natural enemies]. Wellington: Archives New Zealand R1762009.
21. King, C. M., & Powell, R. A. (2007). The Natural History of Weasels and Stoats: Ecology, Behavior and Management, 2nd edition. New York: Oxford University Press.
22. Hawke's Bay Herald 5 June 1890. Rabbit extermination.
23. West Coast Times 18 June 1890. Mr. G Mueller on his trip south and back.
24. Anon. (1890). Annual Report on the Rabbit Nuisance [1889–90]. Appendix to the Journal of the House of Representatives, 1890 H-9.
25. Guthrie-Smith, H. (1969). Tutira. The Story of a New Zealand Sheep Station, 4th edition. Wellington: A. H. & A. W. Reed. Pp. 354–355, 389.
26. King, C. M., Norbury, G., & Veale, A. J. (2017). Small mustelids in New Zealand: invasion ecology in a different world. In D. W. Macdonald, C. Newman, & L. A. Harrington (Eds.), The Biology and Conservation of Musteloids (pp. 257–277). Oxford: Oxford University Press.
27. King, C. M., Innes, J. G., Flux, M., & Kimberley, M. O. (1996). Population biology of small mammals in Pureora Forest Park: 2. The feral house mouse (*Mus musculus*). NZ Journal of Ecology, 20, 253–269.

28. Garvey, P. M., Glen, A. S., & Pech, R. P. (2015). Foraging Ermine Avoid Risk: Behavioural responses of a mesopredator to its interspecific competitors in a mammalian guild. Biological Invasions, 17, 1771–1783.
29. Haworth, D. (2018). Weasel (*Mustela nivalis*) dynamics in South Island beech forests of the Maruia Valley. DOC Research and Development Series, 355, 1–25.
30. Otago Daily Times 12 November 1908. New Zealand's mammalian menace.
31. Thomson, G. M. (1922). P. 72.
32. Harting, J. E. (1894). The Weasel, *Mustela vulgaris*. Zoologist (Third Series), 18, 417–423 and 445–454.
33. Hawke's Bay Herald. 5 June 1890. Rabbit extermination.
34. Veale, A. J., Clout, M. N., Holland, O. J., & Gleeson, D. M. (2015). An invasive non-native mammal population conserves genetic diversity lost from its native range. Molecular Ecology, 24(9), 2156–2163.
35. Hursthouse, C. (1857). New Zealand, or Zealandia, The Britain of the South. London: Edward Stanford.
36. Star, P. (2006). "Nature's Trump Card": Confronting the rabbit problem in southern New Zealand, 1867–1897. ENNZ: Environment and Nature in New Zealand, 1, 3–9.

CHAPTER 11

Lucky Escapes and Nasty Surprises (1884–1920)

> I question if such enormous benefit has ever resulted from the introduction and acclimatisation of any one animal, as that which has attended the Mungoos in Jamaica and the West Indies; and I marvel that Australia and New Zealand do not obtain this useful animal in order to destroy the plagues of Rabbits in those countries. W.B Espeut (1882).[1]

Earlier chapters of this book have summarised the story of the all-too-successful spread of deliberately introduced mustelids that were expected to save the early wool export industry from the rabbit pest. This strategy was regarded as securely based on the fact that mustelids kill rabbits, from which it followed that the more mustelids were brought in, the more rabbits they would kill. Today, amid the enthusiasm generated by the Predator-Free New Zealand 2050 campaign, might be a good time to examine this logic as a clear example of the critical distinction between (a) observed facts about mustelids and (b) the way that facts can be understood only via the pre-existing world pictures and the widely shared assumptions common at the time. As philosopher Mary Midgley points out: facts are not gathered in a vacuum, but to fill gaps in a world picture that already exists.[2]

C. M. King, *Invasive Predators in New Zealand*,
Palgrave Studies in World Environmental History,
https://doi.org/10.1007/978-3-030-32138-3_11

Mongooses, grey foxes, pine martens, wild cats and other predatory species were also seriously proposed for possible importation for the same purpose. None of them succeeded, but the fact that they were suggested at all indicates the strength of the imperative to continue searching for a solution to the largest threat to the colonial economy of the time, labelled with remarkable understatement as 'the rabbit nuisance'. So now is also a good time to review the contemporary arguments for and against these lesser-known predators. Then as always, these arguments largely consisted of two or more parties tossing isolated facts at each other.

By contrast, some apparently harmless herbivorous species, also deliberately introduced with every good intention, have turned out to be wolves in sheep's clothing. Deer, Himalayan tahr, chamois and others have done plenty more damage to native vegetation than expected, but by far the most significant nasty surprise was delivered by the Australian brushtail possum. These are common tree-dwelling, leaf-eating marsupials, clothed not in sheep's wool but in their own soft and luxurious fur. In the unmanaged Australian forests of the mid-nineteenth century, brushtails bred well, caused no visible harm to the native trees or wildlife and supplied valuable pelts for a flourishing fur trade.

The standing native forests in New Zealand, most comprising endemic species very different from those of Australia, were at that time seen to be conspicuously short of harvestable products suitable to provide a sustainable economic return. Importing these harmless herbivores from just across the Tasman seemed a simple and obvious way to establish a profitable fur industry from land unsuitable for sheep, with no expected risk to native fauna. Who could possibly object?

The one thing that both of these stories have in common is the hard lesson that both taught: animals often do not do what we expect them to do. Mongooses were confidently expected to attack rabbits in New Zealand as vigorously as they had the cane-field rats in Jamaica[3]; instead, the liberated mongooses disappeared, and the accelerating spread of rabbits of the 1870s was totally unaffected. Brushtail possums were expected to maintain roughly the numbers and fur harvest potential in New Zealand as was normal in Australia; instead, possums multiplied to some 20 times their average Australian density, rapidly turning from useful sources of income into an expensive curse for foresters, dairy farmers and conservationists which is still with us.[4]

Failed Expectations: Predators That Never Established

During the nineteenth century, almost no known predatory mammal or bird escaped consideration for translocation to New Zealand at one time or another. Some proposals were based on the flimsiest evidence, for example, that the European badger 'is perfectly harmless and inoffensive, and at the same time an indefatigable destroyer of rabbits'.[5] Settlers and their English correspondents who earnestly promoted such a completely unsuitable species (badgers live mainly on earthworms) illustrate not only the extent of their concern about the rabbit invasion, but also their ignorance of the animals and their unfamiliarity with conditions in New Zealand.

One species, the red fox, arrived before the rabbit problem became a serious issue, but was swiftly prohibited. The feral cat was already present, and at the height of the crisis its distribution in the wild was often deliberately supplemented by transporting tame city cats to the back country (Chap. 6).

Of six other predatory species proposed for importation, only one, the mongoose, ever actually arrived in numbers potentially sufficient to launch a population. Proposals for at least five more were considered but rejected. Among the interesting questions to arise from a review of them are: How many of these predatory species might have done any good had they become established? What do these records tell us about how the colonial government and the settlers responded to the rabbit crisis, and could they have acted any differently on what they knew at the time?

From what we know now, we can guess that most of the species considered at the time would either not have settled, or, if they had, would have done no better in solving the rabbit problem than those that did. Still, knowing the background is useful because it suggests that our criticism of what colonial authorities did should become tinged with some appreciation of what they managed to avoid doing. Australia admitted foxes and domestic ferrets from early days, but (on New Zealand advice) refused stoats and weasels (Chap. 10); New Zealand prohibited foxes just in time, and refused several other candidates, but imported huge numbers of ferrets, stoats and weasels for turning out into the wild (Chaps. 8 and 9). It is worth a quick excursion into history to explain these opposite combinations; the debate continues on which is worse.

Red Fox

Hunting the red fox was an indispensable element of the nineteenth-century English country life that wealthy early colonists so much wanted to import into New Zealand. The *Lyttelton Times* of 25 October 1864 refers to three red foxes sent to New Zealand on the ship *William Miles*, consigned to Mr. Charles Prince. Of the three, one vixen got loose and jumped overboard, but the remaining dog and vixen arrived in 'excellent health'.[6] These may or may not have been the same animals for which John Scott Caverhill of Hawkswood (Canterbury) was granted permission to 'retain possession of foxes' on his property provided they be held in captivity. Another report of 1873 commented that 'not long ago', someone imported a pair of foxes into Canterbury, which 'fortunately died without issue'.[7]

If any of these foxes actually were liberated in the Canterbury of the early 1860s, says E.C. Rolls, they did not establish, perhaps because colonising success for foxes depended on the prior spread of rabbits. In Australia, where foxes were imported for traditional foxhunting on horseback well before wild rabbits were common,[8] the originally abundant small native mammals provided sufficient food for foxes until rabbits caught up. By the time the New Zealand authorities were considering their version of the rabbit problem, foxes were already out of favour in Australia. Rolls adds: 'No one benefitted by their coming because by the time there were enough to hunt there was little interest in [traditional, horseback] hunting. … All the wastage has been of wildlife'.[9]

Fortunately, Canterbury's narrow escape prompted the colonial government to use the New Zealand Protection of Animals Act specifically to prohibit red foxes from importation in 1867, while wild rabbits were still relatively scarce. Other potentially undesirable imports were not mentioned at that stage (mink, muskrats and grey squirrels were added later). The need to add more species to the prohibited list led the *Press* to point out

> the inadequacy of the provision made under the Protection of Animals Acts against the importation of noxious animals … It is unlawful for anyone to introduce 'any fox, venomous reptile, hawk, vulture or other bird of prey into the colony' under a penalty of a hundred pounds or imprisonment for six months. He may, however, introduce any beast of prey, or vermin he pleases, with the single exception of the fox. Anyone, therefore, who should take a fancy to turn adrift a few tigers, wolves, or other carnivorous animals is at perfect liberty to do so, so far as the penal statutes are concerned.[10]

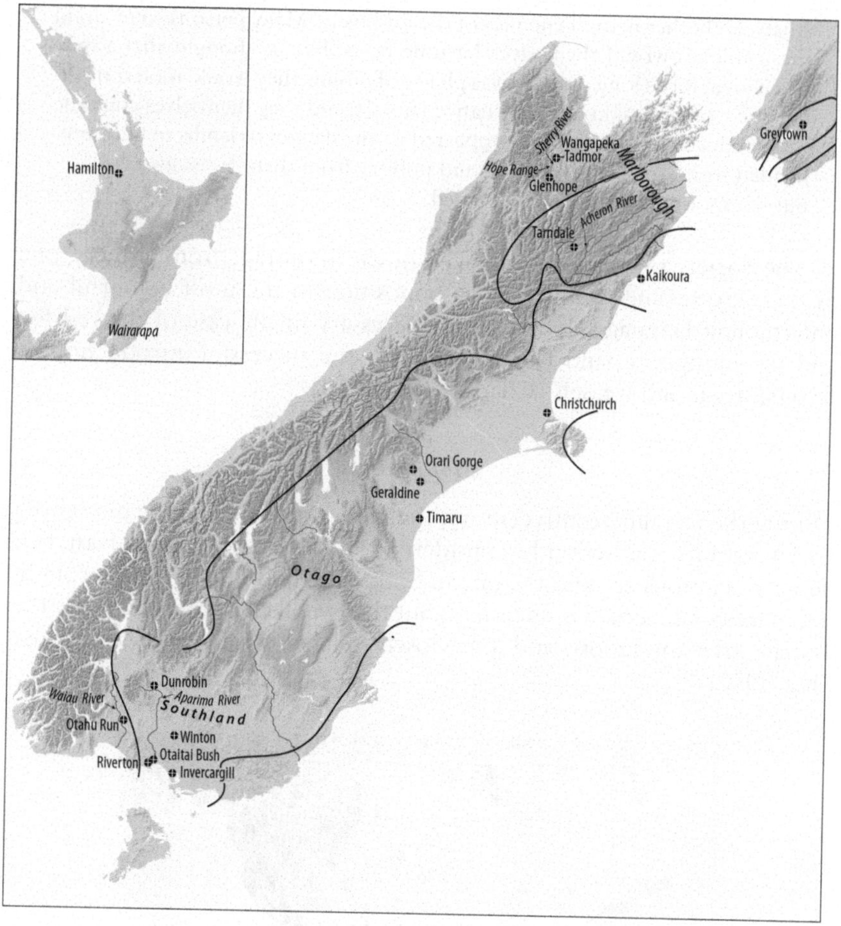

Fig. 11.1 Map showing places mentioned in Chapter 11. (Cartography by Max Oulton from Author's data)

Undaunted, the Acclimatisation Society in Southland (Fig. 11.1) continued to include red foxes on their wish list. At the Society's annual general meeting in April 1880, J.L. Leesmith moved

> That this Society, in view of the alarming increase of the rabbit pest, and the ruinous consequences resulting therefrom, recommend, and will encourage by every lawful means, the introduction into this province of weasels, foxes,

stoats, and other natural enemies of the rabbits ... Many persons were afraid these animals would themselves become pests, but he thought that a mistaken idea, for as long as there was plenty of rabbits they would molest nothing else worth speaking of [i.e., native fauna!], and they themselves could be easily reduced as the rabbits disappeared ... the danger to lambs to be apprehended from foxes, and to game and poultry from them [was discussed] ... but the motion was ultimately carried.[11]

The Society's intentions provoked howls of outrage from farmers, one of them expressing deadly serious opposition in the most colourful and entertaining language. It was all un-necessary in the case of the red fox and the mongoose, and failed to prevent the successful introduction of ferrets, stoats and weasels (Chaps. 8, 9 and 10).

Mongoose

During the late nineteenth century, the idea of introducing the mongoose to kill rabbits was seriously considered, to the extent that at least two actual liberations are known, one in Southland and one in Marlborough. The species concerned is uncertain, but the two likeliest candidates are the Indian grey mongoose and the closely related small Asian mongoose (Fig. 11.2).

Fig. 11.2 Indian grey mongoose. (Wikipedia Commons)

Both are bold, inquisitive predators, weasel-like in shape and habits, with a grizzled, long-haired coat and long tail. They are both native to much of southern Asia. Both are active by day, searching energetically for almost any small prey including rodents, birds, lizards and invertebrates. Their generalist diets and tolerance of human environments give them two important advantages as potential rabbit control agents in agricultural habitats.

The first recorded proposal to bring mongooses to New Zealand arose in Winton, a small town 30 km inland from Invercargill. Winton was on the front line of the expansion of rabbit numbers northward from their early liberation sites on the coast between Invercargill and Riverton (Chap. 7, Fig. 11.1). Southland farmers facing serious damage to their land in the mid-1870s organised many meetings to discuss what to do about the rabbits, even while in other areas of New Zealand, rabbits were still welcome. In December 1876, the Winton and Forest Hill Farmers' Club discussed

> the steps to be taken for the reduction of the numerical strength of the army of rabbits at present laying successful siege to the farming industry … [they discussed] a letter recommend[ing] the introduction of an animal called the mongoose, indigenous to India, and to be obtained there at a shilling per head … the adaptability of the animal for the proposed purpose had been verified by an old Indian pensioner resident here … It was estimated that about £300 would [pay for] one thousand mongoose, including cost of transit, together with a keeper's salary. Several members stated that with the co-operation of other clubs, there would be no difficulty in raising even more money than that.[12]

New Zealand was a favourite destination for retired officers of the East India Company, so the presence of an Indian pensioner to enthuse Southland farmers about such an unfamiliar animal is less surprising than it sounds. Perhaps the unnamed pensioner also provided contact details of knowledgeable people in India to whom the club might turn for advice.

Soon after the meeting, A. McKellar wrote to India on behalf of the Club, asking for information about the mongoose and its potential as a rabbit exterminator. The following detailed extracts from the replies, reprinted in the *Southland Times* in May 1877, give not only the names and affiliations of McKellar's contacts but also some information about the animals critical to understanding what happened next:

From J. G. Charles, Esq, Deputy Commissioner of Police, Calcutta, 9th March, 1877: From my personal knowledge of the Bengal mongoose I believe it might be introduced in Australia and New Zealand with the greatest advantage. It is an animal of very hardy nature, quite able to take care of itself, very fierce and very courageous, killing rats, ... very restless, roving about over a considerable space. It would cause great destruction amongst rabbits, I am quite sure, particularly in destroying the young ones in their burrows. I further believe it would be very unlikely to multiply in such numbers as to become a nuisance.

From Mr H. J. Rainey, [Indian naturalist], March 4th, 1877: As the Indian, or rather the Bengal mongoose ... does prey on the wild hare which abound in these parts, and thereby materially lessens their number, there is every likelihood that these mongooses would prove even more destructive to rabbits ... The rabbits would, probably, be exterminated in this way in a comparatively short space of time ... The mongoose would certainly prove destructive to poultry and game, but ... these ... would be cared for, whilst the rabbits are not, [so rabbits] would be the principal victims.[13]

The Dunrobin run, on the Aparima River 120 km inland from Invercargill (Fig. 11.1), was worked by a pair of brothers, W.H. and C. Basstian. They took over this run in 1859, and named it after a run in Tasmania from which they brought sheep.[14] Like their neighbours at Winton, the Basstians lived close to the earliest liberation sites of rabbits, and were among the first to suffer from rabbit damage. No doubt they all discussed the rabbit problem many times, and agreed that importations of the natural enemy would be the best answer. Christopher Basstian was a Southland and Otago Councillor and chairman of the Wallace County Council,[15] so had plenty of opportunity to influence the public debate. He had no qualms about introducing exotic species: he had already established the first colony of Tasmanian brushtail possums near Riverton in 1858 (see below, p. 267).

Perhaps it is no coincidence that less than six months after the Winton meeting, a brief notice appeared in the *Southland Times*:

Fifteen mongoose from India arrived here by the *Ringarooma* on Sunday, to the order of Mr C. Basstian, of Dunrobin. They have been imported as rabbit destroyers.[16]

Within a week came a blistering reply from a correspondent signing himself 'Fair Play':

> Sir, — In a recent issue of the *Times* I notice that a certain squatter has imported a number of 'Mongoose' for the destruction of rabbits. As it is (I believe) an admitted fact that the mongoose is destructive of poultry, and therefore will be specially obnoxious to small holders, does not its introduction appear to be at least a questionable good, even admitting it to answer the purpose of its introduction. The cockatoos [farmers] have their work to do in dealing with the rabbit nuisance in an equal degree (compared with their holdings) with the run-holder, and it appears very selfish on the part of the man to whom wool is alone the object, to seek to attain it by letting loose a pest and destroyer on his industrious neighbour. Every small settler will agree with me that the sooner the mongoose are got rid of the better. The rabbit encroachment can be fully dealt with by fair and legitimate means without detriment to anyone or the introduction of noxious vermin. I am, &c, Fair Play.[17]

Indeed, there was cause for concern that Basstian's animals might spread, because when a wild mongoose was caught in a rabbit trap at Otaitai Bush in February 1878, it was immediately assumed to have come from Dunrobin.

Nearby, Robert Cuthbertson fought against rabbits on his Southland farm, the Otahu run, for at least six years of accelerating losses (Chap. 7). At the height of his troubles, he wrote to the Colonial Secretary on behalf of the Southland Pastoral Association:

> The Rabbit Nuisance in Southland is assuming gigantic proportions, and that now the farmers acknowledge that they are unable to cope therewith, an admission on their part which goes far to show the extent of the evil … In view of the inability of both farmers and runholders in Southland to cope with the nuisance, by the methods at present in use … the Government of New Zealand should communicate with H M Government in India, to ascertain if the Indian ferret or mongoose would be likely to prove useful in destroying rabbits in this part of the Colony – viz. 45° So[uth] Lat[itude], if they could be got in large numbers, and the cost, and if they could stand the voyage. I have also to suggest that the Govt. might endeavour to import some of these animals to put on the Govt. reserves in Southland.[18]

Talk of bringing the mongoose to New Zealand stimulated 'Settler' in Greytown to forward to the local newspaper a letter from W.B. Espeut, a planter in Jamaica:

> Prior to 1872 I had to contend with an enemy almost as bad—the rat—which destroyed my sugar canes to such an extent as to threaten me with ruin. In that year (1872) I imported from India nine mongooses, and let them loose in my cane-fields. In a few months I observed a marked decrease in the ravages of the rats, and this relief increased rapidly, until, after two years, my loss from rats was nil. For the last five years, I have not spent £9 in destroying rats, or lost £50 in rat-eaten canes; but before 1872 I used to expend between £250 and £300 a year in killing rats, and lost sugar to the value of several hundreds of pounds yearly … I estimate that the island has gained in the nine years since the introduction of this useful animal more than two million pounds sterling increase in production of sugar, cocoa [and other crops formerly ravaged by rats] … I do not think [you] need fear the mongoose; he has done an immense service to me and to all cultivators of the soil of Jamaica, and all the bad that can be said about him is that now and then he will gobble up a chicken or a duckling. But it will take many thousands of ducklings and chickens to equal the loss now caused by rabbits; and a mongoose cannot climb; any careful hen-wife can completely secure her chickens in wire-netted hen-houses or pens.[19]

Espeut's letter to the newspaper was a summary of his more detailed paper published by the Zoological Society of London, which ended with the comment reproduced at the head of this chapter.

This was just the sort of economic argument that wool growers liked to hear, and it may well have had a strong influence in the days when objections to the importations of predators were usually based on the consequences for domestic poultry or introduced game birds, as opposed to those for native fauna. As 'Settler' had put it:

> When we remember that our wool export is declining at the rate of a million annually, whilst our annual interest is increasing, and that runs that used to carry thousands of sheep now only carry a far diminished number, it is time for the small settler to remove his objection to the free introduction of the natural enemy by the Government and, admitting the rabbit evil, at once set about protecting his poultry yard. Wire is cheap, and a small yard can be soon enclosed against ferret or mongoose, if imported.[20]

So the pros and cons of the mongoose idea continued to be debated into the mid-1880s, long after the official programmes of importing stoats and weasels, and releasing locally bred ferrets by the thousand, were both already well under way (Chaps. 8 and 10). The predatory powers of the mongoose were never doubted, but its origin in a tropical habitat was a concern. For example:

> The mongoose, provided it is brought from the colder parts of India, might turn out a success; but the mongoose from the plains of India is not likely to have a numerous progeny in the colder parts of New Zealand.[21]

George Bullen of Kaikoura was one of the witnesses examined by the Joint Committee on the Sheep and Rabbit Acts, which reported in October 1884. He was already known as a staunch supporter of ferrets as rabbit predators, but was also asked if he had any experience of the mongoose. He replied that he had succeeded in importing nine or ten of the small mongooses from the Himalayas. One escaped and was killed in a trap by a neighbour, much to Bullen's regret because 'a mongoose costs a good bit of money'. The rest he let go on the run. He observed that they were very active, and could kill rabbits in the open, but could not say whether they survived in the wild. One member of the Committee pointed out that the mongoose comes from a different climate, and that whether they could stand the cold was still unknown.

If New Zealand's cool climate was the problem, mongooses should have done well across the Tasman, because much of Australia is a lot warmer than anywhere in New Zealand. In 1883–1884, mongooses were released 'by the hundred' into parts of the Murray River basin thick with rabbits, but they did not survive there either. The explanation there seems to be that rabbit trappers, 'the natural enemy of the natural enemy of the rabbit', had 'done away with them all' in defence of their business.

Nevertheless, the issue would not go away. The most serious and thorough official investigation into the potential advantages of importing the mongoose is recorded in a stream of correspondence occupying most of 1888, and preserved in one archives file.[22] The series started on 3 January with a letter to the secretary for land from Arthur Manners of Hull, England. He recited the history of the mongoose in saving the sugar cane industry of Jamaica in much the same terms as Espeut's first-hand account given above, and added:

> The mongoose you will thus see is a much more powerful aid both as regards fecundity and in respect to his destructive power to any of the animals used in England for such purposes (as the weasel, stoat and ferret).

The superintending inspector at the Stock Office, Benjamin Bayly, wrote on the file: 'These animals are already turned out in NZ and are reported to be doing well. Enquiries might be made what the cost of procuring them in India and introducing them would be'. The Colonial Office passed Manners' recommendation to the Agent-General in London, with the request that he make enquiries as to where mongooses are obtainable, and their number, cost, and so on.

The Agent-General, F.D. Bell, wrote to Dr. Philip Sclater FRS, secretary of the Zoological Society of London, on 11 June, asking if the Society had any information that could be conveyed to the New Zealand government as to whether it would be advisable to bring in mongoose, 'either in substitution for or in addition to Stoats and Weasels, a large number of which have been imported there of late'.

Bell informed the minister that Sclater's reply left him in no doubt that the mongoose

> would be very effective toward destroying the rabbit-pest; but the introduction of the animal would on the other hand be a serious danger for the flocks, as the mongoose would kill young lambs not less freely than rabbits. I have therefore been strenuously advised against that step.

However, Sclater also suggested that Bell consult D. Morris, assistant director, Kew Gardens, which he did on 18 July 1888. The reply pointed out the following:

> It may destroy the young rabbits, but its powers of increase are so enormous, and it is difficult to destroy when once established in a country, that it is pretty well certain to prove a greater nuisance to Colonists than even the rabbits themselves ... Where the mongoose is abundant it is almost impossible to raise poultry, as they eat both eggs and chickens. ... I have no doubt they would attack lambs, [possibly] in preference to rabbits if the latter became at all scarce in their neighbourhood ... I have had seven years' experience of the mongoose in Jamaica, and although it certainly destroyed the rats in cane fields, and did some good to the sugar planters, the animal everywhere now is become an unmitigated nuisance in the Colony.

Bell confirmed to Wellington on 11 September 1888:

> It is further advised that such a fierce and rapacious animal as the mongoose should not be introduced into NZ as a natural enemy of the rabbit.

The Colonial Secretary acknowledged this letter with thanks and dropped the subject. However, that was not quite the end of the story.

Mongooses make good pets, and in India tame mongooses are kept to control rodents, much like house cats.[23] There were no regulations to prevent those civil servants and former East India Company staff, who chose to migrate with their households and staff to New Zealand, from bringing their pet mongooses with them. Years later, an unknown number of individuals of, probably, the Indian grey mongooses were imported into New Zealand by a retired Indian army officer, who had a farm and menagerie in the Nelson district. Escaped or released mongooses were sighted in the wild on three occasions at Glenhope in 1912 and in the Clark River valley between 1914 and 1918. There were also unconfirmed reports of mongooses in the Sherry River–Hope Range areas until 1920. They presumably died out, but not before some had dispersed 20 km or more.

Although we cannot now be sure which species of mongoose is referred to in any of the above accounts, the exploits of the small Indian mongoose in Hawaii, Fiji and the West Indies qualify it to join a list of the nine most damaging mammals to have caused problems in other island countries. We therefore have cause to be grateful that we escaped the consequences that have followed successful colonisations of other islands by mongoose—so far.

However, accidents do happen. In November 2014, a New Zealand border inspection officer found three dead animals in a consignment of new tyres imported from Malaysia. They were long bodied and long tailed, a shape which suggested that they could be weasels. But in the photograph published in the *Waikato Times*, they looked quite unlike weasels, but much more like small mongooses. Unfortunately, the carcases were discarded before they could be properly identified. At least we can be relieved that the three animals did not survive the journey, but the incident shows that border security can never be relaxed.

Pine Marten

Among many Scots who emigrated to New Zealand in 1853 was James Menzies, a Highlander born in 1821 in Kinloch Rannoch, familiar with moorland wildlife. For some time he held the Dunalister run on the lower Mataura River, was a member of the Southland Provincial Council, and later became the first superintendent of Southland during the short period 1861–1870 when it was administratively separate from Otago. He was a member of the Legislative Council until his death in 1888, so he was in a position to contribute to high-level discussions of what might be done about the rabbit pest.

The archives of the Sheep and Cattle Branch of the Colonial Office preserve an interesting correspondence[24] set off in mid-1882 by a proposal from Menzies to R. Oliver, of the Legislative Council, stating:

> I think that if the Govt desires to establish in the Colony the natural enemies of the rabbit it could not find a more deadly one than the American Pine Marten ... The Scottish varieties are almost always trapped far away from habitation – on the moors or in rocky dens – if it was possible to get the marten from San Francisco it would keep down the rabbits in the high rough country where men could not act effectively and if it should ever become too numerous its fur would become a valuable article of export, and then its numbers would be kept within reasonable limits.

Oliver accordingly passed a memo to the Colonial Secretary, Thomas Dick, dated 16 July 1882:

> Dr Menzies has strongly advised the introduction of the marten from America and Scotland to aid in extirpating the rabbit pest. He says that in wooded and rocky places the marten would prove a very destructive enemy of the rabbit and that its fur is very valuable so that if it should increase through the plentiful supply of food it could in its turn be got rid of by trapping and other means.

Two days later, Bayly wrote on the file:

> The introduction of the sable or pine marten is a matter that has for some time engaged my attention. For the high wooded country in the South Island especially in the vicinity of the Lakes as well as towards the W coast I believe they would be invaluable and I would strongly recommend that this introduction be attempted. From their hardy nature as well as disposition to keep away from habitations I am under the impression they would be preferable to the ferret.

So Dick sent a formal request to F.D. Bell, the Agent-General in London, asking him 'to ascertain if the Pine Marten can be obtained at San Francisco, or sent out by a vessel from Glasgow'. His letter read:

> It has been represented to the Government that the marten of Scotland would be of much use in aiding to keep down the rabbits, particularly in the wooded portions of the Colony. I have accordingly to request you to be good enough to ascertain whether the marten could be obtained in any

quantities at what price per pair, whether it could be sent out [?by?] vessel; and generally to report whether it is likely that the introduction of this animal is possible and would be advantageous.

Bell replied on 5 October 1882:

I made enquiries in Scotland through Mr [illegible name] respecting the Marten, and I have been informed by him that the dealer who supplied the ferrets shipped last month from the Clyde states 'that the marten cat is almost extinct, that the price for one, if it could be obtained, would be five pounds (£5), and that they are not good rabbiters like ferrets.' I am making further enquiries here in London and will inform you of the result by next mail.

The file ends there, but probably only because the idea appeared expensive and impractical, not because Dick and Bayly guessed the futility of introducing a forest-dwelling, arboreal predator to control numbers of an open-country, burrowing pest.

The subject came up again four years later in the 1886 report of the Joint Committee on Rabbit and Sheep Acts. The Committee was taking evidence from witnesses, one of whom summarised the previous enquiries, including Bell's estimate of £5 a head for Scottish martens and his promise to provide further information. He reported that Bell did not write again and R. Creighton, the New Zealand agent in San Francisco, did not respond to Dick's enquiry. Dr. Menzies pointed out that the marten is getting very scarce in Scotland, but they are common in California, and Creighton ought to be able to get them without difficulty. The Committee resolved, on Menzies' motion, that our 'Agent in San Francisco be instructed to send pine martens'.[25]

When the Committee's resolution was reported in the press, the *Waikato Times* (based in Hamilton, North Island) responded:

It is to be fervently hoped the Government will confine their favours in this direction to the other Island … Martens will be even more destructive [to birds] than stoats and weasels, for the home of the former is among trees; where their food consists principally of birds. … They are many times larger than either the stoat or weasel; the fact that they were commonly called 'marten cats' will convey some idea of their size. … I have been told their skins were worth half a guinea each, but I fear each marten would do more damage than its skin would pay for even at that price.[26]

Fortunately, New Zealand never had to calculate the odds, because, so far as anyone knows, no martens ever arrived.

Patagonian Grey Fox

Donald McCaskill was a British settler in far southern Chile, who had heard about the plagues of rabbits in Australia and New Zealand, and about the prize of £25,000 offered by the New South Wales government in 1887 for an effective means of defeating them.[27] He was among several entrepreneurs to write to the Colonial Secretary in Wellington, offering to sell him a solution to the problem under the guise of a public service. His letter (undated, but received on 8 May 1889) was sent from Punta Arenas to the NZ Livestock Office in Wellington:

> I beg to recommend the use of a species of small animal found in this country, in shape and appearance resembling a fox, but about the size of a domestic cat [not named but probably the Patagonian or South American gray fox, averaging 2.5–5.5 kg]. These animals I know to be destructive to rabbits, and are harmless to lambs. I have carefully watched them and never knew of them attacking lambs … our yield of lambs, where said animals abound, is always 80–90%. If you agree to pay me £300 and a free passage to New Zealand, and send me boxes suitable for their conveyance, I will catch alive 80 or 90 of said animals, feed them … and attend to them during the voyage. I have been in New Zealand and I know something of the losses incurred by the rabbit plague and will be willing to do the country a service.[28]

The Colonial Secretary replied promptly on 28 May: 'Thank you … but the Stock Inspector does not see his way to take advantage of the proposal … and you omitted to name the animal you refer to'. Perhaps he already knew from his opposite number in Australia that so large a prize had stimulated >1000 proposals, of which none were judged worthy of it, so the offer had been withdrawn.

McCaskill had recommended the Patagonian fox for importation on the grounds that it was large enough to catch rabbits but too small to present any risk to lambs. He perhaps did not notice that small predators generally prefer small prey and the Patagonian fox is no exception. One recent study found that its diet consisted mainly of rodents and insects.[29] Unlike the mongoose, this fox is well adapted to a cool climate, so had it been introduced in sufficient numbers it might have become established. But like the common weasel, another cold-adapted specialist on small prey, the shortage of small rodents probably would have kept its numbers very low (Chap. 10), with zero benefit to New Zealand in both cases.

Scottish Wild Cat

T.G. Pudan of Glasgow was another public-spirited citizen concerned about the economic distress of runholders suffering the effects of the rabbit plague in New Zealand. He too saw an opportunity to help and, at the same time, to engage in a profitable business on his own account. He wrote to the Colonial Secretary from Scotland on 1 March 1888, asking, cautiously,

> are you prepared to entertain a proposal for the extermination of rabbits in New Zealand, and if so, on what terms? I have a method that does not involve the use of poison … and would not be noxious to man, horses, cattle, camels, sheep, swine, goats or dogs. I beg to enquire what methods have already been tested.[30]

Bayly replied, of behalf of the Colonial Secretary, on 25 April 1888, stating that 'no special sum is offered by the Colony but … any successful scheme … [would be] sure of ample reward'.

Pudan, anxious about his intellectual property, and probably also interested in the prize, decided to entrust Bayly with a confidential proposition. On 5 July 1888, he wrote:

> [I] acknowledge yours … and enquire if your Government would object to the introduction of wild cats for the purpose? It is well a known fact that these animals are the most destructive of ground game … killing 20 for every one they eat while being [?harmless] to domestic animals. A few hundreds let loose would soon clear the runs … the cats when no longer required for destroying rabbits would be a source of sport for the colonists … the idea is confined to myself and I trust the Government of the Colony may not make it public unless decided to test it … I belong to the Indian Civil Service and am at present on sick leave, due back to India in October … I'd expect compensation for loss of salary if I had to take time off to do this.

On top of the file is a memo from the Live Stock Office with a note from J. Shaw, perhaps Bayly's secretary, dated 8 October 1888: 'Hon Mr Richardson. Shall the writer be informed that the introduction of stoats and weasels is considered preferable to the wild cat'. Beside it is a note from the Colonial Secretary instructing his office to write to Pudan: 'Thanks for letter which shall be treated as confidential, and decline offer. 12 October 1888'.

Tasmanian Native Cat

In 1882, the *New Zealand Herald* heard that Sir George Grey had recently charged some members of the government with contemplating the introduction of the Tasmanian native cat [the eastern quoll] to help with destroying rabbits. An editorial commented that

> the Tasmanian cat, though small, is a fierce creature, and would be dangerous to children. The rabbit is a costly trouble, but it will not do after all to pay too high a price for his removal. ... Really it is time that our zoological experiments were placed under some system of special regulation.[31]

Another newspaper, the London *Graphic*, on reporting the shipment of 100 stoats and weasels from England to New Zealand (Chap. 10), asked: Why don't they bring in native cats from Australia? One obvious answer was that the native cats of Australia and Tasmania could do nothing with the rabbits even in their native country where they presumably had the advantage of long residence. So far as it is possible to tell, no quolls made the trans-Tasman trip.

Burrowing Owl and Little Owl

Helpful suggestions continued to arrive in the Colonial Secretary's mailbox long after the idea of introducing alien predators had proved disastrous. The long endurance of public interest in 'the natural enemy' (Chaps. 7 and 8) is illustrated by a late entry into the debate, concerning predatory birds rather than mammalian carnivores. This one, among the more unusual of such letters, came from S.F. Fraser, addressed to J.M. Ritchie, the chief inspector stock and dated 8 December 1892:

> [D]uring a residence on the Pacific coast of USA I observed the habits of a small species of owl which I believe could be introduced as a useful natural enemy to the rabbit and also to small birds. ... The bird in question preys during the day on the ground squirrel (or sage rat), rears its young in

> burrows under the ground … it is too small to injure lambs and in every respect would be perfectly harmless except as a destructant to young rabbits. Its haunts are the open prairie country and I think as a natural enemy it would prove very useful in the Otago and Canterbury districts. If imported it would have to be done through San Francisco by the mail service boats.[32]

Ritchie took the question seriously enough to write a letter on 23 March 1893 to R. Creighton, New Zealand's agent in San Francisco. He repeated the substance of Fraser's suggestion, and ended with the request that Creighton 'make enquiries and inform me if you think there is any likelihood of this bird being beneficial if introduced'.

Creighton's Californian contacts would have told him at once that Fraser was misinformed about the diet of burrowing owls. They do often live near ground squirrels and use their old burrows, but are too small to prey on them; rather, they live on insects and rodents smaller than ground squirrels.[33] Ritchie no doubt politely declined Fraser's suggestion.

Instead, in 1906 the Otago Acclimatisation Society sponsored the introduction of the little owl, a specialist in catching small birds. Many European species of finches and sparrows had been introduced by homesick colonists, and were soon causing their own form of nuisance in orchards and gardens. Within three years, the arrival of the owls pleased at least one fruit grower, Mr. Everson of Central Otago, who 'speaks highly of the little owls, which completely obviated the small birds pest'.

The Biggest Nasty Surprise: The Brushtail Possum

Brushtail possums are cat-sized climbing marsupials, with an appealing foxy face (hence their specific name *vulpecula*) and a long prehensile tail. They come in various colour forms, ranging from black through shades of brown to clear grey. Possums from Tasmania, the cooler end of Australia, carried the largest, darkest and thickest fur pelts, so comprised more than half of all the known imports; the smaller greys came from mainland Australia. Subsequent releases and frequent deliberate translocations of locally bred progeny helped to mix and spread the stock.

The Dunrobin station in Southland (Fig. 11.1) was run by the Basstian brothers from Tasmania. They were keen on the acclimatisation of useful animals (p. 256), and were familiar with the superior fur-bearing qualities of Tasmanian black possums, so it is hardly surprising to find that in 1858 they were responsible for the first successful release of brushtail possums in the forest behind Riverton. Over the next 40 years, especially during the decade after 1890, many more legal and illegal imports and assisted dispersal programmes were organised. Official government approval was granted to the acclimatisation societies to create a valuable industry, but by 1889 the Southland Acclimatisation Society began to fear that uncontrolled trapping for fur was destroying this important resource. They urged the colonial government to bring possums under the Protection of Animals Act 1880. The predictable effect was to accelerate unofficial efforts to spread possums far and wide.

Possums are browsers, that is, they feed on the leaves, flowers and fruit of trees, as opposed to grazing grass at ground level. Few of the native forest trees of New Zealand are equipped with the toxic compounds in their leaves which protect Australian trees from browsers, so possums in New Zealand made the most of their new opportunities to convert defenceless foliage into lots more possums, all doing the same. The effects on native vegetation were not too noticeable at first, so for a long time possums were still seen as a valuable resource, and protected by landowners against illegal theft in the same way as were livestock.

Settlers near the first possum release area behind Riverton began complaining about browsing damage by possums in orchards as early as 1891, although it is also possible that these complaints were really a disguised argument against restrictions on trapping. For the next 50 years, the conflict of interests between fur trappers and farmers produced a confused series of contradictory legislation, ranging from granting absolute protection to possums in forests in 1913 (which did not stop widespread poaching and export of pelts mislabelled as rabbit skins) to permitting harvest under licence in 1921.[34] But after 1922, the increasing evidence of browsing damage in native forest prompted the government to refuse permission for any new releases. That immediately set off a new round of illegal (unreported) imports by private individuals, which lasted until 1946. Then, as the evidence multiplied as fast as did the possums, the tide of public opinion turned, and possums made a sudden transition from asset to pest. All restrictions on possum trapping were revoked, and efforts to find more and better ways of killing possums have continued ever since.

Brushtail possums have several important advantages in New Zealand that their ancestors did not have in Australia. Not only do they have access to a more nutritive diet; they are also free of many parasites, competitors and predators which still help to limit the numbers of their ancestors. Consequently, the density of possums in New Zealand averages up to 20 times higher than in Australia.[35] That alone multiplies their potential to damage New Zealand native biota, in ways completely unpredictable from their numbers and feeding behaviour at home.

First, intensive possum browsing can quickly cause selective mortality of their preferred tree species, and hence a shift in the species composition of New Zealand native forests.[36] Many native podocarp forests show signs of canopy damage, even to still-standing, totally defoliated skeletons of once vigorous emergent giants.[37] Of course, canopy trees die for other reasons too, and the details in any one case are debatable, but the general effect is well recognised.

Second, while moving about in the canopy, possums often come across roosting or nesting birds, and/or compete with them for dry nest sites. Because possums are primarily browsers, no one connected the increasing numbers of possums with the declining reproductive success of forest birds, until the development of new techniques for routine video monitoring of nests. Then, in the early 1990s, Kerry Brown and his team caught them in the act.[38]

Suddenly, shocking video evidence of possums raiding birds' nests, crunching up the eggs and chewing the heads off the chicks (Fig. 11.3), added another species to the already-far-too-long list of predators of native fauna. Discovery of the unexpected predatory behaviour of possums certainly counts as among the most unpleasant surprises in the history of

Fig. 11.3 Brushtail possum destroying chicks in their nest. (Photograph by David Mudge)

wildlife conservation in New Zealand—and is all the more disastrous now because the very abundance of possums immediately catapulted them, with stoats and rats, to the top of the list of national targets for expensive predator control operations envisaged by Predator-Free New Zealand 2050 (Chap. 12).

Third, as if such extensive damage to forests and birds was not enough bad news already, possums have one more weapon up their furry sleeves. Unlike threats to conservation values, which are dearly held but hard to quantify in money, this one has an all-too easily definable economic cost. Since 1967, possums in New Zealand have become the main wildlife reservoir of the debilitating disease of cattle, bovine tuberculosis (bTB) (Chap. 12). For all these reasons, possums present a far greater threat to the biodiversity and economy of New Zealand than all the mustelids put together.

A Second Nasty Surprise: The European Hedgehog

Hedgehogs (Fig. 11.4) were, like possums, supposed to be harmless to all vertebrate life, and are much loved in the UK (Mrs. Tiggywinkle has a lot to answer for).[39] They were brought to New Zealand by homesick settlers, and also in the hope that they would control garden pests such as slugs and snails. Hedgehogs soon became enormously abundant—much more so than in the UK—but, for reasons unknown, their numbers in both countries have declined in recent years.[40]

Fig. 11.4 Adult European hedgehog. (Photographed in New Zealand by Pat Morris)

Hedgehogs are still common throughout the lowland districts of New Zealand, where they have settled into their role of invertebrate predators all too well, and not only in gardens. In North Island podocarp-broad-leaved forest, the prime habitat of kiwi, one study[41] estimated that roughly 660 g of invertebrates per hectare per night are destroyed by hedgehogs, compared with 9/ha/night by mice, 32/ha/night by possums and 39/ha/night by rats. Wherever hedgehogs are common, such a toll has implications for the food supply of kiwi and other native fauna depending on invertebrates.

Worse, hedgehogs eagerly destroy the eggs and chicks of ground-nesting birds. One five-year study on South Island braided river beds identified hedgehogs as responsible for 19% of the losses of eggs from 172 nests of banded dotterel, black stilt and black-fronted tern nests.[42] Other local studies have attributed even higher losses to hedgehogs, in part because they are often the most abundant predators on open country. An intensive three-year trapping operation to protect native fauna at Macraes Flat, Otago, removed 2294 hedgehogs, 635 feral cats, 423 ferrets, 242 stoats and 42 weasels from 2100 ha.[43] Working out what these figures mean for the native fauna depends on other information not always available, but certainly, the hedgehog is far from an innocent, prickly reminder of home.

Hedgehogs are among several reasons why ecologists and conservationists of the twenty-first century marvel at the diversity of natural enemies proposed to deal with introduced pests, ranging from the plausible to the absurd. The ideas came from well-intentioned people who, long before the development of invasion ecology as a science, struggled to understand why the behaviour of animals in their home countries turned out to be such an unreliable guide to how their descendants would behave in New Zealand. But at least we can be grateful to acknowledge that, from today's perspective, there are no regrets about the many predatory species that were *not* introduced in the hope that they might help to control rabbits.

The same logic helps us to sympathise with the settlers who were so innocently but fatally mistaken in hailing the arrival of the brushtail possum as a valuable resource, and the hedgehog as a harmless reminder of home. They lived in a world that had, as yet, no qualms about killing useful animals by the million, and no concern at all for native species. They had no way to predict that possums and hedgehogs would find life in New Zealand so much to their taste, or that their uncontrollable spread and huge numbers would cost future generations of New Zealanders (not only the human ones) so dearly.

The question to ponder is: How many equally well-meant strategies now being considered to tackle the pest problems of the twenty-first century (Chap. 12) might also produce seriously damaging unforseen results?

Summary

Decades of justifiable regret about the colonial government's rabbit control policies should not disguise relief at the surprising number of natural enemies proposed for introduction that were not officially approved. They include two species which are serious pests of wildlife in other countries. The red fox has been responsible for massive damage to small native marsupials in Australia, but was prohibited from New Zealand in 1867. One of the small Asian mongoose species was twice actually brought to New Zealand by private landholders, in numbers potentially sufficient for establishment. Neither propagule survived, and the government was strongly advised against trying again. Proposals to introduce the pine marten, Patagonian grey fox, Scottish wild cat, Australian eastern quoll and Californian burrowing owl were all rejected. By contrast, the brushtail possum, considered to be a harmless fur-bearing vegetarian, turned out to be a very unpleasant surprise. Possums, many times more abundant than in Australia, seriously damage native trees unused to browsing, destroy the eggs and chicks of native birds, and transmit bovine tuberculosis to cattle. The European hedgehog was another unexpected tragedy. Hedgehogs, now also very abundant, destroy huge numbers of endemic invertebrates needed by kiwi and the eggs of ground-nesting birds.

Notes

1. Espeut, W. B. (1882). On the acclimatisation of the Indian mungoos in Jamaica. Proceedings of the Zoological Society of London, 50(4), 712–714.
2. Midgley, M. (1985). Evolution as a religion. London: Methuen & Co. P. 2.
3. Espeut, W. B. (1882).
4. Cowan, P. E. (2005). Brushtail possum. In C. M. King (Ed.), The Handbook of New Zealand Mammals, 2nd edition (pp. 56–80). Melbourne: Oxford University Press. Third edition in press, Eds C.M. King and D.M. Forsyth. CSIRO Publishing, 2020.
5. For a full list of primary sources underlying this chapter, see King, C. M. (2017). Failed proposals to import the mongoose, pine marten, Patagonian

fox and other exotic predators into New Zealand. Journal of the Royal Society of New Zealand. doi: http://10.1080/03036758.2017.1389755
6. Lyttelton Times 25 October 1864. Town and Country.
7. Daily Southern Cross 12 February 1873.
8. Rolls, E. C. (1969). They All Ran Wild: The Story of Pests on the Land in Australia. Sydney: Angus and Robertson.
9. McDowall, R. M. (1994). Gamekeepers for the nation: the story of New Zealand's Acclimatisation Societies 1861–1990. Christchurch: Canterbury University Press. P. 355.
10. Press 17 May 1878, p. 2.
11. Southland Times 22 April 1880. Acclimatisation Society.
12. Southland Times 4 December 1876. Winton.
13. Southland Times 18 May 1877. Winton.
14. Hamilton, G. A. (1952). History of Northern Southland. Invercargill: Southland Times.
15. Hall-Jones, F. G. (1945). Historical Southland. Invercargill: H. & J. Smith for Southland Historical Committee.
16. Southland Times 7 May 1877. Importation of mongoose.
17. Southland Times 12 May 1877. The rabbit question.
18. Cuthbertson, R. J. (1877). Letter 13 March to CSO from Southland Pastoral Assn Wairakei. Wellington: Archives New Zealand R24266788.
19. Wairarapa Standard 8 April 1882. The mongoose vs the rabbit.
20. Ibid.
21. Marlborough Express 27 February 1884. The rabbit plague.
22. Multiple authors. (1888). Correspondence with Agent General re advice on importing the mongoose. Wellington: Archives New Zealand R1760 817.
23. Lodrick, D. O. (1982). Man and mongoose in Indian culture. Anthropos, 77, 191–214.
24. Agent-General. (1882). Forwarding information re Marten cat. Wellington: Archives New Zealand R24375311.
25. Anon. (1886). Report of the Joint Committee on Rabbit and Sheep Acts. Appendix to the Journal of the House of Representatives, 1886 I, I-05.
26. Waikato Times 10 July 1886. The rabbit nuisance...
27. Rolls, E. C. (1969). P. 149.
28. McCaskill, D. (1889). [Recommends introduction of Patagonian fox]. Wellington: Archives New Zealand R17620845.
29. Zapata, S. C., Travaini, A., Delibes, M., & Martinez-Peck, R. (2005). Food habits and resource partitioning between grey and culpeo foxes in southeastern Argentine Patagonia. Studies on Neotropical Fauna and Environment, 40, 97–103.
30. Pudan, T. G. (1888). [Proposal to introduce wild cats]. Wellington: Archives New Zealand R17 620 820.

31. [Editorial]. New Zealand Herald and Daily Southern Cross. 2 September 1882.
32. Fraser, S. F. (1892). [Recommends introduction of burrowing owl]. Wellington: Archives New Zealand R17620977.
33. Coulombe, H. N. (1971). Behavior and Population Ecology of the Burrowing Owl, *Speotyto cunicularia*, in the Imperial Valley of California. Condor, 73, 162–176.
34. Pracy, L. T. (1974). Introduction and liberation of the opossum (*Trichosurus vulpecula*) into New Zealand. New Zealand Forest Service, Information Series, 45, 1–28.
35. Cowan, P. C. (2005).
36. Campbell, D. J. (1984). The vascular flora of the DSIR study area lower Orongorongo Valley, Wellington, New Zealand. NZ Journal of Botany, 22, 223–270.
37. Holland, E. P., Gormley, A. M., & Pech, R. P. (2016). Species- and site-specific impacts of an invasive herbivore on tree survival in mixed forests. Ecology and Evolution, 6, 1954–1966.
38. Brown, K., Innes, J., & Shorten, R. (1993). Evidence that possums prey on and scavenge birds' eggs, birds and mammals. Notornis, 40, 169–177.
39. Morris, P. (2018). Hedgehog. London: HarperCollins.
40. Brockie, R. E., Sadleir, R. M. F. S., & Linklater, W. L. (2009). Long-term wildlife road-kill counts in New Zealand. New Zealand Journal of Zoology, 36(2), 123–134.
41. Innes, J., Kelly, D., Overton, J., & Gillies, C. (2010). Predation and other factors currently limiting New Zealand forest birds. New Zealand Journal of Ecology, 34(1), 86–114. Table 6.
42. Sanders, M. D., & Maloney, R. F. (2002). Causes of mortality at nests of ground-nesting birds in the Upper Waitaki Basin, South Island, New Zealand: a 5-year video study. Biological Conservation, 106, 225–236.
43. Reardon, J. T., Whitmore, N., Holmes, K. M., Judd, L. M., Hutcheon, A. D., Norbury, G., & Mackenzie, D. I. (2012). Predator control allows critically endangered lizards to recover on mainland New Zealand. New Zealand Journal of Ecology, 36, 141–150.

PART IV

What Now?

CHAPTER 12

Hard Lessons in Pest Management (1976–2019)

> [The settlers] reasoned from visible events to invisible processes. Killing individuals reduced their numbers; killing enough of them would get rid of the problem. ... They spoke of conquest, but they learned to obey. Thomas Dunlap (1999)[1]

The Importance of History

The historian Paul Star summarised the double significance of environmental history as a tool that helps not only those who try to understand the past, but also those responsible for managing the future.[2] Nowhere does that idea apply more cogently than to the current official aspiration, introduced in the Preface (p. xvi), to make New Zealand predator-free by 2050.[3] Two of the three invasive species specifically targeted by this plan (the stoat and the brushtail possum; Chaps. 9, 10 and 11) were established deliberately, repeatedly and after many failures. Yet, now that they are so successful, they have created the very opposite of the economic benefits they were confidently expected to bring (So also did the wild rabbit (Chap. 7)). The third named group, the rats (Chap. 4), arrived without permission or encouragement, but could not have been prevented.

C. M. King, *Invasive Predators in New Zealand*,
Palgrave Studies in World Environmental History,
https://doi.org/10.1007/978-3-030-32138-3_12

So the time is right for a closer look at how these surprising contradictions evolved, not in isolation but as part of the all-important bigger world picture. The named villains are only a subset of the 25 species of introduced mammals regarded as pests in New Zealand.[4] Meanwhile, at least 50% of New Zealand's breeding bird species have already been lost, and 60% of the survivors are critically threatened.[5] How did this sad situation arise? And what can we learn from it?

In 1976, I attended a large meeting in Wellington, at which we discussed the problem of how to understand and manage the problem of rat damage to native fauna in New Zealand's nature reserves.[6] Most working scientists of the time interested in rodent biology and protection of wildlife were there, including those who had watched helplessly while ship rats swept away one of the last surviving intact remnants of our pre-human fauna from Big South Cape Island (Chap. 4). I was about to complete my first five-year survey of the biology of stoats in all ten of the then national parks, and John Innes was just finishing his master's thesis on ship rats. The group of participants made an exhaustive review of what was then known about the effects of introduced predators in nature reserves, especially but not only rats on offshore islands. The original publication reporting the proceedings of the meeting is not widely available, but William Stolzenburg gives an account of it in his book, *Rat Island*, and adds a perceptive review of subsequent events.[7]

Two experienced government scientists, Ian Atkinson and Brian Bell, kicked off the first day by presenting devastating first-hand evidence confirming that the arrival of rats on an island heralds certain doom for naïve endemic birds. The logical prediction, that more extinctions must follow if island birds cannot be protected from rats, was challenged by two of the most senior scientists present, Kazimierz Wodzicki (founder of the Department of Scientific and Industrial Research's Ecology Division) and Sir Robert Falla (founding president of the Ornithological Society of New Zealand). Their reasoning was based on Paul Errington's theory that predators take only the 'doomed surplus' from prey populations, as explained in Chap. 4. Objective data supplied by Tony Whitaker and Graham Ramsay on the effects of rats on formerly abundant populations of tuatara and we¯ta¯ showed clearly why predators invading islands behave differently to those that Errington studied. They only prompted more sceptical responses from the entrenched positions.

On the second day, the group discussed available methods for managing rats and ended with the chairman John Yaldwyn making what William Stolzenburg calls 'an infamous statement of surrender', that 'the possibility for complete extermination of New Zealand rodent populations by conventional control methods, even on islands, must be considered remote. Nothing that has been said this afternoon ... would make me think differently'. This opinion was pretty much unanimous at the time and endorsed by reputable scientists for years afterwards.

That conclusion was certainly true then, except for some unintended eradications that followed other conservation work, but it had to be updated after the invention of second-generation toxins such as brodifacoum in the early 1980s. Brodifacoum is a slow-acting toxin, which defeats the cautious taste-a-bit-to-see approach of rats to new foods, by causing no immediate ill effects. The rat concludes that this new stuff is pretty good and returns to take more; soon, it has taken a lethal dose. Over five years 1981–86, the first six successful eradications of rats using brodifacoum (brand name Talon) were achieved on islands ranging from 2 to 21 ha,[8] but bigger islands were still considered far out of reach.

Breaksea Island

Two far-sighted scientists of DSIR Ecology Division, Rowley Taylor and Bruce Thomas, had been dismayed by the pessimistic outcome of the 1976 meeting. But they saw how the development of brodifacoum, if allied with ultra-precise bait distribution, might make a significant difference. They were the first to figure out that those early 1981–86 results could be repeated on much larger islands, if only brodifacoum could be distributed so rapidly and thoroughly that every rat's home range got a share of it. There was no evidence to justify any such attempt, since it had never been tried before, but they planned to do it anyway.

To prove their point, they chose Breaksea Island, a 170 ha, steep-sided, windswept fragment of rock 2 km off the southwest Fiordland coast, infested for decades by Norway rats. First, they cut a measured network of foot tracks in loops around contours of the island, strung climbing ropes up and down its inhospitable cliffs, and established 743 bait stations every 50 m on every surface, horizontal or vertical. Then, in May 1988 they filled every station with brodifacoum tablets, and checked them daily. As fast as the rats stole the baits and carried them back to their nests, so the team renewed the supply. As each tunnel fell silent, they marked its position on a map. Within three weeks, the last of the rats was dead.[9]

Almost immediately, Fiordland skinks began to move back from their refuges on offshore rocky stacks back on to Breaksea Island; beetles, wētā and spiders thrived in previously impossible numbers; and in due course, threatened native species including South Island saddlebacks, mōhua (yellowhead) and knobbled weevils recolonised, or had been reintroduced there. Fiordland crested penguins are expanding their long-established breeding colonies along the island's coast.

The story of Breaksea Island is one of several that marked a turning point in attitudes to conservation management on islands during the 1980s.[10] It is a classic example of why conservationists, faced with apparently insurmountable odds, should never despair, for reasons best summarised by Arthur C. Clarke, the author of dozens of science fiction novels, including *2001: A Space Odyssey*.

Clarke was not only a gifted writer, but also a rigorously realistic and perceptive, future-oriented scientist. In his fictional writing, Clarke postulated advanced technologies without resorting to flawed engineering concepts or explanations grounded in incorrect science or technology. Further, he avoided taking cues from current trends in research and engineering, which will eventually be outdated. Rather, Clarke based his stories on ultra-advanced technologies limited only by fundamental science.

Clarke formulated three Laws of Prediction, which are:

1. When a distinguished but elderly scientist states that something is possible, he is almost certainly right. When he states that something is impossible, he is very probably wrong.
2. The only way of discovering the limits of the possible is to venture a little way past them into the impossible.
3. Any sufficiently advanced technology is indistinguishable from magic.

The effect of Clarke's first law is amply demonstrated by the decades-long history of rodent eradications on New Zealand islands as it unfolded after that critical 1976 meeting. History has shown that Clarke's first law hit the nail on the head: when established scientists declare that something is impossible, there is a strong chance that they are wrong.

His second law has been amply confirmed by visionary conservationists who have, against all advice from more cautious heads, often successfully challenged the conclusion of the 1976 meeting—starting with Breaksea Island in 1988. Between then and 2010, nearly all uninhabited islands under Department of Conservation (DOC) control have been cleared: 147 populations of 13 species of pest vertebrates have been eradicated from 95 islands, totalling 32,000 ha.[11] The first and easiest stages of the national campaign against conservation pests on islands managed by DOC is effectively complete[12] (excluding the Fiordland islands from which it is not yet possible to eradicate stoats long term; p. 297). Some of these programmes against rodents and cats on islands have been reviewed in Chaps. 4, 5 and 6.

The truth of Clarke's third law will be tested in the future rather than in 1976 or in 1988. For any ordinary person without special training, advanced technology indistinguishable from magic already surrounds us in the modern world. I don't know how smart phones work, but I use mine all the time, grateful that the world is not limited by what I can understand. If he were still here, Clarke would be telling us that, if such magical technology is theoretically attainable, all we need to attain it is the political/social will, and lots of money. Is it possible that the new Predator-Free New Zealand 2050 (PFNZ2050) programme could supply it? Not now, but maybe in the future? Sir Paul Callaghan, visionary founder of New Zealand's innovation agency Callaghan Innovation, likened the idea to the US moon shot: in both, success depended on technology that had not been developed at the time that the project was announced.[13]

From Island Eradications to a National Biodiversity Strategy

Suppose Clarke's third prediction is also right and we have, by 2050, developed that longed-for magical technology, unimaginable now in the same way that clearing uninhabited Breaksea of Norway rats—and even more amazing, subantarctic Campbell Island—was unimaginable to those at the 1976 meeting. Suppose further that this magical technology was economically practicable and without unacceptable side effects, making it feasible to envisage large-scale eradication of small mammal pests on all, or as least critical chunks of, the inhabited mainland.

That is what PFNZ2050 envisages, and it is a whole new ball game, because extending landscape-scale operations to inhabited country depends not only on new technology but also on social licence—the practical and political input of the New Zealand population. Learning to play that game will provide the essential skill we need to seize the opportunity to go beyond the existing evidence and to discover the limits of the possible by venturing a little way past it into the impossible.

The impetus behind PFNZ2050 is energised largely by the way repeatedly successful and well-publicised predator eradications on islands have engendered a deep-seated public desire to extend the same conservation benefits on the mainland. It is strongly encouraged by a dawning realisation that this generation may be the last to 'have a shot at saving the bird species we love before they're exterminated by invasive predators'.[14] But to succeed, we have to recognise that these are not the only environmental changes our native species have to contend with. That means broadening our horizons, both conceptually and geographically.

First, we need to look beyond a common focus on removing predators to the wider concept of restoring biodiversity, that is, not only protecting our threatened species from sudden death, but also reviving habitats and ecosystem processes so as to provide them safe places to *live*.[15]

> Although introduced mammalian predators are one of the gravest threats to the animal subset of New Zealand's native biodiversity, there are many other invasive species, and conservation threats (e.g. freshwater quality, land clearance and climate change), which must also be attended for New Zealand to achieve the sustainable biodiversity goals it aspires to. So, although advocating for greater research on introduced mammalian predator management in inhabited landscapes of New Zealand, we caution that this must be balanced against the priorities of other research investments.[16]

Second, the aspiration to look beyond uninhabited offshore islands to much vaster landscapes, not only the huge areas of rugged back country, but also to the deeply modified lowlands populated with people and domestic stock, presents an unprecedented challenge. The time is long past when official bodies such as DOC and Regional Councils can field enough effort to make a difference on their own. So PFNZ2050 depends on integrating the private endeavours of community conservation groups into a coherent action plan complete with sources of funding and expertise. Some of these groups are very successful—can we simply join them all up? An independent report on PFNZ2050 makes some sensible recommendations:

1. Establish a national and regionally linked institution that will provide visibility, strategic advice and practical support to community conservationists including landowners.
2. Develop a national strategic conservation plan to coalesce and prioritise conservation effort across all players and places.
3. Align public funding of conservation activities with conservation need, to maximise the difference made by that investment.
4. Enhance the funding system by reorienting allocation and distribution to focus more stringently on outcomes and streamlined processes to reduce transaction costs while enhancing accountability for outcomes.[17]

PFNZ2050's plans to extend past successes in uninhabited areas to a future national elimination of rats, mustelids and possums[18] will have to be even more ambitious than was Bayly's campaign against rabbits. Our main advantage is that we can be informed by long experience of past campaigns, both successful and unsuccessful, and by the rapid advance of new developments in more and better ways of killing individual pests and dealing with their compensatory responses. More efficient lethal methods can in theory prevent short-term damage to threatened species, and are potentially important despite their drawbacks[19] because they can buy us time to protect what remains while we search for something better. While we wait, what can history tell us about the pitfalls on the way? Why is it so essential to remember that nature is on the side of the pests? For example, lethal operations often prompt compensatory increases in reproductive rate and reductions in natural mortality of the survivors, and changes in responses of other predators and prey in the same animal community. In particular, the history of New Zealand's long campaigns against rabbits warns us clearly that killing individual pests cannot be enough on its own, however well organised and funded. Why is it so important to take account of how and why New Zealand is a totally different country from anywhere else in the world?

Community Interactions

The New Zealand landscape is dominated by ecological communities which have no equivalent elsewhere. First, the living native species of today are a mixture of very ancient lineages that evolved in a totally different world, thrown together with a long series of new arrivals, almost all non-mammalian (Chaps. 1 and 2). The modern resident land-breeding mammalian fauna of New Zealand is an artificial collection gathered, not quite randomly but certainly without reference to any known rules of ecological assembly identified by northern hemisphere ecologists. It includes a higher proportion of introduced species than that of any other country. Many of them belonged, as Darwin saw (Chap. 4), to 'the advancing legions of plants and animals introduced from Europe', but not only from there. As acclimatisation progressed, selected species of game mammals came from Asia, such as Himalayan tahr, axis, sika and sambar deer,[20] plus hundreds of plants from India[21] and China.[22] The importations were well intended, but proceeded without knowledge of or concern for the potentially disastrous results. The most significant consequences of our unusual mixture of mammals are also the least expected. Here are three of them.

Shortage of Small Mammals

First, the determination of the acclimatisation societies to bring in useful game animals and livestock focused their attention on larger game or domesticated species (deer, wallabies, chamois, tahr, cattle, sheep, horses, dogs), so less than a quarter of New Zealand's resident land-breeding species are small (<1 kg average body weight). By contrast, the terrestrial mammalian faunas assembled by natural processes are dominated by small species. In Britain, for example, almost two-thirds of the resident land-breeding species are small by that definition. There they form the bases of the pyramid of numbers of small mammals, and are the favourite prey of many predators. The various species of voles, wood mice and lemmings teem through the grasslands of northern countries but were never considered important, so are simply absent in New Zealand.

Predators everywhere kill many birds and take their eggs, but in the northern hemisphere the populations of small mustelids are normally limited by their staple mammalian prey, especially field mice, lemmings and voles.[23] Stoats and weasels much prefer these when they are abundant, so in good years for voles and wood mice, nesting birds are practically ignored. Only when these primary prey are scarce do hungry predators switch to whatever else they can catch, and then the birds suffer. For example, one long-running study of small birds nesting in boxes in Wytham Wood, outside Oxford, has observed their nesting success since the late 1940s. In one subset of 13 years of that study, both the proportion of boxes raided by weasels and a standardised density index for woodland rodents could be compared. Ewan Dunn found a significant correlation between weasel predation on eggs and chicks in nest boxes (>50% in 1964; thereafter 11–39%/yr) and the rodent index. From 1976, when the nest boxes were made predator-proof, and throughout a subsequent long period of low numbers of woodland rodents, nest losses dropped to <5%.[24]

When stoats and weasels arrived in New Zealand forests during the late nineteenth century, the limited number of their normal small mammal prey species had unexpected and tragic consequences. *Faute de mieux*, stoats here have to depend largely on rats, rabbits and possums for the bulk of their energy needs. Wild house mice are smaller, less meaty and much less abundant than voles, so are only an unreliable and inadequate substitute and cannot offer anything like an equivalent to the buffering effect once offered by native rodents in the Wytham study. The effects on New Zealand biota after a mast year, a heavy fall of forest seeds, are often very serious.

Masting Cycles

The extensive native forests of southern beeches, and of podocarps and mixed hardwoods, have mutually exclusive geographical distributions across the main islands, but both go through masting (seedfall) cycles much like those of northern oak forests. Heavy seed falls used to benefit the native birds, by supplying welcome pulses of extra food; some, like the kākāpō, cannot breed in non-masting years.[25] But masting cycles now benefit rodents and stoats more than birds.

The normal average post-breeding density of stoats in New Zealand is <2–3/km², while that of ship rats, at least in the North Island bush in late summer, can often be 400/km² or more. Very rarely, and only briefly after a heavy seedfall, stoats can reach a maximum of 10/km², while rats and mice can fill traps faster than they can be cleared. When rats are few, for example, when their numbers are knocked down by successful rat-baiting operations, stoats eat more forest birds,[26] even though mouse numbers usually increase when rats disappear. The surviving birds benefit from removal of stoats by secondary poisoning.

These population irruptions of ship rats and wild house mice are smaller and shorter lived than those observed in northern hemisphere voles and lemmings, but are equally significant drivers of mass predation events. The rapid increases in mice and rats in response to the sudden banquet, followed by stoats doing the same, make a post-seedfall year a critical danger period for native biota. Whole cohorts of juvenile kiwi can be wiped out, even in areas protected by trapping. In the 10,000 ha Okarito kiwi reserve, all 14 radio-tracked kiwi chicks were killed between September and December of the 2002–03 breeding season, even though >800 rats were killed in September and >200 stoats in January alone.[27]

Mice and rats disappear too quickly to support high numbers of stoats for long, and even peak populations of mice seldom approach the numbers required to distract a larger than usual population of stoats from eating roughly the usual number of birds per head.[28] So for at least the last 150 years, the simultaneous short-lived irruptions of rodents and stoats in NZ beech forests have *added to* the losses of forest birds during the post-seedfall breeding season, rather than buffering them as is usual in the Northern Hemisphere. These periods of drastic danger were presumably unknown before wild house mice, ship rats and stoats arrived in the forests. Previously, the sudden pulses of extra food benefitted only the native fauna.

Competitors and Enemies

Predators that make life dangerous for small mustelids in Eurasia, especially falcons, buzzards, large owls, foxes and American mink, are few or absent from New Zealand, although feral cats are everywhere. The native Australasian harrier hawks are common in open country, but are primarily scavengers. Along with differences in prey resources and parasites, this unfamiliar distribution of enemies certainly had immediate consequences for the first mustelids released, affecting their chances of settlement and successful breeding, differently for each of the three species.

Ferrets still live mainly in rabbit habitat, and when rabbits are abundant, they are by far the most common of the three mustelids there, although not the most common predators. That honour belongs to the hedgehogs (Chap. 11) and feral cats (Chap. 6), which are removed by the hundreds every year from the protected nesting areas of native birds.[29] By contrast, stoats and weasels are rare in open country, because neither can compete with ferrets for rabbits, and both are small enough to be vulnerable to attack by the resident ferrets, cats and hawks. Stoats avoid cats and ferrets, even at the cost of going hungry.[30] On the other hand, stoats can beat up weasels, which helps to keep weasels scarce except where stoats have been removed to protect kiwi.

The stoats' key advantage is that they are very agile climbers, and can readily find food and cover in forests, both on the ground and in the canopy.[31] With sharp claws, light weight and good balance, stoats can race up a tree trunk, out along small branches and down again head first. These skills give them access to arboreal resources well out of reach of ferrets, and which are less easily available to cats and harriers. During post-seedfall irruptions, they can destroy many native birds, lizards and insects in the trees, as well as kiwi chicks on the ground.

Ship rats are vastly more abundant than are stoats, so have a proportionately greater chance of being first to encounter a bird's nest. Stoats range widely over much larger areas than rats, but can kill only so many rats a day, always far fewer than the number that can be replaced, so stoats have no chance of preventing rat numbers from escalating, or of bringing them down from their peak, until the rats stop breeding. Stoats do kill lots of rats, but removing stoats makes no difference to rat numbers. On the other hand, knocking down the numbers of stoats on the Fiordland coast is allowing more frequent invasions of Secretary Island by rats.[32] So the subtle links between members of the artificial community of small predatory mammals in New Zealand often have unexpected consequences.[33]

Ship rats kill mice whenever they get the chance, so removing rats predictably benefits mice.[34] Plagues of mice are a common problem inside fenced sanctuaries which have successfully excluded rats, and mice at such numbers are a honeypot for weasels. Mice can also destroy large numbers of the ground invertebrates needed by kiwi. Hedgehogs do the same even at normal density.[35] Ship rats and possums together do the worst damage to forest fauna, but removing possums tends to be followed by more rats.[36]

So, contrary to common assumptions, effective predator management to sustain native fauna must not just remove the worst three species targeted by PFNZ2050 (ship rats, stoats and possums) where that is possible. Rather, it must consider the potential results, for the whole interconnected community of small mammal predators and their prey, of any successful attempt to manage particular bits of it (Fig. 12.1).

Fig. 12.1 Small predators and their prey belong to an interconnected community. Targetting possums, rats and stoats, as envisaged by PFNZ2050, should substantially improve life for native birds in forests, even though all habitats are still at risk from introduced herbivores, while cats, ferrets and hedgehogs line up with teeth at the ready. (Cartoon by S. Marks, courtesy Landcare Research.Co.NZ. From Kararehe Kino, Landcare Research, December 2009, p. 16)

At present, it is impossible to attack the whole community of invasive mammals over one very large area except by concerted community effort, guided by spatially explicit modelling, as in the Cape to City programme covering 26,000 ha in Hawke's Bay.[37] That does not mean that doing the same on an even larger scale will never be possible, because, fortunately, new technologies are developing rapidly. *More* important, we must ask: Do we sufficiently understand our species mix well enough to ensure that our actions will provide the outcomes we want? And when can we be confident enough of our knowledge to press ahead with minimum risk of making a bad situation worse?

Climate Change

Above and beyond all detailed arguments about the problems of the immediate present loom the huge and largely unknown consequences of future climate change. Some ecological interactions are already known to be linked to climatic variables, so changes in them are easy to guess. Masting events (massive synchronised seedfalls) in forests are coordinated by climatic clues linked to summer temperatures, with wide-ranging effects on forest ecology (extra food for seed-eating birds and rodents, but also many extra bird-eating predators). More frequent masting events could be the last straw for some forest bird populations.

Mild winters can reduce cold-weather natural mortality of rats and rabbits, and reduce opportunities for baiting operations in winter (the best time). Sea-level changes would disrupt coastal breeding sites of penguins, seals and sea lions. Warmer conditions would favour range expansions of predators, especially ship rats. Rising tree lines would diminish the total areas of cool refuges above the continuously high predation pressure within most forests, creating a 'thermal squeeze'[38] which could push alpine specialists like the rock wren and kea out of their only habitat. The latitudinal ranges of climate-sensitive species could move northwards, wherever any habitat is left to them. Future funding agencies (local and global) may have to choose between projects concerned with preserving the past (by preventing otherwise inevitable extinctions) as opposed to those focused on the future (by protecting breeding stock and restoring natural habitats). We need to divide our attention between immediate crises and longer-term strategies, without losing sight of either.

Understanding Predators

In general, the numbers of animal populations are most strongly influenced by either (1) top-down processes, which vary the rate of removing existing individuals, or (2) bottom-up processes, which vary the rate of adding new individuals. Manipulating top-down processes to manage a population governed from the bottom-up can seldom produce the intended effect unless the removals exceed the additions.

The effects of a predator on the numbers of its prey, and of a trapper on the numbers of a pest, are the same in principle; both are external agents of mortality, acting as one among several other top-down agents also removing individuals from a population. The main general rule governing the consequences is simple: Can the predator/trapper kill target individuals faster than they can be replaced by breeding or immigration, allowing for other causes of loss? If so, the effect is to *add to* other agents of natural mortality, and the target population will decline unless it can increase its fertility rate enough to compensate. If not, the effect is merely to *substitute for* other agents, and the pest population is merely being sustainably harvested. That is why we still have rabbits, rats, mice and possums despite our persistent efforts to kill them off.

Ferrets and stoats as individuals are classic top-down agents of mortality, whereas rabbit populations are governed largely by variation in productivity controlled from the bottom-up by habitat, fertility and juvenile survival. If ferrets can kill rabbits fast enough, they can impose a new top-down effect on top of the normal bottom-up effects linked mainly to the rabbits' resources and fertility. Unless these equations can be estimated beforehand, it is generally impossible to predict the outcome of any predator–prey relationship or trapping programme just by watching it.

The problem is that watching individuals interacting with each other is all a field observer can see, but *visible events* cannot by themselves tell us anything about the *invisible processes* determining the outcome. The old idea of a 'balance of nature' is an acknowledgement that there are ecological rules; it is not an explanation of them.[39]

Watch the Rules of the Game, Not the Players

Thomas Dunlap, a leading US environmental historian, invites us to consider an analogy in which animals are seen as players in a card game. All the players have to play the hand the dealer gave them, as well as they can within the rules of the game. An observer cannot tell from looking at a single hand what those rules are, or how to guess the winner. Bayly's confidence in his strategy was based on the simple view of nature which he and his generation considered obvious, but it did not help them understand why some of their policies worked and some were ruinous disasters. On the other hand, an observer could deduce the rules of a card game by watching several hands played through a long-enough series of rounds, and then might be able to predict which player has the best chance of winning in that particular ecological landscape. That is the difference made to pest management strategy by organised experimental science, replicate sampling, statistical analyses and computer modelling.

The key information that was never available to Bayly is that the effect of a predator on its prey can never be entirely predictable, because it is a matter of numbers (added to and removed from the prey population) and relative rates (of fertility and mortality on both sides), which are all highly variable with habitat and season, among many other parameters. In the naïve form of this apparently simple equation, multiplication usually beats subtraction, but the results are often more complicated in the real world, especially in the diverse and newly assembled animal communities of New Zealand.

One of the sharpest nineteenth-century observers, Richard Henry, poured scorn on claims that ferrets, stoats and weasels were or could ever have been successful in controlling rabbits. As he put it:

> we brought the stoat from the same place as the rabbit, where they lived and throved together for thousands of years … A rabbit that never saw a ferret … seems to take a ferret's measure in a moment … its stratagem has ruined the hopes of the lords of creation, the Chief Inspector for instance … [his report says] 'shortly after the weasels' release, a diminution of rabbits in the neighbourhood was apparent'. Mr Chief Inspector [concluded] 'The weasels are efficient' … [In reality] Bunny undoubtedly disappears from a locality where there is anything objectionable, and he may well be proud of so easily befooling us.[40]

No, Henry concluded, the real natural enemy of any animal is that which removes its food. The rabbit saga illustrates this idea, and adds the further complication that in stable, long-term co-adapted ecosystems, habitat conditions affect the reproductive capabilities of both parties. Rabbits can take food from the mouths of sheep and easily outbreed them, but mustelids cannot do the same for rabbits. This logic explains why the rabbit can be a devastating natural enemy of the sheep, but mustelids are not the primary natural enemies of the rabbit.

Henry was a hundred years ahead of his time in anticipating the conclusions of contemporary community ecology, showing that predators generally can only follow, not reduce, the numbers of fast-breeding prey. Predation by mustelids (and cats) falls most heavily on young rabbits, and if high numbers of predators are present early in spring, when young rabbits first appear, they can have a substantial effect on a rabbit population. This is a critical point, because long experience and careful, controlled research shows that rates of change in rabbit populations are most sensitive to variations in juvenile mortality. Unfortunately for the supporters of mustelids as agents of biological control, this is likely only under two conditions.

First, where rabbits breed almost year round, they can provide a near-continuous supply of young to maintain predator populations over winter. This condition is now met over much of the North Island, but not in the South Island high country, where the rabbit breeding season is short (September to January), the winter supply of young is sparse, and the numbers of mustelids and cats decline from autumn until late spring (November).[41]

Second, where predator numbers are artificially supplemented before rabbits start breeding in spring, they can be ready in good numbers to attack the young rabbits as soon as they emerge.[42] During the nineteenth century, this condition was sometimes met by repeated, well-timed liberations of locally bred ferrets and cats in places where poisoning had already reduced rabbit numbers (Fig. 12.2).

Neither condition could be met by the imported stoats and weasels, which were always liberated where rabbits were already hugely abundant, and in numbers too small to achieve a significant effect. Hence, the era of deliberate introductions of stoats and weasels to control rabbits in New Zealand was not only a tragic misinterpretation of how animal numbers actually work, but also short, expensive, and mostly unsuccessful.

The Importance of Modelling

Real understanding of this complex topic depends on questioning old assumptions, and venturing into the more abstract field of statistical and theoretical modelling. That in turn means watching not individuals but numbers, assembled under the iron-clad rules of representative population sampling and analyses. Manaaki Whenua Landcare Research (MWLR) is one of several research organisations to take up this challenge. Its regular newsletters keep stakeholders informed on progress in the vital and fast-moving field of pest management ecology.

In March 2019, MWLR published a special issue on modelling in ecology, full of current information on how quantitative models are used in animal pest research.[43] For example, statistical models can determine whether a management programme was successful, and if not, what it would take to improve its chances. Other models are used to predict the power of a planned study to answer the question of interest by estimating the minimum number of field samples needed, so as to avoid wasting resources on inadequate efforts. Theoretical models, although necessarily simplified, can disentangle the community and environmental interactions that influence a given result, for example, to decide whether a proposed eradication can be declared complete, or to estimate whether it might be helped or hindered by future climate change.[44]

One such study was set up by Dean Anderson and his colleagues to ask why it has been impossible, so far, to eradicate stoats from Resolution Island (p. 298). The reasons identified by their model include a highly fecund population, insufficiently attractive traps and a substantial proportion of the population that refuse to enter traps. It concluded that control-to-zero density was feasible, but required greater than a twofold budget increase to intensify the trapping rate relative to population growth.[45]

Generalisations, or, worse, actual policies, which are based on anything less than the rigorous analysis methods now available, may in some places at some times be right by chance. If not, they can lead to catastrophe, with no obvious explanation for the difference. That is why we should have some sympathy for Benjamin Bayly and his colleagues. They had nothing to guide their decisions except simple observation, and they assumed that predation was the only factor controlling rabbit numbers instead of one of many. They found a bottom-up system controlled by food supplies in ways they could not see, and tried to turn it into a top-down one controlled by predators killing rabbits as they watched. It took the advent of twentieth-century population ecology to reduce the risk of repeating their mistakes. The transition has been messy, and is still incomplete in the popular mind.

Biological Control

Bayly's operation was among the earliest attempts to establish a biological control regime, defined as a direct human manipulation of one useful species (an agent or natural enemy) in order to achieve an impact on another, unwelcome species.

One of the most famous successful examples came in 1887, when Californian citrus orchards were almost ruined by the cushiony cotton scale insect. Within a year of the deliberate 1888 introduction of vedalia beetles (predatory ladybirds), both scale insects and beetles were reduced to small numbers of individuals playing hide-and-seek among the orchards.[46]

In that same year of 1888, 1453 wild stoats and weasels were imported into New Zealand (Table 10.1) and organised breeding of domestic ferrets in local studs was producing offspring for release by the tens of thousands (Chap. 8), but no such immediate effect on rabbits was achieved. New Zealand authorities may not have known of the startling success of ladybirds in California, but even if they did, the stark differences between vertebrate and invertebrate predators as biological control agents would have made any hopeful comparison with their own action quite invalid.

These two examples illustrate why, although biological control has long been assumed to be better (both cheaper and more effective) than direct human action, and natural enemies have often been promoted to be the best agents of it, it is also a very risky strategy. A successful operation of this kind requires that predator and prey should have matching population dynamics, dispersal capabilities and operational areas.[47] Even where these conditions can be met, the details of the outcome will be determined by the specific characters of the imported agent and the way it interacts with the target pest.

Ladybirds can match or exceed the reproductive and dispersal rates of scale insects. Their aerial searching and wide dispersal are rapid and efficient; they can kill without risk and have few other prey as attractive; scale insects clump together and have no defences except high fertility, so every individual found by a ladybird is easily killed. When ladybirds are surrounded by more scale insects than they can eat, they simply produce more ladybirds, allowing the prey minimum recovery time.

By contrast, no mustelids can match the breeding rate or running speed of rabbits. Searching on foot is slow and inefficient compared with flying; rabbits can hide in and escape within complex burrow systems and rapidly reinvade a cleared area; they are alert and can fight back, so are risky to kill compared with other equally acceptable native prey, including lizards and birds; mustelids tend to stop to eat after killing one grown rabbit or a nest of infants, and, if no more are in sight, usually prefer to digest their meal in peace and resume hunting only when necessary—meanwhile, the surviving rabbits are left to multiply.

So, most importantly, mustelids do not have the inbuilt responses of a successful control agent, as required to ensure their predatory effects match increases in the target populations. Ladybirds do this naturally, and so do other insects introduced into New Zealand as successful biocontrol agents against weeds, but mustelids can do it only when conditions are already against the rabbits (Fig. 12.2).

Why have rats and mustelids made almost no difference to rabbit numbers, but, at the same time, have had devastating effects on native birds?

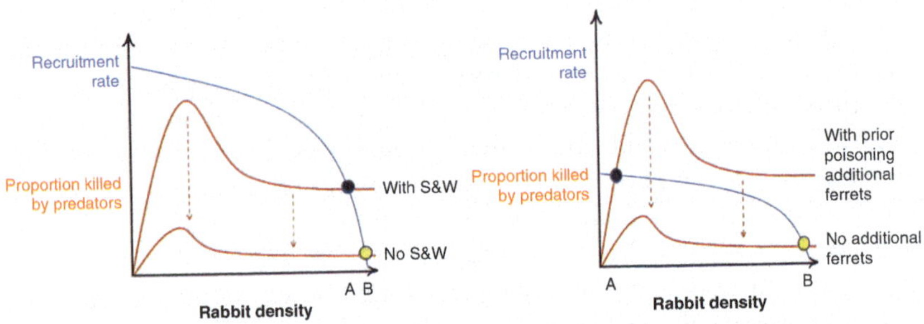

Fig. 12.2 Conceptual model of the per capita relationships between subtraction of rabbits from the wild population by native and introduced predators (lower curves), and the replacement of the rabbits removed by recruitment (breeding and immigration, upper curve) in the southern South Island in the nineteenth century. *Left*: Most landowners simply released ferrets, stoats and weasels among huge numbers of rabbits in the back country, which made very little difference to rabbit density (A, after mustelids were released, is not much different from B, before). *Right*: Bullen's system required knocking down rabbit numbers by poisoning over winter, supplementing ferret numbers in spring, and prohibiting rabbit trapping. Well-timed additional ferret predation, combined with protection for ferrets from traps, increased the per capita predation rate by ferrets to above the recruitment rate of rabbits, so decreasing rabbit densities from B to A. (Modified by Author with permission from Norbury, G. (2011). Mythbusters: Can predators control rabbits in New Zealand? Kararehe Kino, 18, 22)

Because these are clear examples of two quite different predator–prey relationships. Island species that evolved in habitats free of any four-footed mammalian predators, including all of New Zealand's most strongly endemic birds (Chap. 1), typically live long and breed slowly, and have few innate nest defences. Suddenly increased subtraction imposed by a new invader can easily beat multiplication of the residents, and reproduction is overwhelmed by unmanageable losses, as is illustrated by the long list of New Zealand's extinct birds. Lethal control of invasive predators to protect a threatened island fauna is effective and fully justified,[48] up to a limit correlated with the size and location of the island and the number of human residents.[49] Well-managed programmes on uninhabited islands, even very large ones, are now routinely successful.

That is a completely different matter from the reactions of mainland species familiar (in the biological sense) with each other as natural predators and prey. A 'natural enemy' in this sense is the product of an evolved interaction which includes the sort of mutually adaptive population dynamics and flexible behaviour that allow long-term co-existence at landscape scale. The impact of invasive rodents and mustelids on New Zealand's native fauna was overwhelmingly of the disastrous new invader kind; that of mustelids on rabbits, the accommodating old enemy kind.

Learning from History

Uninhabited offshore islands have been the invaluable pioneer locations in eradication history, because they have been the ideal experimental labs. They have all the advantages of being able to deter recolonisation, and any issues connected with social licence are easier to manage. We have learned much from island work. But PFNZ2050 is interested in the inhabited mainland, and history shows that it can also profit from past experience of operations that didn't work, at first or at all. Helpful examples can be found both overseas and at home.

Conventional Weapons

Managing pest populations by direct lethal means (what Caughley[50] called 'frontal assault') has been going on for centuries. In the context of their times, colonial-era land managers first, understandably, searched for more effective and economically feasible means of simply killing more rabbits. Of course, they tried to avoid the ruinous expense of employing men to trap and poison rabbits over huge areas of back country, by making the rabbits contribute to the costs of their own destruction. Hence, the growth of ancillary industries trading in rabbit skins and, later, canned rabbit meat (Chap. 7), but, equally reasonably, those so engaged would always leave a breeding stock for the next year.[51] The problem with any policy of controlling a pest using men paid for their work in products of the pest is that it guaranteed a workforce with a vested interest in doing profitable work that was ineffective in controlling the pest. Only a decommercialisation policy could overcome that unhelpful feedback, and that was not done until 1947.[52]

Coypu

One of the first eradication programmes to overcome the unhelpful effect of operator self-interest was conducted against coypu in eastern England in the 1980s. Coypu are large semi-aquatic South American rodents, imported into the UK for fur farming in 1929. A wild population of coypu became established from escapees in East Anglia, where they caused a lot of damage by burrowing into river banks and chewing up adjacent crops at night. The first organised trapping campaign was declared a failure and abandoned in 1970, and the survivors rapidly recovered to resume their damaging activities, over the remaining 6000 km^2 left to them, for another decade. The one thing that most affected their numbers was a series of cold winters, because coypu are tropical animals with bare feet that easily get frostbitten.

In 1981, a new campaign began, planned to eliminate the problem (familiar in New Zealand experience) that paid trappers are very reluctant to work efficiently enough to put themselves out of a job. The new approach worked (eventually) because the 24 trappers employed on the new programme were promised a bonus of three years' salary each if they got the last one. With such a generous incentive, the trappers applied themselves to the job. Almost 35,000 dead coypu later, and with the help of four cold winters, coypu were eradicated in 1989.[53]

American Mink

New Zealand is not the only country to have discovered how hard it is to eradicate introduced mustelids. American mink are sleek, semi-aquatic mustelids rather like slimmer, water-loving versions of the ferret, and were imported into European captive colonies in the early twentieth century to be bred up for their luxurious fur. Inevitably, many escaped from the

1950s on, and in the absence of competition from the native river otters, their descendants became widespread and ineradicable pests.

In Britain, authorities employed river keepers and game wardens to remove the mink throughout the 1960s, but they were slow to start, and their tools were ineffective. Leg-hold traps had been outlawed in 1958, and mink simply swam on by the undisguised cage traps set on the river banks. The official campaign could not even prevent the spread of mink, so was abandoned in 1970. Mink have since spread throughout the country, threatening nesting waterbirds and devastating populations of the much-loved native water vole.[54]

The good news is that in many places the native otter is now recovering its numbers since the cleaning up of waterways by anti-pollution legislation. The American mink are coming off second best in encounters with its much larger relative.[55] Floating rafts carrying cage traps hidden in enticing tunnels are now proving irresistible to the remaining mink; concentrated, well-informed local programmes are turning the tide against them, although eradication is still well out of reach.[56]

Stoats on Islands

Many native species that can no longer survive on the New Zealand mainland need offshore islands as refuges, and clearing islands of rats is now routine. Islands outside the swimming range of rats are eradication units: rats removed don't come back, so each unit is a separate, finite job. Stoats, however, are different. New Zealand is as far away from their native countries as it is possible to be, both in physical geography and in ecosystem structure, but much of their natural behaviour remains the same,[57] especially their curiosity, intelligence, restless energy and astonishing swimming abilities. Maud Island (309 ha) in the Marlborough Sounds has always been rodent-free and is now restocked with vulnerable native birds, but it is only 900 m from the mainland, just an afternoon's paddle for a stoat that can easily swim 1 km in an hour.[58] Maud has been invaded by stoats three times since 1982; other much larger islands even further off (Kapiti, 5.2 km, in 2009, and Rangitoto, 3 km, in 2010) have also been invaded by stoats.[59] Catching them is very difficult, because they are masters of trap evasion. Worse, females of any age that reach an island are already fertilised.

The chances of success turn out to be closely related to geographic scale. Experience has shown that stoats can be eradicated from small islands distant enough from the shore that the inevitable reinvasion is manageable. Success started with Chalky Island (514 ha) and progressed to Coal (1163 ha) and Anchor Islands (1137 ha). Unfortunately, the huge step up to larger islands has so far been unsuccessful. Secretary Island (8140 ha) and Resolution Island (20,800 ha), off the coast of Fiordland, are both very large and potentially valuable sanctuaries for threatened species, but determined attempts to clear them of stoats have so far been defeated. The hard core of untrappable resident adult females keep producing offspring that soon fall into traps,[60] and occasional swimmers arriving (most often after a beech seedfall on the mainland) do the same. Trapping keeps stoat numbers on the two islands pretty low, but that is not enough. It takes only a very few stoats to make an island uninhabitable for flightless birds. Poison baits cannot be used to supplement trapping because of the risk to kea, an endangered endemic parrot. Population modelling for Resolution (p. 292) suggests that it would take double the investment made so far (millions of dollars spent, 2352 traps checked and reset 24 times from 2008 to 2015, and the tireless dedication of DOC staff) to make much real difference to the outcome. On the other hand, the model should also help improve the probability of achieving at least temporary zero density in the future. Until then, no inshore islands (<5 km out) count as eradication units while stoats remain on the mainland, so eradications on them cannot be permanent.

Plans for stoat eradication programmes on two other large islands (Waiheke and D'Urville), each around the same size as Secretary and Resolution, have learned from previous failures. Both islands are much more accessible for field staff, so they could be equipped with greater trap densities that might make reinvasions more manageable. But stoats are very intelligent and mobile, and the different reactions of the older and more cunning individuals to control devices strongly affect the probability of eradication success. New population models are trying to work out how to improve management protocols to minimise that problem.[61]

Himalayan Tahr

Even the use of very effective tools cannot guarantee success, as illustrated by one long-running direct control operation that could have worked, but didn't. The Himalayan tahr is a handsome mountain goat introduced to the high country of the South Island in 1904–1909. Recreational hunters relish the opportunity of stalking dramatic game animals in a dramatic landscape, and few game animals fit that prescription better than a mature bull tahr in the Southern Alps.

The numbers of tahr increased dramatically to 30–40,000 by 1970. They spread far and wide across the fragile alpine grasslands, damaging the slow-growing native plants with their teeth and the frost-shattered soils with their hooves. But on the open mountain tops, the tahr are also very exposed to fast-moving helicopters carrying sharp-shooters working for the export game meat market, and by 1983, the last few thousand tahr could have been eliminated. Conservationists saw that as a very desirable outcome, but recreational hunters did not; they negotiated a ban on aerial hunting, and a system of monitoring both vegetation and tahr numbers within agreed boundaries.[62] This case is an early example of the critical importance of social licence: even effective pest control technology cannot be used without public approval.

1080

The most effective tool we have for large-scale pest control is the aerial distribution of 1080, an acute toxin. A good 1080 operation can remove possums, rats and mustelids together, at least for a few months at critical bird breeding times. The scientific evidence in favour of this effect is so strong that the Parliamentary Commissioner for the Environment urges that not only should 1080 not be banned, but we should be using more of it.[63]

Possums often live in forest patches adjacent to cattle pastures, and venture out to feed on the grass at night. Cattle sniffing at or licking a possum infected with bovine tuberculosis (bTB) can readily pick up the disease. Confirmation in 1976 of the status of possums as wildlife reservoirs of bTB set off repeated official attempts to reduce the number of infected cattle herds by widespread aerial 1080 baiting of possums, and these efforts have accelerated since the 1990s. The current hugely expensive campaign (costing at least $50 million a year since 2002) has been largely successful in limiting the distribution both of possums and of bTB in livestock. The official threshold of bTB prevalence which will allow New Zealand to declare itself bTB-free is <0.2% of the national herd tested positive. At the end of the 2017–2018 year, only 32 (<0.08%) of nearly 70,000 herds of NZ cattle and farmed deer were infected.[64]

The incidental benefits of this work for conservation are of course enormously welcome, but the programme is designed by agricultural authorities to protect cattle and deer from contact with tuberculous possums. If bTB is ever eradicated, conservation authorities will have to fund the programme themselves. At the moment, 1080 is the only economic method of doing that over the huge land areas required.

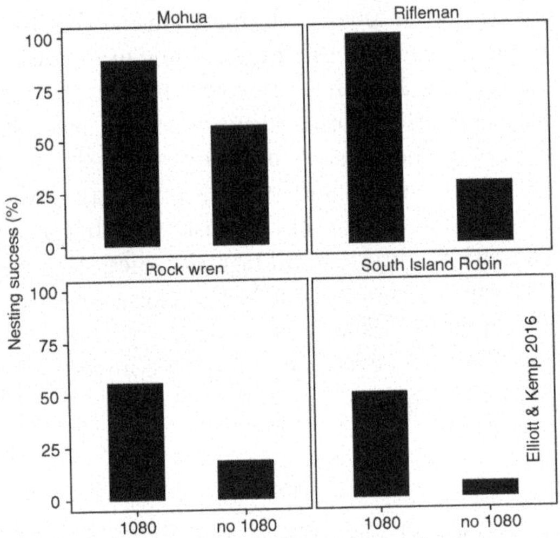

Fig. 12.3 Benefits of a 1080 operation against rats and stoats during the post-seedfall summer of 2014–2015 for the nesting success of four species of small beech forest birds. (From Elliott, G., & Kemp, J. (2016). Large-scale pest control in New Zealand beech forests. Ecological Management & Restoration, 17, 200–209)

The problem is that it is also very controversial, and the ongoing public rejection of evidence against the potential 1080 contamination of water, game meat and forest ecology have exposed a widespread distrust in science. Carefully controlled trials documenting the significant net benefits to whole populations of nesting forest birds (far more important than the risk to individuals; Fig. 12.3) by a large-scale 1080 operation following a massive beech seedfall event[65] make no impression on a passionate anti-1080 lobby unable to look past individual losses to wider population gains.[66]

Evidence demonstrating the benefits of 1080 is even more compelling for brown kiwi, one of our most widely recognised national symbols. Kiwi chicks are acutely vulnerable to stoats until they grow big enough to defend themselves, at about six months old. Hugh Robertson and colleagues have analysed the data from 22 years of monitoring the lives and reproductive success of a total of 142 radio-tagged kiwi in Tongariro Forest.[67] During this programme, 1080 was dropped once every five years over the entire 20,000 ha protected area, and the breeding success of adult kiwi and survival of chicks was monitored throughout the 22 years. In the breeding seasons following the first two years after each drop, which effectively removed most rats and stoats, the survival of chicks to six months old averaged significantly higher (50%, 29%) than in the next two years (17%, 14%), when rats and stoats were back. At least 20 adult kiwis were killed by ferrets over the 22 years: 0 in the first year after a drop, 3 in the second, 5 in the third and 14 in the fourth. On average, through the 4 five-yearly cycles, the kiwi population increased by a healthy 11.55% in the first year after a drop, and by 3.1% in the second year, but decreased by −3.6% in the third year and by −5.9% in the fourth year. So the benefit to breeding kiwi in the first two years of each cycle was largely undone in years 3–4+. The obvious solution is to change to a three-yearly cycle. This has drawbacks of its own, but has been implemented since 2014. Given certain assumptions, the new regime should allow the kiwi population to maintain a steady average recovery rate of 3.6% a year, well above the minimum required. Who then could assert that 1080 does more harm than good for our most iconic native bird?

In practice, the argument has no end in sight, because it is not really about scientific facts but about opposite world views held by the participants. Even the critical differences between 1080 and brodifacoum are often confused. Obviously, only certain things can be seen from any given viewpoint; those looking in from another angle can see only different things, and this applies to many different important issues such as the fluoridation of public water supplies, or vaccination of children. Hence, as philosopher Mary Midgely points out, campaigning activists are usually defending a commitment to a particular view of the world and only secondarily, if at all, the prospects of achieving it.[68]

1080 has more limited applications which might be expected to generate less controversy. For example, continuous trapping on mainland kiwi reserves, however extensive, cannot guarantee the survival of chicks, especially after a mast year (p. 285). Near Whangarei, several years of declining survival rates of kiwi chicks in a well-monitored mainland kiwi sanctuary were attributed to the problem of learned trap-shyness in resident stoats. A quick pulse of 1080 in covered tunnels outwitted them and improved chick survival in both areas.[69] The same team compared the survival of kiwi chicks in areas protected by traps (median survival time 43 days, none to 6 months old) compared with protection by toxins (median 94 days, 33% to 6 months old).[70] No one likes the idea of using poison to kill sentient animals, but in New Zealand at the moment, no other options have much chance of saving our remaining wildlife until we think of something better.

Natural Enemies

The organised introduction of the natural enemies of the rabbit is of course the prime example of a conventional weapon that didn't work. Contemporary observers both in New Zealand and in England pointed out the 'dead certainty', even before the first mustelids arrived, that Bayly's policy would be 'a proposed cure … worse than the disease'.[71] For all the reasons summarised above—primarily that a concentrated focus on natural enemies is far too narrow—that is exactly what happened. All the many vociferous objections and attempts to prohibit the official imports failed until 1889 (Chaps. 7 and 10): the conflicts of interests (ornithologists vs landholders; game-bird hunters vs pastoralists; small farmers vs wealthy runholders; rabbit trappers vs rabbit inspectors) guaranteed that the idea of rabbit control by natural enemies would be a bolted-horse issue from the start.

Long before Bayly's time, farmers were already buying city cats and releasing them on rabbit-infested pastoral runs. Thanks to Bayly, thousands of mustelids followed, so by 1890 ferrets, stoats and weasels were also well established in the wild and breeding freely. They and all other predators of rabbits (wekas and harrier hawks) were still legally protected, in the hope that they might yet turn the scales against the rabbit pest, until an amendment to the Rabbit Nuisance Act 1882 partially removed that protection in 1903.[72]

The sources describing this sad story quote, as simple statements of fact, bald opinions and actions that we now find incredible and reject as false or misleading, but which seemed to be the most reasonable attitudes and the best thing to do at that time. For example, here is an extract from a report written for the Maraura Ensign in July 1898 by Reginald Foster, a former chief stock inspector for Canterbury:

> the danger was not realised until the rabbits commenced to spread northwards like a tidal wave, causing inevitable ruin in their progress … [so] Government introduced stoats and weasels for liberation on the Crown lands … although I, in common with all sportsmen, regret their introduction, I am satisfied that they are, and will continue to be, the main factor in keeping the rabbit pest under [control] on the higher lands in the colony where it is impossible to deal with them effectively by any of the known means of rabbit destruction.

Foster's view of the introductions policy is determined by two unquestioned convictions, which he shared with most of his contemporaries: (1) that the most regrettable consequences of the arrival of stoats and weasels were the losses they caused to sportsmen by their predation on introduced game birds, and (2) that mustelids were fully expected to be capable of controlling rabbit numbers. Few ecologists would now agree with either opinion, but this way of thinking was widely held at the time, in the face of decades of experience to the contrary, and it strongly influenced the nineteenth-century governments' strategy. The only surprise is how many people still think the same.

Now, ferrets themselves are an officially declared noxious species, subject to control campaigns even though ferret numbers are more likely to be determined by rabbit numbers than by human action, especially in the semi-arid pastoral lands where they were first released. Likewise, the numbers of stoats and weasels are strongly influenced by the population fluctuations of forest rodents, especially in mast years.

By 1890, the colonial government had accepted that New Zealand's first experiment in biological control had been a regrettable failure; the passage-paid mustelid experts were simply not up to the job. The first careful analysis of the consequences, by G.M. Thomson in 1922, concluded that by the turn of the century they were 'common in nearly every part of New Zealand, and in some enormously abundant … They have not exterminated the rabbits, in most parts they cannot even keep them in check'.[73]

Later commentators were more forthright. Herbert Guthrie-Smith, a leading runholder and an observant naturalist of Hawke's Bay, bitterly criticised 'the freeing, firstly of rabbits', which he saw as a monumental blunder, 'and secondly of weasels, in the Wairarapa' which he saw not so much a mistake as a deliberate criminal act. Guthrie-Smith reported that the strength of local feeling in Hawke's Bay against those responsible could be judged from the resolution passed 'with ferocious unanimity' to put a bounty of a guinea a head (about NZ$190 in today's money) on every mustelid caught dead or alive. But he was also a realist: he added sadly that 'No paper resolutions, however, can stay a plague once introduced; harm of this sort done cannot be undone'.[74]

Introduced Diseases

Cat flu (feline panleucopaenia) is a serious viral disease of cats, against which most mainland cats are either inoculated or already immune. But on

Marion Island, a 290 km^2 reserve in the southern Indian Ocean populated by millions of breeding seabirds, five cats brought in by lighthouse keepers in 1948 had multiplied to more than 3000 by 1977, and the mortality of petrel chicks was approaching 100%. When conservation managers discovered that the Marion Island cats were all fully susceptible to cat flu, the disease was imported, and the number of cats declined by 29% a year to only 620 by 1982. But the evolution of genetic resistance was signalled by the increasing proportion of cats developing antibodies to the flu, and the rate of decrease in cat numbers slowed. Despite its initial success, the disease failed to eradicate the cats, and switching strategies to trapping and hunting took until 1991 to remove the survivors.[75]

Myxomatosis is a specific and fatal viral disease from Uruguay, which can, in the right conditions, kill all but a few of vast numbers of European rabbits without affecting any other species—not even hares. It requires a blood-sucking vector to transmit the myxoma virus from infected to uninfected rabbits, most efficiently but not only by the specific rabbit flea native to European rabbits.[76] Rabbits transported to both Australia and New Zealand for sport hunting and for meat lost their specific flea on their long journey south in sailing ships, so interrupting this cosy arrangement. Australia did have stickfast fleas, but these are 'the most immobile of all fleas' so did little to help spread the virus. New Zealand rabbits can carry fleas normally found on rats or cats, neither common.

Fortunately for the wool farmers of Australia, the damper parts of their pastoral country have enough mosquitoes to do the job instead.[77] When finally established (after many failures) in Australia in 1951, myxomatosis killed 99% of the rabbits which had plagued Australian pastoralists and their environment for decades. Within two years, Australian wool and meat production was better off by A$68 million and the wool clip by 70 million pounds weight.[78]

Inspired by such a wonderful effect, myxomatosis was brought, first to France and then to the UK in 1953, and soon wiped out more than 99% of the rabbits that had damaged farm production and forestry plantations for centuries. (The man who brought myxomatosis to France was awarded a medal by foresters and taken to court by hunters.)[79] But New Zealand had neither the right kind of flea, nor enough mosquitoes, so attempts to introduce myxomatosis there failed, several times.

Instead, attention turned to a rabbit calicivirus from China, known as rabbit haemorrhagic disease (RHD), which is a less virulent but more humane disease spread by direct contact. It was being trialled on an offshore island in Australia, but escaped; then it was introduced to New Zealand illegally in 1997. Someone nipped across the Tasman, shot a few infected rabbits, mashed up the livers in a blender and smuggled the result home in a carry-on bag. It had good effects at first,[80] but the refusal of the frustrated farmers involved to wait long enough to get an official release of the most virulent possible strain ruined the chances of a much better official operation. The political hand-wringing went on for months, but despite such a flagrant violation of biosecurity legislation, no one was ever prosecuted.[81]

Even where one or other of these introduced diseases worked, any benefit to human affairs did not last, because natural selection is on the rabbits' side. The <1% of rabbits that survived myxomatosis had an enormous breeding advantage, and within a few years they and the disease organisms adapted to each other. By 1995, the population of Australian rabbits was back to some 300 million. By 2010, the survival rate of infected rabbits in the UK increased from 1% to 35%, and the annual cost of rabbit damage to UK agriculture (in lost yield and management costs) was back to £188 million per year.[82] New Zealand rabbits are recovering from RHD, although not yet to previous levels. Numbers are likely to increase as immunity spreads, reviving old arguments about falling back on 1080—still a controversial option, even at the minimal concentrations now used.[83]

Scientists in Australia continue to work on new strains of RHD, and these are being officially (and legally) released in NZ. But the evolutionary arms race will ensure that this battle will continue indefinitely, forwards and backwards, until farmers choose some other way to use the pastures that can no longer feed their sheep.

Common Features of Failed Versus Successful Attempts

Reviewing these case histories suggests a set of simple comparisons.

Failures of pest control operations are usually due to some combination of pessimistic human attitudes and inadequate knowledge of science and technology, including the following. Forgetting that Nature is on the pests' side; ignoring the effects of natural selection; counting the numbers removed, not the numbers remaining, so confusing substitute with additive effects; lack of a coordinated, agreed strategy; lack of flexibility in response to experience; insufficient support from key stakeholders and the general public; inefficient tools; failing to deploy a combination of methods, leaving fertility and immigration unaffected; lack of *post hoc* monitoring. The good news is that we can change all these.

Successes are also usually down to some combination of positive human attitudes and deep understanding of science and new technology, including the following. Indomitable operator confidence, meticulous preparation and sufficient funding; prevention of harvesting and human self-interest in the results; never repeating mistakes, willing to accumulate improvements and combining/switching strategies when necessary; landscape-scale coordination, strong community support; methods that add to natural mortality rather than substitute for it; preventing immigration; targetting fertility; and confirmation of a significant benefit for native species. The even better news is that history can help us identify and build on all these.

Among the best contemporary examples are the growing number of community and urban restoration programmes staffed by indefatigable volunteers willing to keep on tramping round lines of tracking tunnels, traps and bait stations in all weathers, even when the effort feels like (and in a sense is) bailing out the ocean with a teaspoon. Fenced sanctuaries within which most invasive mammals have been eradicated are proving to benefit endemic birds more than introduced species,[84] and a 'halo effect' spreads the offspring of birds nesting inside the fence to adjacent habitats. There are only 14 fenced sanctuaries among a total of 84 sanctuary projects of various types, but community support enables the sustained control of ship rats, possums and stoats to be maintained over at least 100,000 ha of protected but unfenced habitats.[85] Even cities as large as Wellington, with massive Council and public support, are making life for urban pests risky, fearful and short.

Tame Versus Wicked Problems of Managing Invasive Species

D.J. Woodford and his team have usefully distinguished the questions surrounding the management of invasive species as either tame or wicked problems.[86] On this view, tame problems are defined as those with simple or obvious solutions, whereas wicked problems are those for which complete solutions are difficult or impossible to find, largely because of intense conflicts of interest among the major stakeholders. As an example of a wicked conservation problem, Woodford's team analysed the arguments between anglers and conservationists surrounding the management of introduced rainbow trout in New Zealand. Another team applied the idea to explain the persistent opposition to possum control by aerial distribution of 1080.[87] The story of the introduced mustelids is an even better example.

Contemporary ecologists often wonder why such reckless decisions were made, but the historic details of Bayly's programme illustrate how these were, to them, well meant and apparently reasonable environmental management actions. Those responsible were sure they knew best: they simply rejected predictions of a cascade of unintended and undesirable consequences. Serious attempts by private individuals to import ferrets were under way for at least eight to ten years before the government actively supported and extended them. Some runholders, for example, George Bullen of Kaikoura, showed that under certain conditions (Fig. 12.2, right) ferrets could have a significant short-term effect on the numbers of rabbits by cleaning out whole litters of juvenile rabbits. Because Bullen had direct control of his own land, that is, did not have to compromise with other stakeholders, his understanding of his situation was close to Woodford's definition of a tame problem—one that has a simple and obvious solution. Outside these conditions, where no such constraints limited the independent actions of multiple players in an intensely controversial game, the problem of saving the wool industry from rabbits without devastating the native fauna became an intractably wicked problem—one in which disagreements and non-cooperation among stakeholders prevented any real solution.

The history of rabbits and mustelids in New Zealand illustrates the consequences of mistaking a wicked problem for a tame one. The tragedy enacted by Bayly and his colleagues is that they saw the long struggle against rabbits as a tame problem: mustelids kill rabbits, so too many rabbits meant not enough mustelids; the simple and obvious way to have fewer rabbits is to get more mustelids. The conflicts of values of the time between woolgrowers' focus on preserving their profits and their opponents' focus on preserving native species, with the general public as equally argumentative spectators, made the issue a wicked problem; wide agreement was impossible, and, hence, the indiscriminate mass release of imported predators did *not* mean fewer rabbits. It was simply not possible then to understand why not. Fortunately, no one else has tried Bayly's solution on anything like the scale he attempted in New Zealand.

Having said that, it is important to add two necessary cautions. First, we must not use twenty first-century knowledge to criticise decisions made in the totally different atmosphere of the nineteenth century. On the contrary, we have teams of well-informed biologists doing their best to understand past experience in order to learn the valuable principles of pest management applicable today. The most important single key is to recognise that pest populations are determined by the cascading interactions determining the *dynamic balance* between the numbers input (fertility, immigration) versus output (mortality, emigration). Just to complicate matters, all these are locally and seasonally variable, and liable to be shoved off course by unpredictable historical contingency and interactions with other species. So the task of protecting the most critically threatened remaining native fauna long term is a definitely not a tame problem; rather, it is a subset of the much wider issue of biodiversity conservation. In nature, managers cannot change only one thing.

Second, there is no common understanding of what conservation is about. Charles Elton, the widely respected father of invasion biology, defined it way back in 1958 in the language of his time:

> This means looking for some wise principle of co-existence between man and nature, even if it has to be a modified kind of man and a modified kind of nature. That is what I understand by *Conservation*.[88] (original italics)

Not everyone in New Zealand would agree with that view. Opinions about how best to manage wild life around the South Pacific have developed in different directions and now range along a wide spectrum whose extreme points are mutually incomprehensible.

At one end of the spectrum are those who take seriously New Zealand's national identity and statutory commitments to preserving what remains of the *past* by preventing extinctions of the remaining native species. They have a profound longing to 'restore the dawn chorus', including where necessary by large-scale predator control even at enormous cost and potential collateral damage. At the other extreme are those whose focus is on the *future*, who accept that the human-induced transformation of our natural environment is irreversible, and that the best use of the billions allocated to removal of aliens could (except in special, defensible places) be better spent on protecting habitats in which native species can continue to evolve. The debate is brilliantly summarised in a lucid, rational and entertaining book by Keith Thompson, *Where do camels belong?*[89] which should be required reading by all proponents of both sides of the argument. In between are the majority who avoid the spiky bits of both extremes and argue for meeting both demands each in separate locations.

Since the arrival of the first humans in around 1280 AD (Chap. 2), New Zealand ecosystems have become a scrambled mixture of broken native species and introduced elements, rubbing along together in radically changed habitat conditions unfamiliar to both. We cannot return any landscape to the way it once was, any more than we can unscramble an omelette, but we can at least make sure everyone gets a fair share of it. We already distinguish areas where biodiversity protection is the top priority from areas where it is a lost cause. Ultimately, that may be the only practicable approach.

Could Genetic Engineering Be the Technique of the Future?

One feature that all small invasive predators have in common is their breed-fast, die-young lifestyle. High natural mortality means that human efforts to control pests by killing more of them works only if it removes more individuals than will die anyway, and that additional toll is very hard to achieve. Their population numbers depend on rapid replacement of losses, which is normally easy because so many more young are produced than are needed to replace lost adults. Worse, the more rats and stoats we humans remove, the more food and space is left for the remainder, which both help to boost female productivity. On the other hand, depending on frequent good breeding seasons makes their populations vulnerable to bad breeding seasons.

Overcoming these linked effects has in the past required the liberal use of traps, poisons, introduced diseases and the whole arsenal of weapons reviewed above, always costing megabucks and often without achieving the desired effect. In theory, it would not be so much of a challenge if we could attack the equation from the other side and reduce the number of individuals born. Since the whole apparatus of reproduction has evolved to defeat any such risk, natural birth rates of pests are managed by genes selected over uncountable generations for their ability to maximise the potential number of surviving offspring produced in any conditions.

Recent advances in our understanding of how genes work now raise exciting questions and new possibilities for pest managers. We can now not only read the information in the genomic libraries of possums, rats and stoats: we are developing methods of changing them. If reducing fertility is so important, could we edit the reproductive instructions of common pests in ways that would alter the normal laws of inheritance to suit some human agenda? For example, in virtually all vertebrates the sex ratio at birth is nearly 1:1, even in species where only a few dominant males get to breed. But no breed-fast, die-young species can survive long if more than a few generations could be made to produce offspring of only one gender. Since nearly all females breed, but not all males, the key is to reduce the numbers of females.

A new and rather alarming technology is now (since 2012) emerging which is developing a real potential to alter the very foundations of pest management policy in New Zealand.[90] The process uses gene drives, which can, for example, make modified individuals produce only daughterless offspring. With enough of such engineered parents introduced into a natural population of rats, and enough generations with no breeding females, the rat population should eventually die out. Not a single individual would need to meet death by any of the currently unavoidable standard methods which make animal welfare activists shudder. To some, the ethical argument for such a sweeping reduction of animal suffering is almost sufficient justification in itself. If it was ever really feasible, this wonderfully attractive effect could spread across the country at landscape scale, into all the remote places where any human action is impossible.

This sounds, at first, like an ideal answer to those who have long doubted the feasibility of the PFNZ2050 programme. But from another point of view, this technique looks less like an offer of national salvation from intractable pests and more like a modern version of Benjamin Bayly's tragic mustelid introduction programme. Consider the similarities. Rabbits in the nineteenth century had resisted all known methods of control; Bayly sought a self-perpetuating rabbit-killing mechanism; he believed mustelids would supply it. They didn't, and now they cannot be recalled, while rabbits still thrive. Pests in the modern world continue to resist all known methods of control; targeted gene drives offer a permanent, self-perpetuating pest-removing mechanism. If they don't, then in the future we could be left with engineered genes in nature that cannot be recalled, while the pests develop resistance and thrive. Modelling of potential outcomes has found that the chances of eradicating a pest population with a gene drive are uncertain, and mutations conferring resistance are to be expected.[91] Social, legal and political problems are likely to be even more challenging, especially on the international stage.[92]

Thoughtful people at the centre of the debate agree that the benefits of this sort of gene technology are alluring, especially under controlled conditions in the fields of medicine and disease control. But when applied to wild nature, the risk of it going horribly wrong cannot be eliminated, and if it did, the national and international consequences would be incalculable. Of all countries in the world, New Zealand is the one best placed to learn from its own history. We know all too well what happens when policies that seem quite reasonable turn out to have drastic unforseeable effects. Especially, we have learned the hard way why it is so very important that that debate must be in the open and that social licence is negotiated long before deployment, through full public participation, especially with Māori, plus comprehensive international consultation. That is why the Sustainability Council of New Zealand has published a long and thoughtful report[93] discussing all these issues, and responsible geneticists fully acknowledge them.

While the technological challenges of creating controllable gene drives are vast and rapidly advancing, the social license to use such technologies are a different matter. Gene editing technology fits squarely in the realms of magic for most people, and this is unlikely to change in the near future. Even those scientists that do understand it are well aware of the key lessons of history, which demonstrate how easily the complex web of ecosystem interactions can produce unexpected responses to even a simple management operation. An inspiring review of the positive benefits brought by 30 years of conservation genetics for our endemic biota points out that genetics could also make a huge contribution to PFNZ2050, but adds the wry comment that 'biology has a habit of fighting back'.[94]

Beating out a broad social agreement on the potential use of gene drives must count as one of the most significant wicked conservation problems of modern times, and it will be a hot topic for years to come. It will certainly cause a bigger row than the one that followed the illegal introduction of the RHD virus, or even the current repeated uproars over 1080. In the context of this book, we need only emphasise, once again, that future debates on any self-perpetuating mechanism need to include repeated reference to our own history of false beliefs, nasty surprises and irretrievable mistakes.

More than anything, those future debates need to look far beyond the history and the genetics and into the deeper waters of philosophy lurking behind them. We need help to dig up the invisible assumptions that determine our thinking without our conscious assent, a task that requires ruthless dissecting skill and razor-edge logic. No one is better at that job than Mary Midgley. Two passages in her seminal book *The Myths We Live By*, although written in other contexts, need only a few changed words to illuminate our present dilemmas over how to manage gene-drive engineering.[95]

> We know that seventeenth-century mechanists were mistaken in supposing the world to be made of clockwork, and a [twenty-first] century repetition of their overconfidence does not seem likely to work out any better. So questions about biotechnology raise wide issues, not just about the relation of … acts to consequences, but also about where our world pictures come from and what needs to happen when we change them (p. 103).

> From the metaphorical angle, too, the implications of [genetic engineering language] are not encouraging. The idea of improving books by splicing in bits of other books is not seductive because in books, as in organisms, ignoring the context usually produces nonsense … the image constantly suggested by the word 'engineering' [is] a simple analogy with machines. Cogs and sprockets can in principle be moved from one machine to another, since they are themselves fairly simple artefacts, and the working of these machines is more or less fully understood by their designers. Those who use this analogy seem to be claiming that we have a similar understanding of the plants and animals into which we might put new components. But we did not design those plants and animals. This is perhaps a rather important difference. (p. 114)

Of course, no practising geneticists need Midgley's warning, and official gene-drive research is cautious, responsible and internationally monitored.[96] But accidents happen: for example, containment was supposed to prevent the unintended liberation of American mink in the UK, and of RCD in mainland Australia or New Zealand. Worse, science cannot always overrule the economic implications of transformative new technology or policy decisions, especially those loudly advocated by passionate and single-minded lobby groups. In a liberal democracy, public-spirited people often cannot prevent private enterprise from maximising its own profits at the expense of the common good.

As the world learned in 2016, stakeholders can be expected to vote for corporate or national policies that privatise profits and socialise costs. Professor Newton's 1876 warning to the colonial government about the folly of importing stoats and weasels was timely and well argued (Chap. 9), but still rejected by runholders concerned more for personal than national interests. The result was a decision to import a powerful self-perpetuating tool which proved disastrous (Chap. 10), with an uncanny resemblance to its modern equivalent that could one day be incubating in commercial genetics labs.

Where to from Here?

The term 'Anthropocene' summarises a period of cascading and dramatic social, cultural and economic changes around the globe over the last few hundred years. The nineteenth century in particular was a period of uninhibited importation and acclimatisation of exotic animals generally. Colonial governments established *via* imperial expansion from western Europe assumed that they had the power and authority to reshape nature to their own needs. Species were seen as separate bits of nature that could be moved around the landscape like independent entities. The drastic transformation of the New Zealand landscape during the nineteenth century was part of the consequences of this wider movement.[97]

In practice, these simple, unquestioned pioneer expectations were often dumbfounded. Rural settlers had to deal with the capacity of their new environment to 'throw up surprise after surprise',[98] as their confident assumptions of their right to exert control over nature were challenged by contact with reality. They could often establish new species, but not control them, and when things turned out differently, they could not explain why. They did not allow for blizzards and droughts as among the normal conditions to be expected; when early frosts destroyed their crops and lambs, they set their teeth and carried on. When we are tempted to criticise their actions as reckless, we fail to remember that they saw nature only in terms of independent units distributed across an unfamiliar background, not as interacting parts of a dynamic network. They could not understand why consulting overseas experts, familiar only with very different conditions, did not usually help them.

Consequently, in New Zealand as in England but much more rapidly,[99] native ecosystems were disrupted and replaced with 'improved' vegetation and imported animals. New Zealand was not the only example of this process, since species perceived to be useful have been distributed freely around the world since ancient times, but it is certainly one of the most catastrophic. The consequences, especially for the native fauna, remain among the worst known. The whole episode provides a well-documented example of how cheerfully innocent nineteenth-century notions about the balance of nature and the benefits of manipulating natural enemies compare with those of today.

Historical events are best understood from documents written soon enough after the event to be free of later reinterpretations and read in the original words. Chapters 2, 3, 4, 5 and 6 describing the movements of animals and people before about 1860 are not well documented from contemporary sources, so have to be deduced by indirect methods giving little idea of collateral biases. By contrast, we have excellent sources of information about what was going on in the mid-to-late nineteenth century as described in the regional newspapers, and their reports were often very detailed and comprehensive. They are uniquely valuable now, especially to inform our understanding of the motivations behind the deliberate introductions of so many European mammals into New Zealand (Chaps. 7, 8, 9, 10 and 11). More importantly, contemporary reports help us appreciate the differences between what people did then and what we may wish they had done, or would have done in their place. We are fortunate to be able to do that, in order to see our modern dilemmas in context, and to avoid repeating old mistakes.

At the time of writing (October 2019), the dream of national-scale predator control policy for the main, inhabited islands is suspended between a past that seldom worked and a future vision that cannot yet be achieved. At a very deep level, the argument is about values, ethics and national identity.[100] PFNZ2050 is technically impossible at present. Its critics point out, correctly, that predators are not always and everywhere the largest impact on biodiversity; that the policy must necessarily focus on where predators can be removed, not necessarily where biodiversity values are greatest; it will obscure the fact that indigenous habitat cover and quality continues to decline; therefore, it risks diverting effort and resources from higher environmental priorities and better alternatives.[101] On the other hand, those who remember how things have changed since that dismal meeting in 1976 (p. 278) are wary about predicting how technological and social changes could or could not affect the future. Meanwhile, if nothing else so far, PFNZ2050 has galvanised scores of tireless community groups into dogged use of conventional technologies on a heroic scale, while 1080 remains the only effective way to protect what native values remain in huge swathes of the remote back country.

A stable future pest management policy for the inhabited expanses New Zealand must be based on knowledge of how the numbers of animals of all species are powerfully influenced, not only by the flexible parameters of their own population biology, but also by their interactions within the community into which they are adapted by their evolutionary history. For example, the damaging post-seedfall irruptions of stoats in southern beech forests are driven not by extra breeding but by sudden increases in juvenile survival. That in turn can be understood only in the contexts of the extra food supplied by the responses of not only wild house mice and ship rats, but also native seed-eating birds to a sudden pulse in their own food supply. Likewise, restoration (as far as is feasible in a fundamentally dynamic system; Chap. 1) requires a deep knowledge of the environmental conditions to which the surviving native species are already adapted.

Therefore, pest control can be effective only if it combines new, landscape-scale techniques of culling the most pervasive pest populations *and* managing the reproductive capability and juvenile survival rates of the remainder, *and* restoring the natural habitats required by protected species, *and* avoids releasing other problems downstream, *and* is implemented by socially acceptable means. Failures are common, especially when managers cannot anticipate the responses of non-target species, or cannot remove enough pests to make any difference. Over the inhabited landscapes where most people live, both rural and urban, management of introduced predators is increasingly successful, but is also increasingly challenging if all the demands of ecology, economic and ethics are to be met.[102] Yet pest control is only one component of an ideal national conservation strategy.[103]

PFNZ2050 cannot be achieved quickly, so until we find it the old methods are still needed, merely to 'hold the fort' by preventing further damage to what native species we still have. Any national-scale new strategy will take much longer to get results than simple short-term damage prevention, because the solution to any such 'wicked' problem depends not only on technical expertise but also on widespread understanding of how predator populations work, and social approval and willing collaboration in developing potential ways to manage them.

For the future, we can choose optimism (which, according to Winston Churchill, sees opportunity in every difficulty rather than vice versa). We can pick our battles more carefully now, because we have learned those most bitter of lessons, taught over a hundred years by rats, cats and mustelids. President Kennedy was not put off announcing the American moon shot simply because it depended on technology that did not exist at the time. So does Predator-Free New Zealand, but that did not put off Sir Paul Callaghan from advocating it. For New Zealanders, a deep longing to restore Aotearoa's lost dawn chorus has for decades been the widely agreed end: now, clear vision, knowledge of the past, and steely determination are the first, indispensable steps on the way to developing the unknown means.[104]

Summary

The lessons of environmental history are vital, both for understanding the past and for managing the future. New Zealand is showing the world how to avoid past mistakes and to treat predator control as only part of a full-scale national biodiversity strategy. Advanced eradication expertise has cleared most uninhabited offshore islands of introduced predators, allowing recolonisation of exiled native rarities. A public aspiration to extend the same success to the inhabited islands has fuelled a government-supported programme aiming to eradicate rats, possums and stoats by 2050. That is at present possible locally, but cannot be extended to the national scale without as-yet unknown new technology. Meanwhile it has stimulated countless community groups to intensive voluntary action. They use conventional tools to help keep common birds from sliding onto the threatened list. Important questions include: how to achieve additive, rather than substitute, mortality; how to detect learned trap evasion and the number of pests remaining; how to estimate the post-operation replacement rate. For remote landscape-scale operations conducted by professionals, strictly controlled aerial distribution of 1080 toxin is unpleasant and controversial but the only proven hope of retaining at least parts of the mainland native fauna until better techniques become available. New possibilities such as gene drives are among many suggestions presenting classic wicked problems to future generations.

Notes

1. Dunlap, T. R. (1999). Nature and the English diaspora: environment and history in the United States, Canada, Australia and New Zealand. Cambridge: Cambridge University Press, p. 73.
2. Star, P. (2006). "Nature's Trump Card": Confronting the rabbit problem in southern New Zealand, 1867–1897. ENNZ: Environment and Nature in New Zealand, 1, 3–9.
3. Owens, B. (2017). The Big Cull: Can New Zealand pull off an audacious plan to get rid of invasive predators by 2050? Nature, 541, 148–150; Russell, J. C., Innes, J. G., Brown, P. H., & Byrom, A. E. (2015). Predator-Free New Zealand: Conservation Country. BioScience, 65, 520–525.
4. King, C. M. (Ed.) (2005). The Handbook of New Zealand Mammals, 2nd Edition. Melbourne: Oxford University Press. Third edition in press, Eds C.M. King and D.M. Forsyth. CSIRO Publishing, 2020.

5. Tennyson, A. J. D. (2010). The origin and history of New Zealand's terrestrial vertebrates. New Zealand Journal of Ecology, 34, 6–27.
6. Dingwall, P. R., Atkinson, I. A. E., & Hay, C. (Eds.). (1978). The Ecology and Control of Rodents in New Zealand Nature Reserves. Wellington: NZ Department of Lands and Survey Information Series 4.
7. Stolzenburg, W. (2011). Rat Island: Predators in Paradise and the world's greatest wildlife rescue. New York: Bloomsbury.
8. Towns, D. R., & Broome, K. G. (2003). From small Maria to massive Campbell: Forty years of rat eradications from New Zealand islands. NZ Journal of Zoology, 30, 377–398.
9. Taylor, R. H., & Thomas, B. W. (1993). Rats eradicated from rugged Breaksea Island (170 ha), Fiordland, New Zealand. Biological Conservation, 65, 191–198.
10. Towns, D. R., Broome, K. G. (2003); Russell, J. C., & Broome, K. G. (2016). Fifty years of rodent eradications in New Zealand: another decade of advances. New Zealand Journal of Ecology, 40, 197–204.
11. Towns, D. R. (2011). Eradications of vertebrate pests from islands around New Zealand: what have we delivered and what have we learned? In C. R. Veitch, M. N. Clout, & D. R. Towns (Eds.), Island Invasives: Eradication and Management (pp. 364–371).
12. Russell, J. C., & Stanley, M. C. (2018). An overview of introduced predator management in inhabited landscapes. Pacific Conservation Biology, 24, 371–378.
13. https://www.youtube.com/watch?v=noIP5lbuJHk
14. Brown, M., & Maseyk, F. J. F. (2018). Transforming community conservation funding in New Zealand. Retrieved from Palmerston North: Predator Free New Zealand Trust.
15. Biodiversity Collaborative Group. (2018). Draft National Policy Statement for Indigenous Biodiversity. Retrieved from Wellington: Biodiversity (Land and Freshwater) Stakeholder Trust.
16. Russell, J. C., & Stanley, M. C. (2018).
17. Brown, M., & Maseyk, F. J. F. (2018).
18. Owens (2017); Russell et al. (2015).
19. Tompkins, D. M., & Veltman, C. J. (2006). Unexpected consequences of vertebrate pest control: Predictions from a four-species community model. Ecological Applications, 16, 1050–1061.
20. King, C. M. (Ed.) (2005). Handbook, for species accounts.
21. Beattie, J. (2015). Plants, Animals and Environmental Transformation: Indian-New Zealand Biological and Landscape Connections, 1830s–1890s. In V. Damodaran, A. Winterbottom, & A. Lester (Eds.), The East India Company and the Natural World (pp. 219–248). Basingstoke: Palgrave Macmillan.

22. Beattie, J. (2008). Acclimatisation and the 'Europeanisation' of New Zealand, 1830s–1920s. Environment and Nature in New Zealand, 3, 1–12.
23. King, C. M., Norbury, G., & Veale, A. J. (2017). Small mustelids in New Zealand: Invasion ecology in a different world. In D. W. Macdonald, C. Newman, & L. A. Harrington (Eds.), The Biology and Conservation of Musteloids (pp. 257–277). Oxford: Oxford University Press.
24. Savill, P. S., Perrins, C. M., Kirby, K. J., & Fisher, N. (Eds.) (2010). Wytham Woods: Oxford's Ecological Laboratory. Oxford: Oxford University Press. P. 180.
25. Fiddler, A. E., Lawrence, S. B., & McNatty, K. P. (2008). An hypothesis to explain the linkage between kakapo (*Strigops habroptilus*) breeding and the mast fruiting of their food trees. Wildlife Research, 35(1),1–7.
26. Murphy, E., Maddigan, F., Edwards, B., & Clapperton, K. (2008). Diet of stoats at Okarito Kiwi Sanctuary, South Westland, New Zealand. New Zealand Journal of Ecology, 32(1), 41–45.
27. Murphy, E., Maddigan, F., Edwards, B., & Clapperton, K. (2008).
28. White, P. C. L., & King, C. M. (2006). Predation on native birds in New Zealand beech forests: The role of functional relationships between stoats and rodents. Ibis, 148, 765–771.
29. Reardon, J. T., Whitmore, N., Holmes, K. M., Judd, L. M., Hutcheon, A. D., Norbury, G., & Mackenzie, D. I. (2012). Predator control allows critically endangered lizards to recover on mainland New Zealand. New Zealand Journal of Ecology, 36, 141–150.
30. Garvey, P. M., Glen, A. S., & Pech, R. P. (2015). Foraging Ermine Avoid Risk: Behavioural responses of a mesopredator to its interspecific competitors in a mammalian guild. Biological Invasions, 17, 1771–1783.
31. King, C. M., & Powell, R. A. (2007). The Natural History of Weasels and Stoats: Ecology, Behavior and Management, 2nd edition. New York: Oxford University Press.
32. McMurtrie, P. M. (unpubl.).
33. Ruscoe, W. A., Norbury, G., & Choquenot, D. (2006). Trophic interactions among native and introduced animal species. In R. B. Allen & W. G. Lee (Eds.), Biological Invasions in New Zealand (pp. 247–263). Heidelberg: Springer-Verlag; Ruscoe, W. A., Ramsey, D. S. L., Pech, R. P., Sweetapple, P. J., Yockney, I., Barron, M. C., … Duncan, R. P. (2011). Unexpected consequences of control: Competitive vs. predator release in a four-species assemblage of invasive mammals. Ecology Letters, 14(10), 1035–1042.
34. Bridgman, L., Innes, J., Gillies, C., Fitzgerald, N., Rohan, M., & King, C. (2018). Interactions between ship rats and house mice at Pureora Forest Park. New Zealand Journal of Zoology, 45, 238–256.

35. Innes, J., Kelly, D., Overton, J., & Gillies, C. (2010). Predation and other factors currently limiting New Zealand forest birds. New Zealand Journal of Ecology, 34, 86–114.
36. Sweetapple, P. J., & Nugent, G. (2007). Ship rat demography and diet following possum control in a mixed podocarp–hardwood forest. NZ Journal of Ecology, 31, 186–201.
37. Lustig, A., James, A., Anderson, D., & Plank, M. (2019). Pest control at a regional scale: Identifying key criteria using a spatially explicit, agent-based model. Journal of Applied Ecology, 56, in press. doi: https://doi.org/10.1111/1365-2664.13387
38. Walker, S., Monks, A., & Innes, J. (2019). Thermal squeeze will exacerbate declines in New Zealand's endemic forest birds. Biological Conservation, 237, 166–174.
39. Dunlap (1999). P. 141.
40. Henry, R. (1887). The New Zealand Rabbit and its prey. Lyttleton Times, Christchurch. Copy in Auckland Institute Museum, QL737.L32 HEN.
41. Gibb, J. A., & Williams, J. M. (1994). The rabbit in New Zealand. In H. V. Thompson & C. M. King (Eds.), The European Rabbit: The History and Biology of a Successful Colonizer (pp. 158–204). Oxford: Oxford University Press.
42. Parkes, J. (1995). Rabbits as pests in New Zealand: A summary of the issues and critical information. Manaaki Whenua Landcare Research, Contract Report LC9495/141, Christchurch.
43. Gormley, A. M. (Ed.) (2019). Modelling in Ecology. Kararehe Kino 32. (Manaaki Whenua Landcare Research, Lincoln).
44. Veitch, C. R., Clout, M. N., Martin, A. R., Russell, J. C., & West, C. J. (Eds.) (2019). Island Invasives: Scaling up to meet the challenge. Gland, Switzerland: IUCN Species Survival Commission.
45. Anderson, D. P., McMurtrie, P., Edge, K. A., Baxter, P. W. J., & Byrom, A. E. (2016). Inferential and forward projection modeling to evaluate options for controlling invasive mammals on islands. Ecological Applications, 26, 2546–2557.
46. Caltagirone, L. E., & Doutt, R. L. (1989). The history of the vedalia beetle importation to California and its impact on the development of biological control. Annual Review of Entomology, 34, 1–16.
47. Norbury, G., & Jones, C. (2015). Pests controlling pests: Does predator control lead to greater European rabbit abundance in Australasia? Mammal Review, 45, 79–87.
48. Russell, J. C., Jones, H. P., Armstrong, D. P., Courchamp, F., Kappes, P. J., Seddon, P. J., … Tershy, B. R. (2016). Importance of lethal control of invasive predators for island conservation. Conservation Biology, 30, 670–672.

49. Towns, D. R., & Broome, K. G. (2003).
50. Caughley, G. (1977). Analysis of vertebrate populations. London: John Wiley & Sons.
51. Ritchie, J. D. (1892). Annual Report on the Rabbit Nuisance [1891–92]. Appendix to the Journal of the House of Representatives, 1892 H-19.
52. Gibb, J. A., & Williams, J. M. (1994).
53. Gosling, L., & Baker, S. (1989). The eradication of muskrats and coypus from Britain. Biological Journal of the Linnaean Society of London, 38, 39–51.
54. Harris, S., & Yalden, D. W. (Eds.). (2008). Mammals of the British Isles: Handbook, 4th Edition. Southampton: The Mammal Society.
55. Bonesi, L., & Macdonald, D. (2004). Impact of released Eurasian otters on a population of American mink: A test using an experimental approach. Oikos, 106, 9–18.
56. Reynolds, J. C., Short, M. J., & Leigh, R. J. (2004). Development of population control strategies for mink Mustela vison, using floating rafts as monitors and trap sites. Biological Conservation, 120, 533–543.
57. King, C. M., Norbury, G., & Veale, A. J. (2017). Small mustelids in New Zealand: invasion ecology in a different world. In D. W. Macdonald, C. Newman, & L. A. Harrington (Eds.), The Biology and Conservation of Musteloids (pp. 257–277). Oxford: Oxford University Press.
58. King, C. M., Veale, A., Patty, B., & Hayward, L. (2014). Swimming capabilities of stoats and the threat to inshore sanctuaries. Biological Invasions, 16, 987–995.
59. Crouchley, D. (1994). Stoat control on Maud Island 1982–1993. Ecological Management, 2(2), 39–45; Prada, D., Veale, A., Duckworth, J., Murphy, E., Treadgood, S., Howitt, R., … Gleeson, D. M. (2014). Unwelcome visitors: Employing forensic methodologies to inform the stoat (*Mustela erminea*) incursion response plan on Kapiti Island. New Zealand Journal of Zoology, 41, 1–9; Veale, A. J. et al. (2012). Genetic population assignment reveals a long-distance incursion to an island by a stoat (*Mustela erminea*). Biological Invasions, 14, 735–742.
60. Veale, A. J. et al. (2013). Using genetic techniques to quantify reinvasion, survival and in situ breeding rates during control operations. Molecular Ecology, 22(20), 5071–5083.
61. Gormley, A. M. (Ed.) (2019). Modelling in Ecology. Kararehe Kino 32. (Manaaki Whenua Landcare Research, Lincoln).
62. Forsyth, D. M., & Tustin, K. G. (Eds.). (2005). Himalayan tahr. In Handbook of New Zealand Mammals, 2 Edition (pp. 361–373). Melbourne: Oxford University Press.
63. Parliamentary Commissioner for the Environment. (2011). Evaluating the use of 1080: Predators, poisons and silent forests. Wellington: Parliamentary Commissioner for the Environment.

64. OSPRI. (2018). Annual Report, 2017/18. Wellington: OSPRI New Zealand Ltd.
65. Elliott, G., & Kemp, J. (2016). Large-scale pest control in New Zealand beech forests. Ecological Management & Restoration, 17, 200–209.
66. Hansford, D. (2016). Protecting Paradise: 1080 and the Fight to Save New Zealand's Wildlife. Nelson: Potton & Burton.
67. Robertson, H. A., Guillotel, J., Lawson, T., & Sutton, N. (2019). Landscape-scale applications of 1080 pesticide benefit North Island brown kiwi (*Apteryx mantelli*) and New Zealand fantail (*Rhipidura fuliginosa*) in Tongariro Forest, New Zealand. Notornis, 66, 1–15.
68. Midgley, M. (1992). Science as salvation. London: Routledge. P. 94.
69. Robertson, H. A., Craig, E., Gardiner, C., & Graham, P. J. (2016). Short pulse of 1080 improves the survival of brown kiwi chicks in an area subjected to long-term stoat trapping. New Zealand Journal of Zoology, 43, 351–362.
70. Robertson, H. A., Colbourne, R. M., Graham, P. J., Miller, P. J., & Pierce, R. J. (2011). Experimental management of Brown Kiwi *Apteryx mantelli* in central Northland, New Zealand. Bird Conservation International, 21, 207–220.
71. Otago Daily Times, 3 May 1880. Acclimatisation with a vengeance.
72. Wells, P. (2009). The Fall and Fall in the Legal Status of Mustelids in New Zealand. Environment and History, 15, 343–368.
73. Thomson, G. M. (1922). The naturalisation of animals and plants in New Zealand. Cambridge: Cambridge University Press. P. 72.
74. Guthrie-Smith, H. (1969). Tutira. The Story of a New Zealand Sheep Station, 4th Edition. Wellington: A. H. & A. W. Reed. P. 354.
75. Bester, M. N., Bloomer, J. P., van Aarde, R. J., Erasmus, B. H., van Rensburg, P. J. J., Skinner, J. D., Howell, P. G., & Naude, T. W. (2002). A review of the successful eradication of feral cats from sub-Antarctic Marion Island, Southern Indian Ocean. South African Journal of Wildlife Research, 32, 65–73.
76. Fenner, F., & Ross, J. (1994). Myxomatosis. In H. V. Thompson & C. M. King (Eds.), The European rabbit: The history and biology of a successful coloniser (pp. 205–239). Oxford: Oxford University Press.
77. Rolls, E. C. (1969). They all ran wild. Sydney: Angus and Robertson. P. 177.
78. Rolls (1969) p. 184.
79. Thompson, H. V., & King, C. M. (Eds.). (1994). The European Rabbit: The History and Biology of a Successful Colonizer. Oxford: Oxford University Press.
80. Norbury, G., Heyward, R., & Parkes, J. (2002). Short-term ecological effects of rabbit haemorrhagic disease in the short-tussock grasslands of the South Island, New Zealand. Wildlife Research, 29, 599–604.

81. Parliamentary Commissioner for the Environment. (1998). The rabbit calcivirus disease (RCD) saga; a biosecurity/bio-control fiasco. Wellington: Parliamentary Commissioner for the Environment.
82. Williams, F., Eschen, R., Harris, A., Djeddour, D., Pratt, C., Shaw, R. S., ... Murphy, S. T. (2010). The Economic Cost of Invasive Non-Native Species on Great Britain. CABI Project No. VM10066, Wallingford, UK.
83. Nugent, G., Twigg, L. E., Warburton, B., McGlinchy, A., Fisher, P., Gormley, A. M., & Parkes, J. P. (2012). Why 0.02%? A review of the basis for current practice in aerial 1080 baiting for rabbits in New Zealand. Wildlife Research, 39, 89–103.
84. Bombaci, S., Pejchar, L., & Innes, J. (2018). Fenced sanctuaries deliver conservation benefits for most common and threatened native island birds in New Zealand. Ecosphere, 9, 02497. https://doi.org/10.1002/ecs2.2497
85. Innes, J., Fitzgerald, N., Binny, R., Byrom, A., Pech, R., Watts, C., ... Burns, B. (2019). New Zealand ecosanctuaries: Types, attributes and outcomes. Journal of the Royal Society of New Zealand. doi: http://10.1080/03036758.03032019.01620297
86. Woodford, D. J., Richardson, D. M., MacIsaac, H. J., Mandrak, N. E., van Wilgen, B. W., Wilson, J. R. U., & Weyl, O. L. F. (2016). Confronting the wicked problem of managing biological invasions. NeoBiota, 31, 63–86.
87. Green, W., & Rohan, M. (2011). Opposition to aerial 1080 poisoning for control of invasive mammals in New Zealand: risk perceptions and agency responses. Journal of the Royal Society of New Zealand, 42, 185–213.
88. Elton, C. (1958). The Ecology of Invasions by Animals and Plants. London: Methuen. P. 145.
89. Thompson, K. (2014). Where do camels belong? The story and science of invasive species. London: Profile Books.
90. Dearden, P. K., Gemmell, N. J., Mercier, O. R., Lester, P. J., Scott, M. J., Newcomb, R. D., ... Penman, D. R. (2017). The potential for the use of gene drives for pest control in New Zealand: a perspective. Journal of the Royal Society of New Zealand, 48, 225–244.
91. Prowse, T. A. A., Cassey, P., Ross, J. V., Pfitzner, C., Wittmann, T., & Thomas, P. (2017). Dodging silver bullets: Good CRISPR gene-drive design is critical for eradicating exotic vertebrates. Proceedings of the Royal Society B-Biological Sciences, 284, 20170799.
92. Espeut, K. M., & Gemmell, N. J. (2017). Conservation demands safe gene drive. PLOS Biology, 15, e2003850.
93. A Constitutional Moment.: Gene Drive and international governance. www.sustainabilitynz.org

94. Wallis, G. P. (2019). Thirty years of conservation genetics in New Zealand: What have we learnt? Journal of the Royal Society of New Zealand. doi: http://10.1080/03036758.03032019.01586735
95. Midgley, M. (2004). The myths we live by. Abingdon: Routledge.
96. Campbell, K. J., Saah, J. R., Brown, P. R., …., Shiels, A. (2019). A potential new tool for the toolbox: assessing gene drives for eradicating invasive rodent populations. In Veitch, C. R., Clout, M. N., Martin, A. R., Russell, J. C., & West, C. J. (Eds.), Island Invasives: Scaling up to meet the challenge (pp. 6–14). Gland, Switzerland: IUCN Species Survival Commission.
97. Crosby, A. W. (1986). Ecological Imperialism: The Biological Expansion of Europe, 900–1900. Cambridge, UK: Cambridge University Press; Dunlap, T. R. (1999). Nature and the English diaspora: environment and history in the United States, Canada, Australia and New Zealand. Cambridge: Cambridge University Press; Peden, R. (2011). Making Sheep Country: Mt Peel station and the transformation of the tussock lands. Auckland: Auckland University Press; Pawson, E., & Christensen, A. A. (2014). Landscapes of the Anthropocene: From dominion to dependence? In J. Frawley & I. McCalman (Eds.), Rethinking invasion ecologies from the environmental humanities (pp. 64–83). London: Routledge.
98. Holland, P. (2013). Home in the howling wilderness. Auckland: Auckland University Press. P. 139.
99. Pawson, E., & Brooking, T. (Eds.). (2002). Environmental histories of New Zealand. Melbourne: Oxford University Press.
100. Wallach, A. D., Bekoff, M., Batavia, C., Nelson, M. P., & Ramp, D. (2018). Summoning compassion to address the challenges of conservation. Conservation Biology. doi: https://doi.org/10.1111/cobi.13126
101. Linklater, W., & Steer, J. (2018). Predator Free 2050: A flawed conservation policy displaces higher priorities and better, evidence-based alternatives. Conservation Letters, 11, e12593. doi: https://doi.org/10.1111/conl.12593
102. Gormley, A. M. (Ed.) (2019). Modelling in Ecology. Kararehe Kino 32. (Manaaki Whenua Landcare Research, Lincoln).
103. Russell, J. C., & Stanley, M. C. (2018). An overview of introduced predator management in inhabited landscapes. Pacific Conservation Biology. doi: https://doi.org/10.1071/PC18013
104. Russell et al. (2015).

Appendix: List of Species Mentioned in the Text

NB: This list omits many species known in New Zealand, but not mentioned in the book. For more information, go to Tennyson, A. J. D. (2010). The origin and history of New Zealand's terrestrial vertebrates. New Zealand Journal of Ecology, 34, 6–27 and references therein.

English name, if given	Scientific name	Māori name, if known
Native birds, living or very recently extinct		
Bellbird	*Anthornis melanura*	kōmako
Creeper, Brown	*Mohoua novaeseealandiae*	pīpipi
Dotterel, banded	*Charadrius bicinctus*	pohowera
Duck, Blue	*Hymenolaimus malacorhynchos*	whio
Falcon, New Zealand	*Falco novaeseelandiae*	kāeaea
Fantail	*Rhipidura fuliginosa*	tīwaiwaka
Harrier, Australasian	*Circus approximans*	kērangi
Heron, White-faced	*Ardea novaehollandiae*	matuku
Kākā, North Island	*Nestor meridionalis septentrionalis*	kākā
Kākāriki, see Parakeet		
Kākāpō	*Strigops habroptilus*	kākāpō
Kea	*Nestor notabilis*	kea
Kingfisher	*Todiramphus sanctus vagans*	kōtare
Kiwi, Little Spotted	*Apteryx owenii*	kiwi pukupuku
Kiwi, North Island Brown	*Apteryx australis mantelli*	kiwi
Kōkākō, North Island	*Callaeas cinerea wilsoni*	kōkako

(*continued*)

C. M. King, *Invasive Predators in New Zealand*,
Palgrave Studies in World Environmental History,
https://doi.org/10.1007/978-3-030-32138-3

(continued)

English name, if given	Scientific name	Māori name, if known
Morepork	*Ninox novaeseelandiae*	ruru
Muttonbird, see Shearwater, sooty		tītī
Owl, Laughing	*Sceloglaux albifacies*	whēkau
Parakeet, Red-crowned	*Cyanoramphus novaezelandiae*	kākāriki
Parakeet, Orange-fronted	*Cyanoramphus malherbi*	kākāriki
Parakeet, Yellow-crowned	*Cyanoramphus auriceps*	kākāriki
Penguin, Yellow-eyed	*Megadyptes antipodes*	hoiho
Petrel, Black	*Procellaria parkinsoni*	tāiko
Petrel, Cook's	*Pterodroma cookii*	tītī
Petrel, Diving	*Pelecanoides urinatrix*	kuaka
Petrel, Grey-faced	*Pterodroma macroptera gouldi*	
Petrel, Storm	*Fregetta maorianus*	
Pigeon, New Zealand	*Hemiphaga novaeseelandiae*	kererū
Pipit, Campbell Island	*Anthus novaeseelandiae aucklandicus*	
Pipit, New Zealand	*Anthus novaeseelandiae*	hīoi
Plover, Spur–winged	*Vanellus miles*	
Prion, Broad-billed	*Pachyptila vittata*	pararā
Pūkeko	*Porphyrio porphyrio melanotus*	pūkeko
Rail, Banded	*Gallirallus philippensis assimilis*	moho pererū
Rifleman	*Acanthisitta chloris*	kōhurehure
Robin, North Island	*Petroica australis longipes*	tātāwai
Robin, Stewart Island	*Petroica australis rakiura*	tātāwai
Saddleback, North Island	*Philesturnus rufusater*	tīeke
Saddleback, South Island	*Phiesturnus carunculatus*	tīeke
Shearwater, Sooty	*Puffinus griseus*	tītī
Silvereye, White eye	*Zosterops lateralis*	pihipihi
Snipe, Campbell Island	*Coenocorypha aucklandica*	hākuwai
Snipe, Stewart Island	*Coenocorypha iredalei*	hākuwai
Stilt, Black	*Himantopus novaeseelandiae*	tūarahia or kaki
Takahē, South Island	*Porphyrio hochstetteri*	takahē
Teal, Campbell Island	*Anas nesiotis*	
Tern, Black-fronted	*Chlidonias albostriatus*	tarapirohe
Tern, fairy	*Sternula nereis*	tara iti
Piopio, North Island	*Turnagra tanagra*	piopio
Piopio, South Island	*Turnagra capensis*	piopio
Tomtit, North Island	*Petroica macrocephala toitoi*	hōmiromiro
Tomtit, South Island	*Petroica macrocephala macrocephala*	ngirungiru
Tūī	*Prosthemadera novaeseelandiae*	tūī
Warbler, Grey	*Gerygone igata*	kōriroriro
Weka, Māori hen	*Gallirallus australis*	weka

(*continued*)

(continued)

English name, if given	Scientific name	Māori name, if known
Buff	*Gallirallus australis hectori*	weka
Wren, Lyall's	*Traversia lyalli*	
Wren, Bush	*Xenicus longipes*	
Wren, Rock	*Xenicus gilviventris*	hurupounamu
Yellowhead	*Mohoua ochrocephala*	mohua
Native birds, extinct by early nineteenth century		
Adzebill, South Island	*Aptornis defossor*	
Coot	*Fulica prisca*	
Eagle, Haast's	*Harpagornis moorei*	
Goose, South Island	*Cnemiornis calcitrans*	
Harrier, Eyles'	*Circus eylesi*	
Merganser, Southern	*Mergus australis*	
Moa, North Island giant	*Dinornis novaezealandiae*	
Owlet-nightjar	*Aegotheles novaezealandiae*	
Penguin, waitaha	*Megadyptes waitaha*	
Raven, New Zealand	*Corvus antipodum*	
Snipe, Little Barrier Island	*Coenocorypha barrierensis*	
Takahē, North Island	*Porphyrio mantelli*	
Wren, Stout-legged	*Pachyplichas yaldwyni*	
Resident introduced birds		
Blackbird	*Turdus merula*	
Owl, Little	*Athena noctua*	
Pheasant, Ring-necked		
Song thrush	*Turdus philomelus*	
Other native fauna		
Flax snail	*Placostylus ambagiosus*	
Frog, Hochstetter's	*Leiopelma hochstetteri*	
Gecko, Green	*Naultinus elegans punctatus*	
Gecko, Forest	*Mokopirirakau granulatus*	
Skink, Common	*Cyclodina omatum*	
Skink, Copper	*Cyclodina aenum*	
Skink, Striped	*Oligosoma striatum*	
Skink, Tree	*Oligosoma striatum*	
Tuatara	*Sphenodon punctatus*	
Velvet worm	*Peripatus, Peripatoides*	
Wētā	70 species in 2 families of endemic Orthoptera	
Mammals (New Zealand natives and introduced or proposed)		
Badger, European	*Meles meles*	
Bat, Greater short-tailed	*Mystacina robusta*	
Bat, Lesser short-tailed	*Mystacina tuberculata*	

(*continued*)

(continued)

English name, if given	Scientific name	Māori name, if known
Bat, Long-tailed	*Chalinolobus tuberculatus*	
Cat, Feral	*Felis catus*	
Cat, Tasmanian native	*Dasyurus viverrinus*	
Cat, Wild	*Felix sylvestris*	
Chamois	*Rupicapra rupicapra*	
Dog, Polynesian, see kuri		
Dolphin, Hector's	*Cephalorhynchus hectori*	
Dolphin, Maui's	*Cephalorhynchus hectori maui*	
Ferret/Polecat, Feral	*Mustela furo*	
Fox, Patagonian grey	*Lycalopex griseus*	
Fox, Red	*Vulpes vulpes*	
Fur seal, New Zealand	*Arctocephalus forsteri*	
Goat, Feral	*Capra hircus*	
Hedgehog	*Erinaceus europaeus occidentalis*	
Kuri, Polynesian dog	*Canis familiaris*	
Kiore, Polynesian Rat	*Rattus exulans*	
Marten, American	*Martes americana*	
Marten, Scottish Pine	*Martes martes*	
Mink, American	*Martes americana*	
Mongoose, Grey	*Herpestes edwardsi*	
Mongoose, Small Asian	*Herpestes javanicus auropunctatus*	
Mouse, House	*Mus musculus*	
Muskrat	*Odonatra zibethicus*	
Otter, Sea	*Enhydra lutris*	
Pig, Feral	*Sus scrofa*	
Possum, Australian brush-tail	*Trichosurus vulpecula*	
Rabbit	*Oryctolagus cuniculus cuniculus*	
Rat, Norway	*Rattus norvegicus*	
Rat, Ship	*Rattus rattus*	
Rat, Polynesian	*Rattus exulans*	
Sea lion, Hooker's	*Phocarctos hookeri*	
Sheep	*Ovis aries*	
Squirrel, Grey	*Sciurus carolinensis*	
Stoat	*Mustela erminea*	
Tahr	*Hemitragus jemlahicus*	
Weasel	*Mustela nivalis vulgaris*	
Other species mentioned		
Coypu	*Myocastor coypus*	
Ladybird	*Rodolia cardinalis*	
Scale insect	*Icerya purchasi*	
Owl, Burrowing	*Athene cunicularia*	
Swamp hen, Eastern	*Porphyrio melanotus*	

Index

C. M. King, *Invasive Predators in New Zealand*, Palgrave Studies in World Environmental History, https://doi.org/10.1007/978-3-030-32138-3

D

S

The manufacturer's authorised representative in the EU is Springer Nature Customer Service Centre GmbH, Europaplatz 3, 69115 Heidelberg, Germany. If you have any concerns regarding our products, please contact ProductSafety@springernature.com

Printed and bound by CPI Group (UK) Ltd, Croydon, CR0 4YY
23/09/2026
02221544-0008